THE FUNDAMENTALS OF
X-RAY AND RADIUM PHYSICS

EIGHTH EDITION

THE FUNDAMENTALS OF X-RAY AND RADIUM PHYSICS

By

JOSEPH SELMAN, M.D., FACR, FACP

Medical Director
Radiation Therapy Center
East Texas Medical Center Cancer Institute

CHARLES C THOMAS • PUBLISHER
Springfield • Illinois • U.S.A.

Published and Distributed Throughout the World by

CHARLES C THOMAS • PUBLISHER
2600 South First Street
Springfield, Illinois 62794-9265

First Edition, 1954
Second Edition, 1957
Third Edition, 1961
Fourth Edition, 1965
Fifth Edition, 1972
Sixth Edition, 1977
Seventh Edition, 1985
Eighth Edition, 1994

ISBN 0-398-05870-9

Library of Congress Catalog Card Number: 93-10676

With THOMAS BOOKS *careful attention is given to all details of manufacturing and design. It is the Publisher's desire to present books that are satisfactory as to their physical qualities and artistic possibilities and appropriate for their particular use. THOMAS BOOKS will be true to those laws of quality that assure a good name and good will.*

Printed in the United States of America
SC-OK-3

Library of Congress Cataloging-in-Publication Data

Selman, Joseph.
    The fundamentals of x-ray and radium physics / by Joseph Selman.—
8th ed.
        p.   cm.
    Includes bibliographical references and index.
    ISBN 0-398-05870-9
    1. X-rays.   2. Radium.   3. Radiography.   I. Title.
    [DNLM: 1. X-Rays.   2. Radium.   3. Radiation, Ionizing.
    4. Radiography.   5. Health Physics.   WN 110 S468f 1993]
    QC481.S45   1993
    005.4'469—dc20
    DNLM/DLC
    for Library of Congress   93-10676
                                                                      CIP

Dedicated to my wife

PREFACE

S ince the last edition, the American Society of Radiologic Technologists has issued a revised Curriculum Guide, which includes new instruction requirements in computed tomography, magnetic resonance imaging, and ultrasound imaging. Computed tomography had already been described in the Seventh Edition, and now the Eighth Edition takes up the other two subjects.

A short section on logarithms has been added to Chapter 1 because of its application to ultrasound imaging and radiographic density. Other additions include digital fluoroscopy (digital fluorography); high-frequency generators, especially for mammographic units; high-frequency control of filament current; and the gamma camera. Obsolete devices such as valve tubes and field emission radiographic equipment have been deleted. The section on intensifying screens has been updated.

Recently recommended changes in exposure limits for radiation workers have replaced the MPD formula $5(N - 18)$, so this section has been revised, as has the exposure limit for the general public. In addition, the ALARA concept has been expanded.

Throughout this new edition, concepts have been further clarified, old figures revised, and new figures added where deemed necessary.

Special thanks are due Joe Burrage and Larry Spitler (RT) of the Gilbert X-Ray Company of Texas, Willis Preston (Preston Imaging, Tyler, Texas), and Jim Murray (RT) for information on newer types of radiologic imaging equipment; and to Eastman Kodak Company, DuPont, 3M, GE Medical Systems, Bennett X-Ray Corporation, LORAD Medical Systems, Philips of North America, Acoma Medical Imaging Inc., and NuTech Inc. (Tyler, Texas) for technical information and advice about their equipment and supplies.

Again, I wish to thank my artist, A. Howard Marlin, for his patience and cooperation in creating the excellent new figures and revision of the old.

Finally, I wish to thank Payne Thomas and Michael Thomas of Charles C Thomas, Publisher, for giving me the opportunity to revise this book for the seventh time since its First Edition in 1954.

JOSEPH SELMAN, M.D.

CONTENTS

THE FUNDAMENTALS OF
X-RAY AND RADIUM PHYSICS

Chapter 1

SIMPLIFIED MATHEMATICS

ALL OF THE PHYSICAL SCIENCES have in common a firm basis in mathematics. This is no less true of radiologic physics, an important branch of the physical sciences. Clearly, then, in approaching a course in radiologic physics you, as a student technologist, should find your path smoothed by an adequate background in the appropriate areas of mathematics.

We shall assume here that you have had at least the required high school exposure to mathematics, although this may vary widely from place to place. However, realizing that much of this material may have become hazy with time, we shall review the simple but necessary aspects of arithmetic, algebra, and plane geometry. Such a review should be beneficial in at least two ways. First, it should make it easier to understand the basic principles and concepts of radiologic physics. Second, it should aid in the solution of such everyday problems as conversion of radiographic technics, interpretation of tube rating charts, determination of radiographic magnification, and many others that may arise from time to time.

The discussion will be subdivided as follows: (1) arithmetic, (2) algebra, (3) ratio and proportion, (4) geometry, (5) graphs and charts, and (6) large and small numbers. Only fundamental principles will be included.

ARITHMETIC

Arithmetic is calculation or problem solving by means of definite numbers. We shall assume that you are familiar with addition, subtraction, multiplication, and division and shall therefore omit these operations.

Fractions

In arithmetic, *a fraction may be defined as one or more equal parts of a unit.* For example, $\frac{1}{2}$, $\frac{1}{3}$, and $\frac{2}{5}$ are fractions. The quantity below the line is called the *denominator;* it indicates the number of equal parts into which the unit is divided. The quantity above the line is the *numerator;* it indicates the number of equal parts taken. Thus, if a pie were divided into three equal parts, the denominator would be 3; and if two of these parts were taken, the numerator would be 2. Hence, the two segments would represent $\frac{2}{3}$ of the pie.

In a more general sense, fractions represent *the division of one quantity by another.* This extends the concept of fractions to expressions in which the numerator is larger than the denominator, as in the fraction $\frac{5}{2}$.

If the numerator is smaller than the denominator, as $\frac{3}{5}$, we have a *proper fraction.* If the numerator is larger than the denominator, as $\frac{5}{3}$, we have an *improper fraction,* because $5 \div 3 = 1\frac{2}{3}$, which is really an integer plus a fraction.

In *adding* fractions, all of which have the *same* denominator, we add all the numerators first and then place the sum over the denominator:

$$\frac{2}{7} + \frac{3}{7} + \frac{6}{7} + \frac{5}{7} = \frac{2 + 3 + 6 + 5}{7} = \frac{16}{7}$$

$$\frac{16}{7} = 2\frac{2}{7}$$

Subtraction of fractions having identical denominators follows the same rule:

$$\frac{6}{7} - \frac{4}{7} = \frac{6 - 4}{7} = \frac{2}{7}$$

If a series of fractions is to be added or subtracted, and the denominators are *different,* then the *least common denominator* must be found. This is the smallest number which is exactly divisible by all the denominators. Thus,

$$\frac{1}{2} + \frac{2}{3} - \frac{3}{4} = ?$$

The smallest number which is divided exactly by each denominator is 12. Place 12 in the denominator of a new fraction:

$$\frac{}{12}$$

Divide the denominator of each of the fractions in the old equation into 12, and then multiply the answer by the numerator of that fraction; each result is then placed in the numerator of the new fraction:

$$\frac{6 + 8 - 9}{12} = \frac{5}{12}$$

When numbers are *multiplied, we take their product.* In multiplying fractions, we take the product of the numerators and place it over the product of the denominators,

$$\frac{4}{5} \times \frac{3}{10} = \frac{4 \times 3}{5 \times 10} = \frac{12}{50}$$

The resulting fraction can be reduced by dividing the numerator and the denominator by the *same* number, in this case, 2:

$$\frac{12}{50} \div \frac{2}{2} = \frac{6}{25}$$

which cannot be further simplified.

Note that when the numerator and the denominator are both multiplied or divided by the same number, the value of the fraction does not change. For instance,

$$\frac{3}{5} \times \frac{2}{2} = \frac{6}{10}$$

is the same as $\frac{3}{5} \times 1 = \frac{3}{5}$

When two fractions are to be divided, as $4/5 \div 3/7$, the fraction that is to be divided is the *dividend,* and the fraction that does the dividing is called the *divisor.* In this case, $4/5$ is the dividend and $3/7$ the divisor. The rule is to invert the divisor (called "taking the reciprocal") and multiply the dividend by it:

$$\frac{4}{5} \div \frac{3}{7}$$

$$\frac{4}{5} \times \frac{7}{3} = \frac{28}{15} = 1\frac{13}{15}$$

Percent

A special type of fraction, **percent**, is represented by the sign % to indicate that the number standing with it is to be divided by 100. Thus, 95% = 95/100. We do not use percentages directly in mathematical computations, but first convert them to fractions or decimals. For instance,

150 × 40% is changed to
150 × 40/100 or 150 × 2/5
or 150 × 0.40.

All these expressions equal 60.

Decimal Fractions

Our common method of representing numbers as multiples of ten is embodied in the **decimal system**. A **decimal fraction** has as its denominator 10, or 10 raised to some power such as 100, 1000, 10,000, etc. The denominator is symbolized by a dot in a certain position. For example, the decimal 0.2 = 2/10; 0.02 = 2/100; 0.002 = 2/1000, etc. Decimals can be multiplied or divided, but care must be taken to place the decimal point in the proper position:

$$\begin{array}{r} 2.24 \\ \times\ 1.25 \\ \hline 1120 \\ 448 \\ 224 \\ \hline 2.8000 \end{array}$$

Note that we add the total number of digits to the right of the decimal points in the numbers being multiplied, which in this case turns out to be four. We then point off four places from the right in the answer to determine the correct position of the decimal point. The decimal system is used everywhere in science and in the vast majority of countries in daily life.

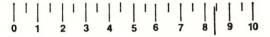

Figure 1.1. With this calibrated scale we can estimate to the nearest tenth. Thus, the position of the pointer indicates 8.4 units, the 0.4 being the last significant figure.

Significant Figures

The precision (reproducibility of results) of any type of measurement is limited by the design of the measuring instrument. For example, a scale calibrated in grams as shown in Figure 1.1 allows an estimate to the nearest tenth of a gram. Thus, the scale in Figure 1.1 reads 8.4 grams. The last figure, 0.4, is estimated and is the *last significant figure* — that is, it is the last meaningful digit. Obviously, no greater precision is possible with this particular instrument. To improve precision, the scale would have to show a greater number of subdivisions.

Now let us see how significant figures are used in various operations. For example, in addition:

$$
\begin{array}{lr}
\text{item 1} & 98.26 \;\text{grams} \\
\text{item 2} & 1.350 \;\text{g} \\
\text{item 3} & \underline{260.1 \quad\text{g}} \\
& 359.710 \;\text{g}
\end{array}
$$

Notice that three digits appear after the decimal point in the answer. But in item (3) there is only one digit 1, after the decimal point; beyond this, the digits are unknown. Therefore, the digits after the 7 in the answer imply more than is known, since the answer can be no more precise than the least precise item being added. In this case, the answer should be properly stated as 359.7. Thus, in addition and subtraction the answer can have no more significant figures after the decimal point than the item with the *least number of significant figures after its decimal point.*

A different situation exists in multiplication and division. Here, the total number of significant figures in the answer equals that in the items having the *least total number of significant figures.* For example, in

$$25.23 \text{ cm}$$
$$\times 1.21 \text{ cm}$$

$$2523$$
$$5046$$
$$2523$$

$$30.5283 \text{ cm}^2$$

1.21 has fewer significant figures—three in all. Therefore the answer should have three significant figures and be read as 30.5 (dropping the 0.0283).

In general, to *round off* significant figures, we observe the following rule: if the digit following the last significant figure is equal to or greater than 5, the last significant figure is increased by 1; if less than 5, it is unchanged. The rule is applied in the following examples:

45.157 is rounded to 45.16
45.155 is rounded to 45.16
45.153 is rounded to 45.15

where the answer is to be expressed in four significant figures.

ALGEBRA

The word *algebra,* derived from the Arabic language, connotes that branch of mathematics which deals with the relationship of quantities usually represented by letters of the alphabet—Roman, Greek, or Hebrew.

Operations. Mathematical operations with *letter symbols* are the same as with *numerals,* since both are symbolic representations of numbers which, in themselves, are abstract concepts. For example, the concept "four" may be represented by 4, 2^2, 2×2, $2 + 2$, or $3 + 1$; or by the letter x if the value of x is specifically designated to represent "four." In algebra, just as in arithmetic, the fundamental operations include addition, subtraction, multiplication, and division; and there are fractions, proportions, and equations. Algebra provides a method of finding an unknown quantity when the relationship of certain known quantities is specified.

Algebraic operations are indicated by the same symbols as arithmetic:

+ (plus) add
− (minus) subtract
× (times) multiply
÷ (divided by) divide
= equals

To indicate *addition* in algebra, we may use the general expression

$$x + y \qquad (1)$$

The symbols x and y, called *variables*, may represent any number or quantity we choose. Thus, if $x = 4$, and $y = 7$, then, substituting these values in equation (1),

$$4 + 7 = 11$$

Similarly, to indicate *subtraction* in algebra, we may use the general expression

$$x - y$$

If $x = 9$ and $y = 5$, then

$$9 - 5 = 4$$

Notice that algebraic symbols may represent whole numbers, fractions, zero, and negative numbers, among others. Negative numbers are those whose value is less than zero and are designated as $-x$. In algebraic terms, we may add a positive and a negative number as follows:

$$x + -y$$

If $x = 8$ and $-y = -3$, then

$$8 + -3$$

is the same as

$$8 - 3 = 5$$

The + sign is omitted in the designation of positive numbers, being reserved to indicate the operation of addition.

On the other hand, we may subtract a negative quantity from a positive one:

$$x - -y$$

If $x = 4$ and $-y = -6$, then

$$4 - -6$$

is the same as

$$4 + 6 = 10$$

Multiplication in algebra follows the same rules as in arithmetic. However, in the multiplication of letter symbols the $\times$ sign is omitted, $x \times y$ being written as xy. If $x = 3$ and $y = 5$, then substituting in the expression xy,

$$3 \times 5 = 15$$

Division in algebra is customarily expressed as a fraction. Thus, $x \div y$ is written as x/y. If $x = 3$ and $y = 5$, then

$$3 \div 5 = \frac{3}{5}$$

When two negative quantities are multiplied or divided, the answer is positive. Thus, $-x \times -y = xy$; and $-x/-y = x/y$. When a positive and a negative quantity are multiplied or divided, the answer is negative; thus, $x \times -y = -xy$, and $x \div -y = -x/y$.

In solving an algebraic expression consisting of a collection of *terms* we must perform the indicated *multiplication and division first,* and then carry out the indicated addition and subtraction. An example will clarify this:

$$ab + c/d - f = ?$$

Suppose $a = 2, b = 3, c = 4, d = 8$, and $f = 5$. Substituting in the preceding expression,

$$2 \times 3 + \frac{4}{8} - 5 = ?$$

Performing *multiplication and division first,*

$$6 + \frac{4}{8} - 5 = ?$$

Then, performing addition and subtraction,

$$6\frac{4}{8} - 5 = 1\frac{4}{8} = 1\frac{1}{2}$$

A parenthesis inclosing a group of terms indicates that all of the terms inside the parenthesis are to be multiplied by the term outside the parenthesis. This is simplified by performing all the indicated operations in correct sequence—inside the parenthesis first—and then multiplying the result by the quantity outside the parenthesis. For example,

$$6 (8 - 4 + 3 \times 2) =$$

$$6 (8 - 4 + 6) =$$
$$6 \times 10 = 60$$

Equations. The simpler algebraic equations can be solved without difficulty if fundamental rules are applied. You can easily verify these rules by substituting numerals. In the equation

$$a + b = c + d \tag{2}$$

$a + b$ is called the **left side**, and $c + d$ the **right side.** Each letter is called a **term.** If any quantity is added to one side of the equation, the same quantity must be added to the other side in order for this to remain an equation. Similarly, if any quantity is subtracted from one side, the same quantity must be subtracted from the other side. To simplify the concept of the equation we may picture it as a see-saw as in Figure 1.2. If persons of equal weight are placed at each end, the board will remain horizontal—the equation is balanced. If a second person is now added to one end of the see-saw, a person of similar weight must be added to the other end in order to keep the board level.

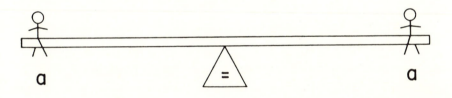

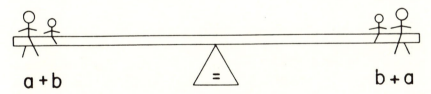

Figure 1.2. Analogy of algebraic equation to a see-saw.

Return again to the simple equation:

$$a + b = c + d$$

in which each term is a variable. If any three of the variables are known, the fourth can be found. Suppose, a is 3, b is 4, c is 1, and d is unknown. Substituting these values in the equation,

$$3 + 4 = 1 + d$$

How can d be found? Simply rearrange the equation so that d is alone, that is, *the only term in its side.* In this case, subtract 1 from both sides of the equation. However, a mathematical short cut can be used: *a term may be transposed from one side of an equation to the other side, provided it is given the opposite sign.* Following this rule, 1 becomes a minus 1 when moved to the left side. Thus,

$$3 + 4 - 1 = d$$
$$6 = d$$
$$\text{or,} \quad d = 6$$

Usually, in solving algebraic equations, we rearrange the terms before substituting their numerical values. In the equation

$$a + b = c + d$$

to find d, transpose c to the left side and change its sign.

$$a + b - c = d$$
$$\text{or,} \quad d = a + b - c$$

(Reversing both sides of an equation does not alter its equality.)

In algebraic equations in which terms are multiplied or divided, analogous rules apply. For example, equation

$$x = y/z$$

may be solved for y by multiplying both sides by z,

$$xz = yz/z$$
$$xz = y$$
$$\text{or,} \quad y = xz$$

The same result may be obtained by moving z from the denominator of the right side to the numerator of the left side. Thus, we have the short-cut rule for cross-multiplication: *if the denominator of one side of an equation is moved, it multiplies the numerator of the other side and, conversely, if the numerator of one side is moved, it multiplies the denominator of the other side.*

Suppose that in the equation $x = y/z$, x and y are known; then z is solved as follows: move z into the numerator of the opposite side as a multiplier,

$$xz = y$$

and move x into the denominator of the right side as a multiplier, and

$$z = y/x \qquad (3)$$

The above rule can be readily tested. Suppose that y is 12, and x is 3. Substituting in equation (3),

$$z = 12/3$$
$$z = 4$$
$$4 = 12/3 \qquad (4)$$

If we wish to move 3, we must place it in the numerator of the left side of equations (4),

$$4 \times 3 = 12 \qquad (4a)$$

Note that numerical equations (4) and (4a) balance.

Now, referring again to equation (4), suppose we wish to move 12. We must place it in the denominator of the left side

$$4/12 = 1/3$$

Again, it is evident that the equation balances.

RATIO AND PROPORTION

A *ratio* is a fixed relationship between two quantities, simply indicating how many times larger or smaller one quantity is relative to another. It has essentially the same meaning as a fraction. One symbol that expresses a ratio is the colon (:). Thus, $a:b$ is read "a is to b." Or, 1:2 is read "1 is to 2." In modern mathematics ratios are usually represented as fractions:

$a:b$ is the same as a/b

1:2 is the same as $1/2$

As noted above, the fraction $1/2$ indicates that the numerator is $1/2$ as large as the denominator. Similarly, $2/3$ indicates that the numerator is $2/3$ as large as the denominator.

The meaning of ratio is important to the technologist, because it underlies the concept of *proportion,* defined as *an expression showing that two given ratios are equal.* Thus, we may have an algebraic proportion,

$$a/b = c/d \tag{5}$$

which is read "a is to b as c is to d." The same idea can also be presented numerically. For example,

$$3/6 = 4/8$$

If any three terms of a proportion are known, the fourth may easily be determined. Suppose in proportion (5) a is 2, b is 4, d is 8, and c is to be found. Then,

$$2/4 = c/8$$

Moving 8 to the numerator of the left side (crossmultiplying),

$$\frac{2 \times 8}{4} = c$$

$$c = 16/4 = 4$$

There are three general types of proportions that are of interest in radiography.

1. *Direct proportion.* Here, one quantity maintains the same ratio to another quantity as the latter changes. For example, the statement "a is proportional to b" means that if b is doubled, a is automatically doubled; if b is tripled, a is tripled, etc. In order to represent this mathematically, let us assume that the quantity a_1 exists when b_1 exists; and that if b_1 is changed to b_2, then a_1 becomes a_2. Thus,

$$a_1/b_1 = a_2/b_2$$

The numbers below the letters are called "subscripts" and have no significance except to label a_2 as being different from a_1. Such a direct proportion is solved by the method described above.

2. *Inverse proportion.* Here, one quantity varies in the opposite direction from another quantity as the latter changes. Thus, the statement "a is inversely proportional to b" means that if b is doubled a is halved, if b is tripled, a is divided by 3, etc. Such a proportion is set up as follows:

$$\frac{a_1}{1/b_1} = \frac{a_2}{1/b_2}$$

Crossmultiplying,

$$a_1/b_2 = a_2/b_1 \tag{6}$$

A numerical example should help clarify this. Suppose that when

a_1 is 2, b_1 is 4; if a is inversely proportional to b, what will a_1 become if b_1 is changed to 8? In this case, a_2 is the unknown, and b_2 is 8. Substituting in equation (6),

$$2/8 = a_2/4$$
$$a_2 = 8/8 = 1$$

Thus, a is halved when b is doubled.

3. *Inverse Square Proportion.* See pages 333–337 (Fig 18.9).

PLANE GEOMETRY

Plane geometry is that branch of mathematics which deals with figures lying entirely in one plane; that is, on a flat surface. Some of the elementary rules of plane geometry will now be listed.

1. A *straight line* (in classical geometry) is the shortest distance between two points. It has only one dimension—length.

2. A *rectangle* is a plane figure composed of four straight lines meeting at right angles. Its opposite sides are equal. The sum of the lengths of the four sides is called the *perimeter.* The area of a rectangle is the product of two adjacent sides. Thus, in Figure 1.3 the perimeter is the sum $a + b + a + b = 2a + 2b$. The area is $a \times b$ or ab.

3. A *square* is a special rectangle in which all four sides are equal, as in Figure 1.4.

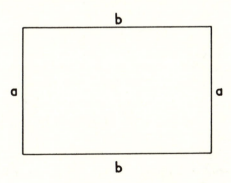

Figure 1.3. Rectangle.
Perimeter $= a + b + a + b$
$= 2a + 2b$

4. A *triangle* is a figure made up of three straight lines connecting

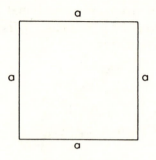

Figure 1.4. Square.
Perimeter = 4a
Area = $a \times a = a^2$

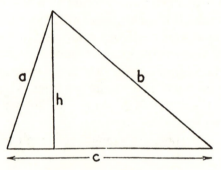

Figure 1.5. Triangle.
Perimeter = $a + b + c$
Area = $\frac{1}{2}ch$

three points that are not in the same line and lie in the same plane. The perimeter of a triangle is the sum of the three sides. The area of a triangle is one half the base times the altitude (the altitude is the perpendicular distance from the apex to the base). This is shown in Figure 1.5.

5. A *circle* is a closed curved line which is everywhere at an equal distance from one point called the center, lying in the same plane as the center. A straight line from the center to any point on the circle is the *radius.* A straight line passing through the center and meeting the circle at two points is the *diameter,* which is obviously equal to twice the radius. The length of the circle is its *circumference,* obtained by multiplying the diameter by a constant,

π (pi). Pi always equals about 3.14. The area enclosed by a circle equals π times the square of the radius. Figure 1.6 shows these relationships.

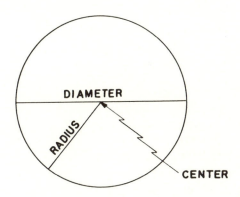

Figure 1.6. Circle.
Circumference = π × diameter = πd
Area = π × radius² = πr^2

Similar Triangles

Of great importance in radiology, and closely related to ratio and proportion, is the proportionality of *similar triangles.* Such triangles have identical shape, although they differ in size. *In similar triangles the corresponding sides are directly proportional and the corresponding angles are equal.* Thus, similar triangles represent the geometric equivalent of direct proportion.

Figure 1.7 shows two similar triangles. The corresponding sides (those opposite the equal corresponding angles) are proportional, so that

$$a/A = b/B, \text{ or}$$
$$a/A = c/C, \text{ or}$$
$$b/B = c/C$$

A line drawn across any triangle parallel to its base produces two similar triangles, one partly superimposed on the other, as shown in Figure 1.8. Line *BE* has been drawn parallel to the base *CD.* Then triangle *ABE* is similar to triangle *ACD* and their

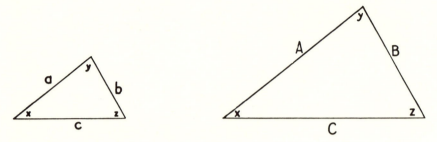

Figure 1.7. Similar triangles. The corresponding angles are *equal;* that is, $x = x$, $y = y$, $z = z$. The corresponding sides are proportional; thus, $a/A = b/B = c/C$.

corresponding sides are proportional. This may be simplified by separating the two triangles as in Figure 1.7.

A thorough comprehension of similar triangles is essential to the understanding of photographic and radiographic projection. Suppose an object is placed in a beam of light originating at a point source and allowed to cast a shadow or image on a surface (see Figure 1.9). The image has to be larger than the object. Knowing the distances of the object and image, respectively, from the light source, as well as the object size, we can determine the image size from the similarity of triangles ABC and ADE. We do this by setting up the following proportion:

$$\frac{\text{image size}}{\text{object size}} = \frac{\text{image distance}}{\text{object distance}}$$

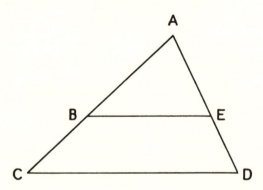

Figure 1.8. Similar triangles. *BE* is parallel to *CD*. Triangle *ABE* is similar to triangle *ACD*. Therefore,

$$AB/AC = AE/AD = BE/CD.$$

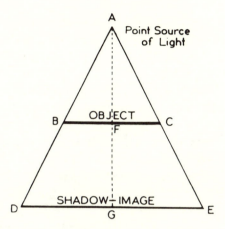

Figure 1.9. Projection of the shadow image of an object by a point source of light. $DE/BC = AG/AF.$

Substituting the known values in the equation, we can predict the size of the image by solving the equation. Note that if any three of the values are known, the fourth can be readily obtained from the same equation. This principle also applies in radiography, as will be shown in Chapter 18.

GRAPHS AND CHARTS

For practical purposes, graphs and charts may be regarded as diagrams representing the relationship of two quantities, one of which depends on the other. The dependent factor is called the *dependent variable.* The factor which changes independently is called the *independent variable.*

When data are accumulated, showing how a dependent variable changes with a change in the independent variable, they can either be compiled in a table, or represented by a graph. However, tables do not give intermediate values as do graphs.

The construction and interpretation of a graph may be exemplified as follows: the optimum developing time for x-ray film increases as the temperature of the developing solution decreases (not inversely proportional, however). Table 1.1 reproduces the data actually obtained with a particular developer. To chart this information on graph paper, first plot the temperature (independent variable) along the horizontal axis of the graph. Then plot the developing time (dependent variable) on the vertical axis, as shown in Figure 1.10. In mathematics, the dependent variable is said to be a *function* of the independent variable. To graph the tabulated data, take the first temperature listed in the table and locate it on the horizontal axis. Trace vertically upward from this point to the horizontal line corresponding to the correct developing time, and mark the intersection with a dot. Repeat this for all the values in the table, and then draw a line that best fits the dots. This constitutes a *line graph* or *curve.*

Having constructed a graph, how do we read it? Suppose that we wish to determine the correct developing time for a temperature of 18.5 C (65 F). Locate 18.5 C (65 F) on the horizontal axis and trace vertically upward to the intersection with the curve; from this point of intersection trace horizontally to the vertical axis on the left, where the correct developing time can be read. This is shown in Figure 1.10, where the tracing lines are formed by broken lines with arrows. The horizontal tracing line meets the vertical axis at approximately $3\frac{1}{2}$ minutes, which is the correct developing time at this temperature. Thus, a properly constructed graph gives the developing times for temperature readings that did not appear in the original table, a process called *interpolation.*

TABLE 1.1
TIME–TEMPERATURE DEVELOPMENT DATA

Temperature		Developing Time in Minutes
C	F	
17	62	$4\frac{1}{2}$
18.5	65	$3\frac{1}{2}$
20	68	3
21	70	$2\frac{3}{4}$
22	72	$2\frac{1}{2}$
24	75	2

LARGE AND SMALL NUMBERS

Because ordinary notation is too cumbersome to express the extremely large numbers often encountered in physics, we resort to the *exponential system* in powers of 10. For example, the

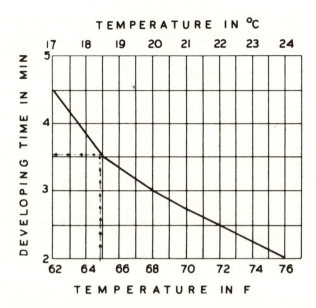

Figure 1.10. Method of using a time-temperature development curve. Select the correct temperature on the horizontal axis, in this case 18.5 C (65 F). Follow vertically, as shown by the arrows, to the curve. Then follow horizontally to the time axis where the correct developing time in this case is seen to be 3½ min.

number 10^3 means that 10 is multiplied by itself three times: $10 \times 10 \times 10$. In the quantity 10^3, 10 is the *base* and 3 the *exponent*, and is read as "ten to the third power." In general,

$$10^1 = 10$$
$$10^2 = 100$$
$$10^3 = 1000, \text{ etc}$$

Note that the exponent of 10 indicates the number of zeroes required to express the number in ordinary notation.

Let us now apply this principle. The velocity of light in air is 30,000,000,000 cm per sec. Place a decimal point after 3; there are ten zeroes to the right of the decimal point. Therefore, this number may be simply expressed as 3.0×10^{10} cm per sec, read as "three point oh times ten to the tenth." A more complicated number such as 4,310,000,000,000,000 may be simplified in the same way by placing a decimal point after the first digit, 4, and counting the number of *places* to the right, in this case fifteen. The corresponding number in the powers-of-ten system is 4.31×10^{15}.

Suppose, instead, we are to convert 6.3×10^6 to *ordinary notation.* Count six places to the right of the decimal point, filling out the places with zeroes. Since the first place is already occupied by a digit, 3, five zeroes must follow. Therefore, the number is 6,300,000. In general, *any number followed by 10^6 is expressed in millions.*

The same system can be applied to *extremely small* numbers, by the use of negative exponents. This is based on

$$10^{-1} = \frac{1}{10} = 0.1$$
$$10^{-2} = \frac{1}{100} = 0.01$$
$$10^{-3} = \frac{1}{1000} = 0.001, \text{etc.}$$

From this you can see that a negative exponent indicates the number of decimal places to the *left* of a digit, *including the digit.* Thus, 0.004 would be expressed as 4×10^{-3}. To obtain this result, count the number of places to the right of the decimal point, including the first digit,

$$\overset{\cdots}{0.004}$$

The counted number of places (three in this instance) then becomes the negative exponent of 10; thus, 4×10^{-3}. A more complicated number such as 0.00000000000372 would be simply expressed as 3.72×10^{-12}, read as "three point seven two times ten to the minus twelfth."

The simplified system of notation facilitates the multiplication and division of large and small numbers. To multiply such numbers, we multiply the digits and make the exponent of 10 in the answer equal to the algebraic sum of the exponents of 10 in the multipliers. Thus,

$$4 \times 10^3 \times 2 \times 10^2 = 4 \times 2 \times 10^{(3+2)} = 8 \times 10^5$$
$$3 \times 10^6 \times 2 \times 10^{-4} = 3 \times 2 \times 10^{(6-4)} = 6 \times 10^2$$

To divide numbers in this system, we divide the digits and derive the exponent of 10 in the answer by subtracting algebraically the exponent in the divisor from the exponent in the dividend:

$$\frac{6 \times 10^5}{2 \times 10^3} = \frac{6 \times 10^{(5-3)}}{2} = 3 \times 10^2$$

$$\frac{9 \times 10^7}{3 \times 10^{-2}} = \frac{9 \times 10^{(7+2)}}{3} = 3 \times 10^9$$

Note in the above examples that when a power number is

moved from the denominator of a fraction to the numerator, or *vice versa*, we simply change the sign of the exponent. Thus,

$$\frac{8}{3 \times 10^{-5}} = \frac{8 \times 10^5}{3}$$

LOGARITHMS

In several areas of imaging, we shall have to use quantities known as *logarithms.* For our purposes, only the basic aspects of logarithms will be necessary. In fact, the preceding section included powers-of-ten, which are logarithms (abbreviated *log* [*sing.*) and *logs* [*pl.*]).

As an example of the use of logs, consider ,the intensity of sound. It so happens that the human ear, in comparing a sound 100 times louder than another, perceives it as being only *twice* as loud; 100 = 10^2, so the exponent is the "twice." If the sound is 1000 times the first, the ear perceives it as 3 times louder; 1000 = 10^3, so now 3 represents the "three times."

It turns out that 2 is the logarithm of 100, stated as log 100 = 2; similarly, log 1000 = 3. In other words, logs are simply the exponent to which 10 must be raised to equal a given number. Note that 10 is the *base* of the system of *common logs.* (There is another system—the Naperian, with a different base.) Conversely, the antilogarithm (antilog) of 2 is 100. The following examples will help summarize simple logs:

10^0 = 1	log 1 = 0	antilog 0 = 1
10^1 = 10	log 10 = 1	antilog 1 = 10
10^2 = 100	log 100 = 2	antilog 2 = 100
10^3 = 1000	log 1000 = 3	antilog 3 = 1000

You can see that in each case, for example, log 100 = 2 because 10 has to be raised to the second power to equal 100.

The same applies to negative exponents:

10^{-1} = 0.1	log 10^{-1} = -1	antilog -1 = 0.1
10^{-2} = 0.01	log 10^{-2} = -2	antilog -2 = 0.01
10^{-3} = 0.001	log 10^{-3} = -3	antilog -3 = 0.001

You can see that in each instance, for example, log 0.01 = -2 because 10 has to be raised to the -2 power to equal 0.01.

In the past, logs and antilogs were extremely useful in multiplication and division, especially of large and small numbers, because, to multiply powers of 10, one simply adds the exponents; and to divide powers of 10, one subtracts exponents. In a chain multiplication with a mixture of positive and negative exponents, all to the base 10, one simply adds the exponents algebraically and uses the resulting exponent as a power of 10, as in the preceding section.

Special tables of logarithms are available to find the logs to the base 10 (ie, common logs) for numbers that are *not* integer powers of 10; for example, let us multiply 43.61 × 25.23 by using logs.

$$\log 43.61 = 1.6396$$
$$\log 25.23 = 1.4017$$

3.0413 total value of exponent

Applying this new exponent to 10,

$$10^{3.0413} = \text{antilog of } 3.0413 = 1100$$

Actually multiplying the original numbers 43.61 × 25.23 = 1100.2. If a 5-place log table had been used, the answer would have shown the .2; this example has been given simply to show how logarithms work. Manual calculations with the aid of log tables have been supplanted by pocket calculators having a wide range of complexity. Even the multiplication problem just presented can be solved by the use of a simple pocket calculator.

QUESTIONS AND PROBLEMS

1. Reduce the following fractions:

    (a) $4/8$ (b) $9/12$ (c) $10/150$ (d) $4/5$

2. Solve the following problems:

    (a) $3/5 + 2/5 + 4/5 =$
    (b) $2/3 + 3/4 + 3/7 =$
    (c) $1/2 + 2/3 + 4/5 =$

3. Mr. Jones has 100 bushels of potatoes and sells 75 percent of this crop. How many bushels does he have left?

4. Divide as indicated:

    (a) $4/5 \div 3/5$ (b) $4/9 \div 7/16$ (c) $8/9 \div 3/5$

5. If $a = b/c$, what is b in terms of a and c? What is c in terms of a and b?

6. Solve the following equations for the unknown term:

   (a) $x + 4 = 7 - 3 + 8$
   (b) $6 + 3 - 1 = 4 + 1 - a$
   (c) $x/3 = 7/9$
   (d) $4/y = \dfrac{6 + 3}{10}$
   (e) $\dfrac{x + 4}{8} = 7 + 2 - 3$

7. Solve the following proportions for the unknown quantity:

   (a) $a/7 = 2/21$ (c) $4/6 = 7/y$
   (b) $3/9 = x/15$ (d) $3/5 = 9/x$

8. The diameter of an x-ray beam is directly proportional to the distance from the tube target. If the diameter of the beam at a 20-in. distance is 10 cm, what will the diameter be at a 40-in. distance?

9. The exposure of a radiograph is directly proportional to the time of exposure. What will happen to the exposure if the time is tripled?

10. What is the area of a circle having a diameter of 4 cm?

11. What is the area of a rectangular plot of ground measuring 20 ft on one side and 30 ft on the adjacent side?

12. Using the time-temperature curve in Figure 1.10, determine the proper developing time when the temperature is 64 F.

13. Convert 3.6×10^6 to the ordinary number system.

14. Change 424,000 to the exponential system.

15. Solve $320,000 \times 1,200,000 \div 240,000$ using the exponential system.

16. What is log 10^5 to base 10 (common logs)?

17. State the antilog of 4.

18. Multiply $10^5 \times 10^6$ in the powers-of-ten system. Now state the log of the answer.

19. Divide 10^8 by 10^5 and state the log of the answer.

20. What is antilog $(4 + 5)$?

21. Find the log of 1000×10000.

Chapter 2

PHYSICS AND THE UNITS OF MEASUREMENT

Physics, as an exact science, requires a precise vocabulary in which each term has a clear and definite meaning. Not only does this simplify the learning process, but it also facilitates the organization of concepts and their accurate communication to others.

In order to appreciate the position of physics within the framework of science, we may use the term *natural science* to include the systematic study of the universe and its contents. Figure 2.1 shows the subdivision of natural science into its major categories. Here you can readily see that while physics is listed as one of the physical sciences, it actually underlies all of them. In fact, it is ultimately an important contributor even to the biologic sciences.

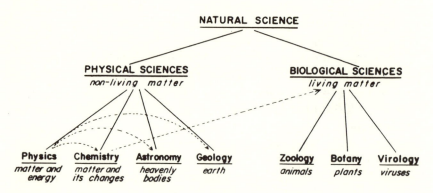

Figure 2.1. Subdivisions of natural science. Notice that physics directly or indirectly plays a part in all branches of science (dashed lines).

What, then, is physics? It may be defined as that branch of science which deals with matter and energy, and their relation to each other. It includes mechanics, heat, light, sound, electricity, and magnetism, and the fundamental structure and properties of matter. For our purpose, we shall be interested in those aspects of

26

physics pertaining to the origin, nature, and behavior of x rays and related types of radiation; that is, *radiologic physics.*

Thus far, we have used the word science rather freely without definition. *Science is organized and classified knowledge.* Natural processes go on continually in the universe, with or without our knowledge. But these do not constitute a science until we have applied the scientific method to them. The *scientific method* comprises the systematic collection of facts, the study of their interrelationship, and the drawing of valid conclusions from the resulting data. Inherent in the scientific method is the assumption that nature is orderly and that sooner or later we can discover the rules by which she behaves.

Not only do we use the scientific method to describe and classify natural occurrences, but we also seek to correlate them by deriving certain principles known as *laws.* Scientific laws are based on human experience, being derived by observation of natural phenomena, by laboratory experiments, or by the application of mathematics. Such laws state as clearly and simply as possible, according to our present state of knowledge, that certain events will always follow in the same order, or that certain facts will always be related in the same way. For example, we know by experience that all objects fall to the ground; this observation has been expanded into the law of gravitation, which applies to the attraction between all bodies of matter.

Scientists are not satisfied with merely discovering the laws of nature. They also attempt to correlate various laws in order to determine their conformity to a *general pattern.* Thus, wherever possible, natural laws are so tied together that their operation can be expressed as a general concept, often on a mathematical basis. Such a broad, unified concept of the laws underlying certain natural phenomena is called a *theory.* Not only do theories provide a better insight into nature but they also suggest new lines of scientific research, since it is common experience that the solution of one problem in science ultimately gives rise to a host of new problems.

Standard Units

An exact science must include not only the observation and classification of natural occurrences, but also their *measurement.*

In other words, physics as an exact science deals with both the "how" and the "how much" of physical phenomena. In order for these measurements to have the same meaning everywhere, they must be expressed in certain standard units. A *unit* is a quantity adopted as a standard of measurement by which other quantities of the *same* kind can be measured, such as the meter as an interval of distance, or the second as an interval of time. *Standard units* are the basic units in any particular system of measurement; they are used mathematically to describe natural phenomena, derive laws, and predict future events within the realm of nature. For example, by the application of science and mathematics, astronomers can foretell precisely the time of an eclipse of the sun or moon.

Units should have a size compatible with that which is being measured. Thus, we measure the length of a baby in centimeters, but we use a larger unit, the meter, to measure the height of a building; and a still larger unit, the kilometer, to measure the distance between cities.

The standard units employed in physics consist of two general types, the simpler *fundamental units* dealing with *length, mass* and time, and the more complicated *derived units*, obtained by various combinations of the fundamental units.

Fundamental Units

The fundamental units, arbitrarily selected and named, have been so standardized that a given unit has the same meaning everywhere. We now have two widely used systems of measurement, the *English* and the *metric.*

Since the metric system is used internationally in the exact sciences, it will be emphasized here, but the equivalents in the English system will also be indicated because we still use this in everyday life. The metric system is also known as the MKS (meter-kilogram-second) or CGS (centimeter-gram-second) system. These will gradually be replaced in science by the International System of Units, known as the *SI system,* which provides for the interconversion of units among all branches of science. The SI system is based on the metric system, and insofar as the fundamental units are concerned, the two systems are identical.

1. **Length.** The unit of length in the metric system is the *meter,* originally defined as the distance between two scratches on a bar

of platinum, kept at Sèvres, France, at the International Bureau of Weights and Measures. (It was redefined in 1960 as 1.65×10^6 times the wavelength of the orange-red radiation of krypton 56.) Roughly equal to a yard, the meter (m) can be conveniently subdivided into smaller units such as a centimeter (0.01 or 10^{-2} m) and millimeter (0.001 or 10^{-3} m); or it can be treated as a multiple such as a kilometer ($1,000$ or 10^3 m). In x-ray physics, we even have occasion to use very small fractions of these units; thus, the angstrom is $1/10,000,000,000$ or 10^{-10} m (or 10^{-8} cm). The following common equivalents should prove useful:

1 meter = 100 cm (centimeters) = 1000 mm (millimeters)
1 Å (angstrom) = $1/100,000,000$ cm = 10^{-8} cm = 0.1 nm
km (kilometer) = 1,000 m (meters) = about $3/5$ mile
1 in. = 2.54 cm

2. **Mass.** Representing the quantity of matter in a body (or inertia) its *mass* is determined by weighing, a procedure which measures the force with which the earth attracts the body at some particular geographic location. Thus, the more massive the body, the greater the gravitational force and, therefore, the greater its weight. As implied in the above definition, the weight of a body may vary from place to place, but equal masses will have the same weight under identical conditions. The unit of mass is the *kilogram,* which is the weight of a standard piece of platinum-iridium (International Kilogram Prototype) kept at the International Bureau of Standards. All other kilograms are more or less exact copies of this standard unit. A more convenient unit is the gram, which is $1/1,000$ of a kilogram.

1 kg (kilogram) = 1000 or 10^3 g (grams)
1 kg = 2.2 lb
28 g = 1 oz

3. **Time.** This is a measure of the duration of events. We are all aware of the occurrence of events and the motion of objects, but to measure duration with relation to our senses alone is inaccurate. The standard unit of time is the *second,* defined as $1/86,400$ of a mean solar day. In other words, the second is a definite fraction of the average time it takes for the earth to make one rotation on its axis.

Derived Units

Of the many derived units, only a few examples of the more common ones will be given here. Those that apply specifically to radiologic physics appear later in the text.

1. **Area** is the measure of a given surface, and depends on length. Thus, a square or rectangle has an area equal to the product of two sides. The area of a circle equals the radius squared times π. Figures 1.3, 1.4, 1.5, and 1.6 in Chapter 1 explain area in detail. In the metric system, area is represented by square meters for larger surfaces and square centimeters for smaller ones. Square centimeters are abbreviated either as sq cm or cm^2.

2. **Volume** is a measure of the capacity of a container, and is also derived from length. The volume of a cube equals the product of three sides. In metric units, volume may be expressed in cubic centimeters (cc) or milliliters (ml). One liter exactly equals 1,000 ml, and is very slightly larger than 1,000 cc—approximately one quart.

3. **Density** is the mass per unit volume of a substance, and may be expressed in kg per cubic meter (kg/m^3).

4. **Specific gravity** has *no units.* It is the ratio of the density of any material to the density of water. The density of water is 1.

5. **Velocity** is speed in a given direction, and can be expressed in cm per sec, or km per sec, or some other convenient units. Speed is commonly expressed in miles per hr (English system).

6. **Temperature** is a measure of the average energy of motion of the molecules of matter. Two systems are commonly employed today. In the metric system temperature is expressed in degrees Celsius (centigrade), but still the Fahrenheit system is the one generally used in the United States. The following data exemplify the differences between these systems:

CELSIUS   0 C = freezing point of water
               100 C = boiling point of water

FAHRENHEIT   32 F = freezing point of water
              212 F = boiling point of water

Conversion from one system to the other can be achieved by reference to tables, or by use of the following equations:

$$C = 5/9 \ (F - 32)$$
$$F = 9/5 \ C + 32$$

Prefixes Applied to SI Units

To simplify the designation of quantities of various sizes, the SI introduced a number of prefixes for standard and derived units. Table 2.1 gives the prefixes for fractions and multiples of units employing the power-of-ten concept. Note that a unit with a smaller prefix always has a larger associated number, than a unit with a larger prefix to express the same value. Thus, 1000 mm = 1 m. (Obviously, 1000 is larger than 1, but mm is 1000 times smaller than m.) These prefixes are of interest to radiologic technologists to a limited degree, but do apply more often in radiotherapy physics.

TABLE 2.1
POWERS-OF-TEN PREFIXES FOR SUBDIVISIONS AND
MULTIPLES OF DEFINED SI UNITS

Subdivisions			*Multiples*		
		$(10^0 = 1)$			
deci-	(d)	10^{-1}	deka-	(da)	10^1
centi-	(c)	10^{-2}	hecto-	(h)	10^2
milli-	(m)	10^{-3}	kilo-	(k)	10^3
micro-	(μ)	10^{-6}	mega-	(M)	10^6
nano-	(n)	10^{-9}	giga-	(G)	10^9
pico-	(p)	10^{-12}	tera-	(T)	10^{12}
femto-	(f)	10^{-15}	peta-	(P)	10^{15}
atto-	(a)	10^{-18}	exa-	(E)	10^{18}

QUESTIONS AND PROBLEMS

1. Discuss the terms science; law; theory.
2. Why are standard units necessary?
3. Define natural science; physics; radiologic physics.
4. Which branch of natural science underlies all the others?
5. How were standard units in the metric system established?
6. With what do the fundamental units deal?
7. Discuss the metric system?
8. Explain the relation between mass and weight. Define the unit of mass.
9. What is the unit of time?
10. State the approximate equivalent of 1 meter in the English system? One gram? One liter?

11. The temperature of a solution is 68°F. Find the equivalent temperature in centigrade units.
12. A patient is found to have a temperature of 40°C. What is his temperature in Fahrenheit units?
13. Find the temperature value which is the same in both the °C and °F systems. (There are two methods, graphic and algebraic.)

Chapter 3

THE PHYSICAL CONCEPT OF ENERGY

Force

ALL MATTER is endowed with a property called *inertia*. This may be defined as the tendency of a resting body (of matter) to remain at rest, and the tendency of a body moving at constant speed in a straight line to continue its state of motion. To set a resting object in motion, we must push or pull it; that is, we must apply a *force*. Conversely, to stop a moving object or change its direction, we must apply an opposing force. It should be pointed out that the *mass of a body is a measure of its inertia:* the greater the mass, the greater the external force needed to change its motion.

Work and Energy

Whenever a force acts upon a body over a distance, *work* is done. For example, to lift an object, we must apply a force to overcome the force of gravity. This applied force multiplied by the distance through which the object is lifted equals the work done. If the object is lifted twice the distance, then twice as much work will be done. Thus,

$$\text{work} = \text{force} \times \text{distance}$$

In a physical sense, work results from the expenditure of *energy*. What, then, is energy? *Energy may be defined as the actual or potential ability to do work.* It is obvious that work and energy must be measured in the same units, since the work done must be equal to the available energy.

There are two main types of *mechanical* energy, that is, the energy involved in the operation of various types of machines:

1. **Kinetic Energy.** Every moving body can do work because of its motion. The energy of such a moving body is called *kinetic energy.* The kinetic energy (K.E.) of a body is expressed by

$$K.E. = \tfrac{1}{2}mv^2$$

where m is the mass of the body in kg, and v is the velocity in m per sec. The unit of work is the joule (jool).

2. **Potential Energy.** A body may have energy because of its position or its temporarily deformed state. For example, a car parked on a hill will start to roll down once its brakes have been released. It is logical to assume that at the instant before the brakes were released the car possessed stored energy, which becomes energy of motion, or K.E., as the car rolls downhill. A wound-up clock spring and a stretched rubber band (temporarily deformed) are further examples of stored or potential energy (P.E.).

Thus, there are two types of mechanical energy—kinetic energy (energy of movement) and potential energy (stored energy).

Law of Conservation of Energy

What is the relationship between potential energy and kinetic energy? This is readily demonstrated by a simple example. Consider a rock poised on top of a cliff (see Figure 3.1). The rock has a certain amount of stored or potential energy. Where did this energy originate? A definite amount of work would have to be done to lift the rock to its position on top of the cliff against the force of gravity, such work being "stored" in the rock as *potential energy.* Now suppose the rock is shoved over the edge of the cliff; as it falls, its potential energy is gradually converted to kinetic energy. As the potential energy decreases, the kinetic energy increases correspondingly, so that at the instant the rock reaches the ground its initial potential energy has been completely changed to kinetic energy.

initial energy		final energy
stored in rock	=	of rock in motion
(P.E.)		(K.E.)

Notice that the rock has different values of potential energy along its course; in other words, its potential energy gradually decreases as it falls toward the foot of the cliff. If the potential energy at any point above the ground is called an *energy level,* then it follows that the energy levels decrease at successively lower points. This important concept will be applied again later in the discussion of the energy levels of the electron shells in the

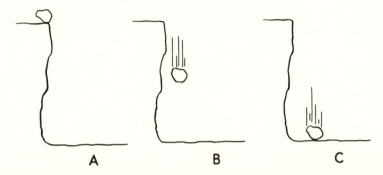

A   B   C

Figure 3.1. The relationship of potential energy to kinetic energy. In *A* the rock is at the top of the cliff, *external work* being required to lift it there against the force of gravity. This work is stored in the rock as *potential energy*. In *B* and *C* the rock is falling and the potential energy is thereby converted to *kinetic energy*. The external work = the potential energy = the kinetic energy.

atom. As exemplified in Figure 3.2, the *difference between any two energy levels equals the amount of potential energy that has been changed to kinetic energy. The rock will not of its own accord move from a lower to a higher energy level; external work is required.* This is another way of saying that *matter will move only from a higher to a lower energy level in the absence of outside work.*

It should be emphasized that the work required to lift the rock to the top of the cliff equals the potential energy of the rock at the summit of the cliff; and this, in turn, equals the kinetic energy attained by the rock at the instant it reaches the ground. At the instant the rock strikes the ground its kinetic energy is converted to heat.

The foregoing discussion exemplifies the *Law of Conservation of Energy*, which states that energy can be neither created nor destroyed, the total amount of energy in the universe being constant. However, energy in one form can be changed into an equivalent amount of energy in other forms.

What are the so-called *"forms of energy?"* They may be listed simply as mechanical, thermal (heat), light, electrical, chemical, atomic, molecular, and nuclear. These can be changed from one form to another. For example, an electric motor converts electrical energy to mechanical energy which is almost equal to the initial electrical energy, any difference being due to the waste of some of the energy due to the production of heat.

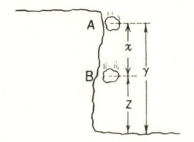

Figure 3.2. Potential energy at A − potential energy at B = kinetic energy at B; or, energy level at A − energy level at B = kinetic energy at B. If potential energy at A = y, and potential energy at B = z, then the difference in energy levels is y − z = x. Thus, x represents the amount of potential energy converted to kinetic energy when the rock drops from A to B.

$$\frac{\text{electrical}}{\text{energy}} = \frac{\text{mechanical}}{\text{energy}} + \frac{\text{heat}}{\text{energy}}$$

Other examples of the transformation of energy include: the conversion of chemical to electrical energy in a battery; the conversion of heat to mechanical energy in a steam engine; and the conversion of chemical to mechanical energy in a gasoline engine. Each of these transformations obeys the Law of Conservation of Energy; any energy disappearing in one form appears in some other form, and is never destroyed.

The most remarkable transformation of all, that of matter and energy, was formulated by Einstein in his famous equation

$$E = mc^2$$

in which E is energy, m is mass, and c is the speed of light in a vacuum. According to this equation, mass and energy are mutually convertible, one to the other. Thus, *matter cannot be destroyed, but can be changed to an equivalent amount of energy, and vice versa.* In fact, an extremely small amount of matter can give rise to a tremendous amount of energy. On this basis, we may expand the conservation law as follows: the sum total of mass and energy in the universe (or in a closed system) remains constant.

QUESTIONS

1. What is the physical concept of work?
2. Define energy; force.

3. Discuss kinetic energy and potential energy. State the equation for kinetic energy and identify the terms and their units.
4. Name the various forms of energy.
5. Define the law of conservation of energy and explain with the aid of an example.
6. What is the source of electrical energy in a battery?
7. Suppose that 1000 energy units of heat are applied to a steam turbine. What is the maximum number of energy units of electricity that can be obtained, neglecting the loss of energy due to friction?
8. Discuss the concept of energy levels.
9. Explain the significance of $E = mc^2$.

Chapter 4

THE STRUCTURE OF MATTER

S PECULATION ON THE BASIC STRUCTURE OF MATTER has occupied the mind of man since ancient times. In fact, a Greek philosopher, Democritus, in about 400 B.C., actually suspected that matter is composed of invisible and indivisible particles, which he called *atoms* (from Greek *atomos* = uncut).

One of the methods of studying something is to break it down, a process called *analysis*. For instance, the most satisfactory way to ascertain the mechanism of a clock is to take it apart, study its individual components, and find their relationship to one another. If this has been done properly it should then be possible to reassemble the component parts into a perfectly operating clock. Such a recombination of parts is called *synthesis*.

SUBDIVISIONS OF MATTER

Let us now apply the process of analysis, based on logic and experiment, to the structure of matter. By definition, *matter is anything which occupies space and has inertia.*

Matter is usually found in nature in impure form, as a *mixture* of indefinite composition. An example is *rock salt,* a mixture of certain minerals (sodium chloride, calcium carbonate, magnesium chloride, calcium sulfate, etc) that can be separated in pure form. One of these constituents is ordinary salt — sodium chloride. If a piece of pure salt is now broken into smaller and smaller fragments, there will eventually remain the tiniest conceivable particle that is still salt. This particle, much too small to be visible, is called a *molecule* of salt. *A molecule is the smallest subdivision of a substance having the characteristic properties of that substance.*

The degree of attraction among the molecules of a given body of matter determines its *state: solid, liquid,* or *gas.* Molecular attraction is strongest in solids, relatively weak in liquids, and weakest in gases.

How can a *molecule* be further subdivided? This cannot be done by ordinary physical means, such as crushing, because the salt molecule is made up of two *atoms* — one of the element sodium and one of the element chlorine — held together by a strong electrochemical force called a *bond* (see Figure 4.1). *The atom is the smallest particle of an element that has the characteristic properties of that element.* Diameters range from about 0.1 to several tenths nm for various atoms (1 nm = 10^{-9} meter or 1 billionth of a meter).

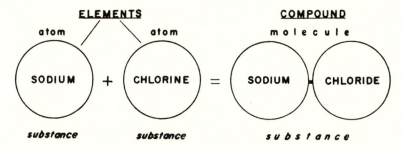

Figure 4.1. Atoms of certain elements are held together by strong electrochemical forces called *bonds,* to form compounds. Elements and compounds are both *substances* because they have a definite, constant structure.

Two new terms have been introduced in the above discussion — *substance* and *element.* These will now be defined. *A substance is any material that has a definite, constant composition,* such as pure salt. (Wood and air are not substances because their composition varies; they are examples of mixtures.) Substances may be *simple* or *complex.* The simple ones, called *elements, cannot be decomposed to simpler substances by ordinary means.* Examples of elements are sodium, iron, lead, oxygen, hydrogen, and chlorine. There are in all ninety-two such naturally occurring elements (see Table 4.1 on pages 46–47).

The complex substances are *compounds,* formed by the *chemical union of two or more elements in definite proportions.* Thus, as we have just seen, salt is a *compound* of equal numbers of atoms of the elements sodium and chlorine in chemical combination (see Figure 4.1); every molecule of salt contains one atom of sodium (Na) and one atom of chlorine (Cl). Thus, the compound is called sodium chloride, symbolized by NaCl. Note that *all elements and compounds are also substances* since they have a definite composition. Compounds may consist of combinations of many atoms, ranging into the thousands.

The commonly occurring gases such as oxygen (O_2), nitrogen (N_2), and hydrogen (H_2) occur in nature in the form of molecules consisting of two atoms. Hence they are known as *diatomic* (two-atom) gases.

Figure 4.2 illustrates the structure of matter down to the atomic level.

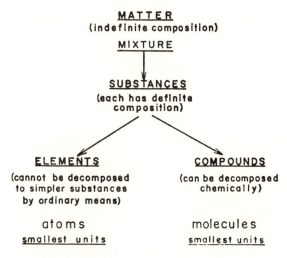

Figure 4.2. Schematic analysis of matter.

In summary then, rock salt is an indefinite *mixture* of *substances* each of which has a definite composition. Pure salt is both a *compound* and a *substance* because it has a definite composition, chemically combined atoms of sodium and chlorine in equal numbers. The smallest particle of salt (as well as other compounds) is a *molecule,* which can be separated chemically into simpler substances—*elements.* Thus, each molecule of salt consists of an atom of sodium and an atom of chlorine (see Figure 4.1). The *atom* is the smallest fragment of a particular element that is still recognizable as such. Atoms cannot be further subdivided by ordinary chemical or electrical methods, but can be broken down into smaller particles by special high energy, atom-smashing machines such as the nuclear reactor and the cyclotron.

Atomic Structure—the Electrical Nature of Matter

Let us now examine the makeup of the atom itself. An atom is not a homogeneous particle, but consists of even smaller particles that are present in the atoms of all elements. Although scientists and mathematicians have given us a highly satisfactory concept of atomic structure, it remains tentative and is undergoing almost continuous modification.

Modern theory has it that the atom is fundamentally *electrical* in nature. Note that the structure of the atom may be represented by a diagram (not an actual picture) to help correlate certain experimental observations that have been treated mathematically.

In 1913 Niels Bohr, a Danish physicist, first proposed the most widely held theory of atomic structure. Although it has been radically revised since his time, his early model of the atom adequately serves to explain the phenomena that we shall encounter in radiologic physics in this course. We shall therefore represent the atom as a submicroscopic system analogous to our solar system in which planets revolve around a central sun (see Figure 4.3). In the center of the atom lies the positively charged core known as the *nucleus,* containing almost the entire mass of the atom. Revolving around the nucleus are the much lighter *orbital electrons,* each carrying a single *negative charge.* Each electron moves continuously in its own *orbit* or path around the nucleus. However, the electrons whose orbits are at a particular distance from the nucleus are grouped together and designated as belonging to a particular *shell* or *energy level.* These shells are identified by letters of the alphabet, the innermost being called the *K* shell, the next the *L* shell, and so on to *Q* (see Figure 4.3). The atoms of various elements differ in the number of electrons that are normally present in the shells.

There is a specific maximum number of electrons that can occupy a particular shell, listed as follows:

K shell	2 electrons
L shell	8 electrons
M shell	18 electrons
N shell	32 electrons

A shell may contain fewer than the maximum allowable number of electrons; thus, starting with the *M* shell, the outermost shell does not have to contain its maximum complement of elec-

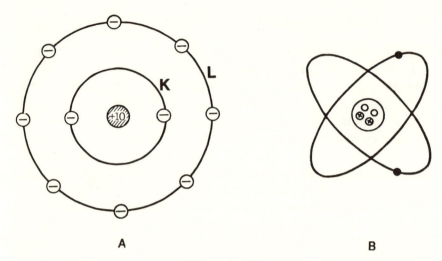

A   **B**

Figure 4.3. *A* shows a greatly simplified atomic model, according to the Bohr Theory, with the nucleus at the center surrounded by some of the *shells* of the electrons. The *K* shell can hold no more than two electrons, *L* no more than eight. The number of electrons that additional shells (up to *Q*) can hold has been determined. The electrons revolve very rapidly in their orbits and, at the same time, spin on their axes. Actually, each electron moves in its own path or *orbital* as in *B,* where the nucleus is shown to contain, in this case, two protons (⊕) and two neutrons (○).

trons before electrons begin appearing in the next higher shell. In fact, an atom tends to be chemically stable when there are eight electrons or an *octet* in the outer shell. For example, the element potassium has two *K* electrons, eight *L* electrons, eight *M* electrons, and one *N* electron, despite the fact that the *M* shell can hold a maximum of eighteen electrons.

In a neutral atom the total number of orbital electrons exactly equals the total number of positive charges in the nucleus.

All of the atoms of a given element have the same total number of positive nuclear charges, but the atoms of different elements have different total nuclear charges. *An atom of a given element maintains its identity only if its nuclear charge is unaltered.* If the nuclear charge is altered, as by irradiation with subatomic particles, the element may be transmuted (changed) into an entirely different element.

Nuclear Structure. Of what does the *nucleus* of an atom consist? We know that the nucleus contains two main kinds of particles or *nucleons* — protons and neutrons. The proton carries a single positive charge (equal and opposite to the negative charge of an electron) and has an extremely small mass, about 1.6×10^{-24} gram, but this is still about 1,828 times the mass of an electron. Ordinarily, protons exist only in the nucleus and can be liberated from the nucleus only under special conditions. Protons of all elements are identical, but **each element has its own characteristic number of nuclear protons.** It must be emphasized that the protons alone give the nucleus its positive charge.

The other important constituent of the nucleus, the **neutron,** is a particle having about the same mass as a proton but carrying no charge; that is, the neutron is electrically **neutral.** (It is incorrect to speak of the neutron as having a neutral charge; rather, it is characterized by absence of charge.) Thus, protons and neutrons compose virtually the entire mass of the nucleus but **only the protons contribute positive charges to the nucleus.** Other nuclear particles have been discovered but will not be considered here.

In summary, we can state that the atoms of various elements consist of the same **building blocks:** protons, neutrons, and electrons. Atoms of various elements differ from one another by virtue of different combinations of these same building blocks. The structure of an atom may be outlined as follows:

1. *Nucleus* — contains one or more:
   a. **Protons** — elementary positive particles with mass about 1.67×10^{-24} g, and diameter about 10^{-13} cm.
   b. **Neutrons** — elementary neutral particles having virtually the same mass as the proton.
2. *Shells* — represent energy levels containing one or more:
   a. **Electrons** — elementary negative particles with mass about 9.11×10^{-28} g, and diameter about 4×10^{-13} cm.

The number of electrons in the shells must equal the number of protons in the nucleus of a **neutral** atom.

Atomic Number

What determines the identity of an element and makes it distinctly different from any other element? The answer lies in the number of nuclear **protons.** In a neutral atom the number of orbital electrons exactly equals the number of nuclear protons (see

Figure 4.3). This number is characteristically the same for all atoms of a given element and is different for the atoms of different elements. *The number of protons or positive charges in the nucleus of an atom denotes its atomic number.* Thus, each element has a particular atomic number and, because of the corresponding specific number and arrangement of orbital electrons, its own distinctive chemical properties.

Mass Number

What determines the mass of an atom? We have already indicated that nearly the entire mass resides in the nucleus and, more specifically, in the nuclear protons and neutrons. These particles comprise the units of atomic mass. Therefore, the *total number of protons and neutrons in the nucleus of an atom denotes its mass number* or *atomic mass* (see Figure 4.4). Protons and neutrons are often called *nucleons.*

NUCLEUS

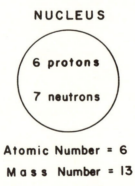

6 protons

7 neutrons

Atomic Number = 6

M a s s Number = 13

Figure 4.4. Nuclear composition (simplified).

Isotopes

If we were to examine a sample of a particular element, we would find that it consists of two or more kinds of atoms that differ in mass number, even though they must all have the same atomic number. For example, the familiar element oxygen is a mixture of atoms of mass number 16, 17, and 18, although all these oxygen atoms have the same atomic number — 8. Such atoms of the same element having different mass numbers are called *isotopes*. Any particular kind of atom, having a specific number of protons and neutrons, is called a *nuclide*.

Hydrogen, the simplest element, has three isotopes, as shown in Figure 4.5. All three hydrogen nuclides have atomic number 1, but they have different mass numbers — 1, 2, and 3. The reason for the difference is immediately evident from the figure; *they differ only in the number of nuclear neutrons.*

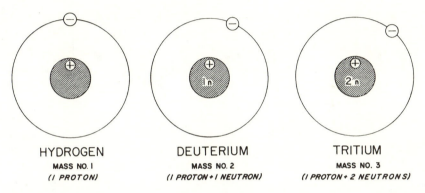

HYDROGEN	DEUTERIUM	TRITIUM
MASS NO. I	MASS NO. 2	MASS NO. 3
(I PROTON)	*(I PROTON + I NEUTRON)*	*(I PROTON + 2 NEUTRONS)*

Figure 4.5. Isotopes of hydrogen. These particular isotopes have the same atomic number, 1, but have different mass numbers. They are called *isotopes* of the same element, hydrogen. Their different mass numbers depend on the different numbers of neutrons in the nucleus. Ordinary hydrogen (mass no. 1) is sometimes called *protium*. Tritium (pronounced trit'-ē-um) is an artificial isotope and is radioactive.

The preceding discussion may be generalized as follows: *isotopes may be defined as atoms that have the same number of nuclear protons (equal to the atomic number of the element) but different numbers of nuclear neutrons.* This definition explains the difference in the mass numbers of isotopes.

Note that the atomic number and the mass number *together* specify the nuclear composition of any given atom, thereby designating a particular *nuclide;* the atomic number alone specifies the number of nuclear protons, and hence identifies the element.

Atomic weight refers to the mass of any atom relative to the mass of an atom of carbon 12 isotope taken as 12. The *atomic mass unit (amu)* is therefore defined as one-twelfth the mass of the

TABLE 4.1
ATOMIC PERIODIC TABLE

The numbers above each element represent atomic weight.
The numbers below each element represent atomic number.
Elements of particular interest to radiologic technologists are in dark type.
Transuranic elements (that is, beyond uranium) are man made.

Period	Group I	Group II	Group III	Group IV	Group V	Group VI	Group VII	Group VIII		
0	1.008 *Hydrogen* *(H)* 1									4.003 *Helium* *(He)* 2
1	6.940 Lithium (Li) 3	9.02 Beryllium (Be) 4	10.82 Boron (B) 5	12.01 *Carbon* *(C)* 6	14.008 Nitrogen (N) 7	16.00 *Oxygen* *(O)* 8	19.00 Fluorine (F) 9			20.183 Neon (Ne) 10
2	22.997 Sodium (Na) 11	24.32 Magnesium (Mg) 12	26.97 *Aluminum* *(Al)* 13	28.06 *Silicon* *(Si)* 14	30.98 Phosphorus (P) 15	32.06 Sulfur (S) 16	35.457 Chlorine (Cl) 17			39.994 Argon (A) 18
3	39.006 Potassium (K) 19	40.08 *Calcium* *(Ca)* 20	45.10 Scandium (Sc) 21	47.90 Titanium (Ti) 22	50.95 Vanadium (V) 23	52.01 *Chromium* *(Cr)* 24	54.93 Manganese (Mn) 25	55.85 *Iron* *(Fe)* 26	58.94 *Cobalt* *(Co)* 27	58.69 *Nickel* *(Ni)* 28
	63.57 *Copper* *(Cu)* 29	65.38 Zinc (Zn) 30	69.72 Gallium (Ga) 31	72.60 Germanium (Ge) 32	74.91 Arsenic (As) 33	79.00 Selenium (Se) 34	79.92 Bromine (Br) 35			83.7 Krypton (Kr) 36
4	85.48 Rubidium (Rb) 37	87.63 Strontium (Sr) 38	88.92 Yttrium (Y) 39	91.22 Zirconium (Zr) 40	92.91 Niobium (Nb) 41	96.0 *Molybdenum (Mo)* 42	99 *Technetium* *(Tc)* 43	101.7 Ruthenium (Ru) 44	102.9 Rhodium (Rh) 45	106.7 Palladium (Pd) 46
	107.88 *Silver* *(Ag)* 47	112.41 Cadmium (Cd) 48	118.70 Indium (In) 49	121.77 *Tin* *(Sn)* 50	127.6 Antimony (Sb) 51	126.93 Tellurium (Te) 52	126.92 *Iodine* *(I)* 53			131.3 Xenon (Xe) 54
	132.9 *Cesium* *(Cs)* 55	137.4 *Barium* *(Ba)* 56	Rare Earths 57–71	178.6 Hafnium (Hf) 72	180.9 Tantalum (Ta) 73	183.9 *Tungsten* *(W)* 74	186.3 *Rhenium* *(Re)* 75	190.2 Osmium (Os) 76	193.1 *Iridium* *(Ir)* 77	195.2 *Platinum* *(Pt)* 78

| 5 | 197.2 Gold (Au) 79 | 200.6 Mercury (Hg) 80 | 204.4 Thallium (Tl) 81 | 207.2 Lead (Pb) 82 | 209.0 Bismuth (Bi) 83 | 210 Polonium (Po) 84 | 211 Astatine (At) 85 | 222 Radon (Rn) 86 |
| 6 | 224 Virginium (Vi) 87 | 226.05 Radium (Ra) 88 | Actinide° Series 89–103 | | | | | |

°Actinide Series

227 Actinium (Ac) 89	232 Thorium (Th) 90	231 Protoactinium (Pa) 91	238.1 Uranium (U) 92
(237) Neptunium (Np) 93	(242) Plutonium (Pu) 94	(243) Americium (Am) 95	(247) Curium (Cm) 96
(249) Berkelium (Bk) 97	(251) Californium (Cf) 98	(254) Einsteinium (Es) 99	(257) Fermium (Fm) 100
(256) Mendelevium (Md) 101	(254) Nobelium (No) 102	(257) Lawrencium (Lw) 103	

(Number in parenthesis is mass number of most stable nuclide.)

carbon 12 nucleus. Because elements consist of mixtures of isotopes, the atomic weight is an average and is therefore almost never a whole number. However, since atomic weight, and mass number, are so nearly equal, these terms are often used interchangeably.

The Periodic Table

The preceding discussion will be made clearer by actual examples. All of the elements can be arranged in an orderly series, called the *periodic table*, from the lightest to the heaviest (on the basis of atomic weight), or from the lowest atomic number to the highest (see Table 4.1). For instance, ordinary *hydrogen* is the simplest element, since it has only one proton and one orbital electron. The atomic number of ordinary hydrogen is therefore 1, and so is its mass number (see Figure 4.6).

The second element in the periodic series is *helium*, with two protons and two neutrons in the nucleus and two electrons in the *K shell* (see Figure 4.7). The atomic number of this element is 2 and the mass number is 4.

Lithium is the third element. Its atomic number is 3, and one of its isotopes has a mass number of 7 (see Figure 4.8). Note that the first shell has the maximum number of electrons, two, and the third electron is in the next shell. Thus, the *K* shell of lithium has two electrons, and the *L* shell, one.

Chemical Behavior

In tabulating the elements in order of increasing atomic number (periodic table), we find that they fall into *eight vertical groups* and *seven horizontal periods* (see Table 4.1). It turns out that the *vertical groups* represent families of elements with surprisingly similar chemical properties. In other words, they participate in chemical reactions in a similar fashion. This relationship of the various elements was first discovered in 1870 by Mendeleev, the eminent Russian chemist. The *periods* (horizontal rows), on the other hand, consist of elements with the same number of electron shells, but with different chemical properties.

Why do the elements in any vertical group have similar chemical properties? These depend on the number of electrons in their *outermost shells.* For instance, lithium (Li), sodium (Na), and

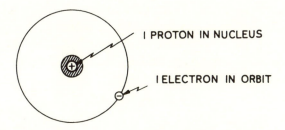

Figure 4.6. Atomic model of ordinary hydrogen.

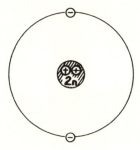

Figure 4.7. Atomic model of helium; two protons in the nucleus and two electrons in K shell.

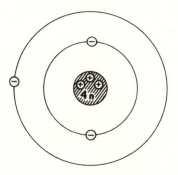

Figure 4.8. Atomic model of lithium.

potassium (K), shown in Table 4.1, belong to the same family—Group I. Their atomic structure is shown in Figure 4.9. The neutrons are omitted here because they do not influence chemical behavior. The diagram shows that although these elements have different numbers of shells, they have one thing in common: *the outermost shell of each has one electron* which it can readily give up, leaving the atom with eight electrons in its outermost shell— the stable *octet* (eight) configuration. This determines the similarity of their chemical behavior.

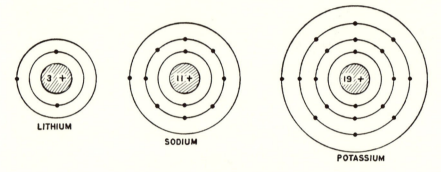

Figure 4.9. Elements with valence of +1.

Consider now vertical group VII: fluorine (F), chlorine (Cl), and bromine (Br). As shown in Figure 4.10, these elements have different numbers of shells, but are similar in that they all have one less electron in the outermost shell than the number of electrons needed to saturate it to a chemically stable octet configuration. Therefore, these elements resemble each other in their chemical properties.

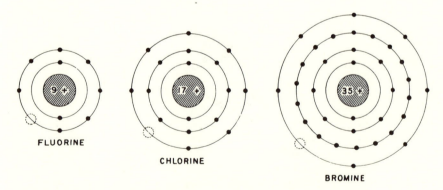

Figure 4.10. Elements with valence of −1.

The number of electrons in the outermost shell determines their combining ability or *valence*. The family of elements represented by lithium, sodium, and potassium has one electron in the outermost shell, and the valence is +1, because these elements can *give* that electron to an element which needs an electron to saturate its outermost shell and form an octet. On the other hand, the elements in the fluorine, chlorine, bromine family, as described above, have one less electron in the outermost shell than the number needed to form an octet. Their valence is −1. These elements can *accept* an electron from the lithium family. The atom that gives up the electron now has one excess positive charge, whereas the atom that accepts the electron has one excess negative charge. Such charged atoms are called *ions* which, because of their opposite charges, firmly attract each other to form a *compound*. The attraction between the two ions is a *chemical bond*, this type being called an *ionic bond* (see Figure 4.11).

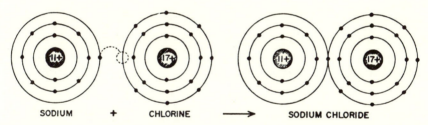

SODIUM   +   CHLORINE   ⟶   SODIUM CHLORIDE

Figure 4.11. Combination of sodium and chlorine chemically to form sodium chloride. The chemical bond is *ionic*, common to the formation of almost all inorganic compounds.

Another type of bond is the *covalent bond* which involves the *sharing* of outer orbital electrons. This is exemplified by the water molecule which consists of two atoms of hydrogen bonded to one atom of oxygen by a sharing of the hydrogen valence electron as shown in Figure 4.12. With this type of bond a shared electron divides its time orbiting around the hydrogen nucleus and the oxygen nucleus, slightly more around the latter. Therefore, the oxygen side of the water molecule is slightly negative with respect to the oxygen side, so we call this *polar covalent bonding*.

Suppose a particular element normally were to have its outer shell saturated with electrons, or in octet configuration. It could not enter into a chemical reaction, since it would be unable to give up, accept, or share an electron. Referring again to Table 4.1,

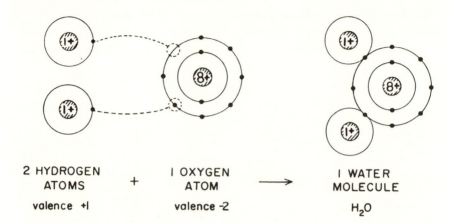

2 HYDROGEN   **I OXYGEN**   **I WATER**
ATOMS + **ATOM** ⟶ **MOLECULE**
valence +I   valence -2 H_2O

Figure 4.12. Chemical combination of hydrogen and oxygen to form water, by sharing of two electrons—that is, a *covalent bond.* Note polarization of molecule—positive on left, negative on right.

notice that there is such a family of elements—Group O—known as the *inert elements.* They include helium, neon, argon, etc, as shown in Figure 4.13. (It has been possible, recently, to bring about chemical union of inert elements in very small quantities.)

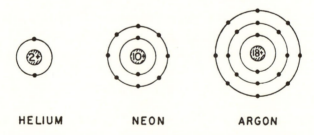

HELIUM   **NEON**   **ARGON**

Figure 4.13. Three members of the family of inert elements, with valence of 0. The outer shell in each case is completely filled with electrons. Therefore, these elements do not ordinarily enter into chemical reactions.

Valence explains why elements always combine in *definite proportions.* Thus, sodium chloride is always formed by the union of one atom of sodium and one atom of chlorine. Water is always formed by the combination of one atom of oxygen with two atoms of hydrogen.

We may conclude from the foregoing discussion that *the chemical properties of an element depend on its valence,* which in turn depends on the number of electrons in the outermost shell. On the other hand, protons and neutrons do not participate in chemical reactions, although they go along for the ride, as it were.

Ionization

If an electron is removed from one of the shells of a neutral atom, or if an electron is added to a neutral atom, the atom does not lose its identity because it retains the same charge on its nucleus so that its atomic number is unaltered. However, the atom is now electrically charged. Thus, if an electron is *removed*, the atom becomes *positively charged* because one of its nuclear charges is now unbalanced and the atom has an excess of positive charge. If an electron is *added* to a neutral atom, the atom becomes *negatively charged.* Such charged atoms will drift when placed in an electric field, as will be discussed later, and are called *ions* (from Greek *ienai* = to go). The process of converting atoms to ions is termed *ionization.*

Note that *ionization is produced exclusively through the addition or removal of orbital electrons;* the protons do not participate in this process, their role being entirely passive.

1. **Exposure of Matter to X Rays or Gamma Rays.** These may remove electrons from the atoms of matter lying in their path. Such atoms become positively charged, that is, positive ions. The removed electrons may ionize other atoms, combine with neutral atoms to produce negative ions, or recombine with positive ions to form neutral atoms.

2. **Exposure of Matter to a Stream of Electrons.** The fast-moving electrons in an electron beam or the electrons released from atoms by x rays as in (1) may interact with and remove orbital electrons from other atoms. An atom, on losing an electron, now has an excess positive charge, but it is soon neutralized by picking up a free electron. Ionization by x rays or fast electrons is of extremely short duration—the ion become neutralized almost instantaneously.

3. **Spontaneous Breakdown of Radioactive Nuclides.** The radiation emitted by the radium series, including charged particles and gamma rays, can produce ionization. The same applies to other radioactive nuclides.

4. Exposure of Certain Elements to Light. When light strikes the surface of certain metals such as cesium or potassium, electrons are liberated from the atoms and emitted from the surface of the metal. This principle is utilized in the photoelectric cell, which is the main component of the x-ray phototimer.

5. Chemical Ionization. If two neutral atoms such as sodium and chlorine are brought together, the sodium atom with valence +1 gives up its outermost or valence electron to the chlorine atom to fill its outermost shell. The sodium atom becomes a positive sodium ion (deficiency of one electron), while the chlorine becomes a negative chlorine ion (excess of one electron). When the compound sodium chloride is dissolved in water its ions separate so that the water now contains sodium ions (Na+) and chlorine ions (Cl−). If the poles of a battery are connected to electrodes and these are immersed in the solution, the Na+ ions move toward the negative electrode (cathode) as *cations,* while the Cl− ions move toward the positive electrode (anode) as *anions,* as shown in Figure 4.14. On arriving at the *cathode,* each Na+ ion picks up an electron and becomes a neutral sodium atom. Each Cl− ion gives up its extra electron to the *anode* and becomes a neutral chlorine atom. These effects of an electric current are called *electrolysis.*

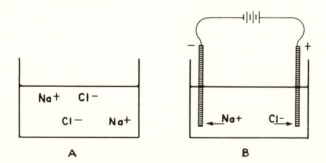

A   B

Figure 4.14. Chemical ionization and electrolysis. In *A* is shown a solution of ordinary salt (sodium chloride = NaCl) whose ions have separated to form equal numbers of Na+ and Cl− ions. In *B* a pair of electrodes has been immersed in the solution; the Na+ ions move toward the negative electrode (cathode) and the Cl− ions drift toward the positive electrode (anode), a process called *electrolysis.*

6. Thermionic Emission. Heating a metal to incandescence (glowing hot) entails the release of electrons from its surface.

These electrons are the loosely bound, outermost orbital electrons of the atoms of the metal; heating the metal imparts sufficient kinetic energy to the electrons to facilitate their escape from the surface, analogous to the evaporation of water molecules from a body of water. The ejection of electrons in this manner is called *thermionic emission,* or *thermionic effect;* it is essential to the operation of an x-ray tube.

Ionization is an extremely important process. Many chemical reactions take place between ions in solution. Ionization of air by x rays underlies the modern measurement of the exposure rate of an x-ray beam. Ionization of body tissues indirectly through preliminary release of electrons underlies the fundamental mode of action of x and gamma rays in therapy.

QUESTIONS AND PROBLEMS

1. Define element; compound; substance.
2. State briefly the Bohr concept of atomic structure.
3. Draw a model of the hydrogen atom and label the parts.
4. What constitutes the nucleus of an atom?
5. Define orbit; shell; octet.
6. How does the mass of a proton compare with that of an electron?
7. A neutral atom has twelve electrons revolving in the orbits around the nucleus. What is its atomic number?
8. An atom has a nucleus containing six protons and two neutrons. What is its atomic number (Z)? Mass number (A)?
9. Define isotope, and give an example. Define nuclide.
10. How do elements combine chemically in fixed or definite proportions?
11. What is an ion? By what methods can ionization be produced?
12. Explain the similarity of chemical behavior of certain elements.
13. Discuss two types of chemical bonds. Give an example of a compound formed by each type.
14. Where are the following located in the atom: proton, neutron, electron? Why are they called the building blocks of matter?

Chapter 5

ELECTROSTATICS

Definition

THE TERM *electrostatics* is defined as *the branch of physics that deals with stationary or resting electric charges.* Another name for resting charges is *static electricity.*

Electrification

We have previously pointed out that all atoms contain electrons in motion about a nucleus. If one or more of these electrons are removed, the atom is left with an excess positive charge. Should the removed electron become attached to a neutral atom, then the latter will become negatively charged. Thus, *there are only two kinds of electric charges, positive and negative.*

The same principle holds for a body of matter (made up of atoms) from which electrons may be removed or to which electrons may be added, a process called *electrification.* So an electrified or charged body has either an excess or a deficiency of electrons, being negatively charged in the first instance and positively charged in the second.

Methods of Electrification

How can we electrify a body? We can do so in three ways:

1. **Electrification by Friction.** You have probably observed after walking over a woolen carpet that touching a metal door knob will cause a spark to jump between your hand and the metal. Again, you may have noted that in combing your hair a crackling sound is heard as the comb seems to draw the hair to it. These are examples of *electrification by friction,* that is, the removal of electrons from one object by rubbing it with another of a different kind. This phenomenon was known to the ancient Greeks who observed that

56

amber, when rubbed with fur, attracts small bits of straw or dried leaves. Since the Greek word for amber is *elektron,* the origin of the root word for *electricity* is obvious.

On the basis of the two kinds of charges developed by friction, we conventionally designate one *negative* and the other *positive.* When a glass rod is rubbed with silk, some electrons are removed from the rod, leaving it positively charged; this is *positive electrification.* At the same time, the silk acquires an excess of electrons and is negatively charged, a process called *negative electrification.*

If amber is rubbed with fur, the amber picks up electrons and becomes negatively charged, thereby undergoing negative electrification, while the fur undergoes positive electrification. It should be emphasized that friction is the simplest and most fundamental method of electrification.

2. **Electrification by Contact.** When a body charged by friction touches an uncharged object, the latter acquires the same charge. This is called *electrification by contact.* If the first object is negatively charged, it will give up some of its electrons to the second object, imparting to it a negative charge. If, on the other hand, the first object has a positive charge, it will remove electrons from the second object, leaving it positively charged. *We can conclude that a charged object confers the same kind of charge on any uncharged body with which it comes in contact.*

3. **Electrification by Induction.** Matter can be classified on the basis of its electrical conductivity, that is, how readily it allows the flow of electrons. The materials in the first group are called *nonconductors* or *insulators* because electrons do not flow freely in them; examples are plastics, hard rubber, and glass. The materials in the second class are called *conductors* because they allow a free flow of electrons; they are exemplified by *metals,* such as silver, copper, and aluminum. Electrification by induction utilizes metallic conductors. Another class, intermediate in conducting ability, comprises the *semiconductors* (see page 225).

Surrounding every charged body there is a region in which a force is exerted on another charged body. This region or zone is called an *electric field.* An uncharged metallic object experiences a shift of electrons when brought into the electric field of a charged object, a process called *electrification by induction.* Note that *only the electrons move.* In Figure 5.1, if the negatively charged rod, *B,* is brought *near* the originally uncharged metal body, *A,* but not touching it, the negative field near *B* will repel electrons on *A* distally, leaving the end of *A* nearest *B* positively charged. The reverse is also true; if *B* were a positively charged rod, the end of *A* nearest *B* would become negatively charged. As *B*, in either case, is withdrawn from the electric field of *A*, the electrons on *A* are redistributed to their original position and *A* is no longer charged.

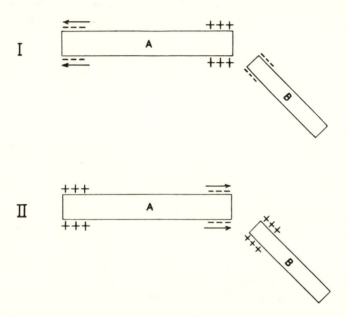

Figure 5.1. Temporary electrification by induction. *A* is a metallic rod. *B* is a charged body. In I electrons on *A* have been repelled to the left by the negative charge on *B*. In II the electrons on *A* have been attracted to the right by the positive charge on *B* (see Figure 5.3).

It is possible to charge a piece of metal by induction so that it will *retain its charge* even after the charging body has been removed. In Figure 5.2, *A* is shown first as it is affected by the electric field of *B*. Connecting the negative end of *A* temporarily to a water pipe (ground) causes the excess electrons to pass down to ground. If the ground connection is broken and *B* is *then* removed, *A* remains positively charged. It must be emphasized that *in electrification by induction, the charged body confers the opposite kind of charge on the metallic body which is placed in its field.*

What is meant by ground? Since the earth is essentially an infinite reservoir of electrons, any charged body can be neutralized

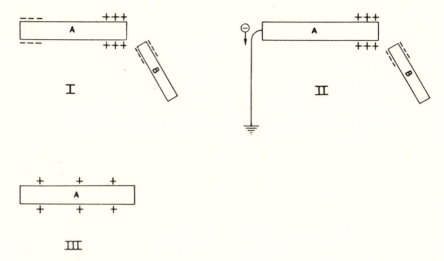

Figure 5.2. Semipermanent electrification by induction. *A* is a metallic rod. *B* is a charged body.

(same number of positive and negative charges) if it is grounded, that is, connected to wet earth by a conductor. Thus, grounding a positively charged body causes electrons to move up from the ground to neutralize the body. On the other hand, grounding a negatively charged body drives its excess electrons to ground, again neutralizing the body. In both instances, the charged body is neutralized by being brought to the same electrical potential energy level as ground. *Ground potential equals zero.* The symbol for ground is ⏚ .

Laws of Electrostatics

There are five fundamental laws of electrostatics:

1. **Like Charges Repel Each Other; Unlike Charges Attract Each Other.** As noted previously, every charge has around it an electric field of force. This field consists of lines of force which start at a positive charge, pass through space, and end on a negative charge (see Figure 5.3). There is a resulting force of attraction between such oppositely charged bodies due to the electric field of force. If, on the other hand, we have two similarly charged bodies, their electric lines of force repel each other as in Figure 5.4 and so there is a repelling or separating

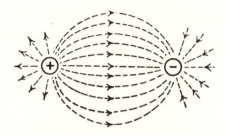

Figure 5.3. Electric fields. The electric lines of force are assumed to start on a positive charge, pass through space, and converge on a negative charge. A negative charge, such as an electron, placed in the field would drift toward the positive charge.

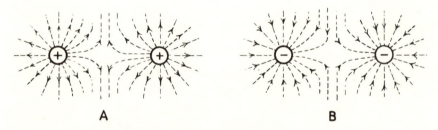

Figure 5.4. When like charges are brought near each other, their electric fields exert a force of repulsion or separation between the like charges.

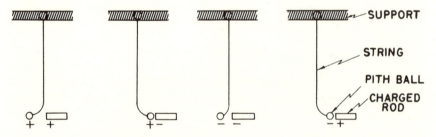

Figure 5.5. Like charges repel. Unlike charges attract.

force between them. Figure 5.5 shows this law demonstrated experimentally.

2. **The Electrostatic Force Between Two Charges is Directly Proportional to the Product of Their Quantities, and Inversely**

Proportional to the Square of the Distance Between Them. The force may be one of attraction or repulsion, depending on whether the charges are different or alike. For example, if we have two unlike charges and the strength of one is doubled, the force of attraction between them is doubled. If both are doubled, the force of attraction between them is 2 × 2 = 4 times as great. If the distance between the original charges is doubled, the force between them is 1/2 × 1/2 = 1/4 as great. If the distance is halved, the force is 2 × 2 = 4 times as great.

3. **Electric Charges Reside Only on External Surfaces of Conductors.** When a hollow metal ball is charged, all of the charges are on the outside surface, while the inside remains uncharged, as is evident from Figure 5.6. This is explained by the mutual repulsion of like charges which tend to move as far away from each other as possible. The greatest distance obviously exists between them if they move to the outer surface of the object.

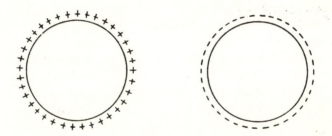

Figure 5.6. Electric charges tend to distribute themselves on the outer surface of a hollow metallic sphere.

4. **The Concentration of Charges on a Curved Surface of a Conductor Is Greatest Where the Curvature Is Greatest.** This is shown in Figure 5.7. The greatest possible concentration of charges occurs on a point which represents an extremely high degree of curvature; in fact, charges become so crowded on a *point* that they easily leak away.

5. **Only Negative Charges (Electrons) Can Move in Solid Conductors.** The positive charges are actually charged atoms which do not drift in solid conductors.

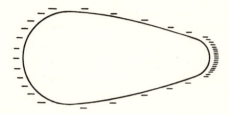

Figure 5.7. Electric charges tend to concentrate in region of sharpest curvature.

Electroscope

An electroscope detects the presence and sign of an electric charge. Under suitable conditions, it can be calibrated to determine the size of a charge. A simple electroscope is shown as a diagram in Figure 5.8.

We can charge an electroscope either by contact or by induction.

1. **Charging an Electroscope by Contact.** Figure 5.9 shows how this is done. If a *negatively* charged rod is allowed to touch the metal knob, as in A, electrons will move from the rod to the knob, stem, and gold leaves of the electroscope. Since the leaves now have the same negative charge, they repel each other and spread apart. The electroscope is now charged *negatively*. On the other hand, if a *positively* charged rod is brought into contact with the metal knob of the electroscope, some electrons will move from the electroscope to the rod, leaving the electroscope with a *positive* charge; the leaves will diverge just as when they were negatively charged.

2. **Charging an Electroscope by Induction.** In Figure 5.10, a negatively charged rod is brought *near*, but not touching, the metal knob of the electroscope. The rod's negative electrostatic field repels electrons from the knob down to the leaves, causing them to separate. If the knob is then grounded while the rod remains near it, the leaves will collapse because they have lost their charge. If the ground connection is next broken and the rod is *then removed*, there will be a deficiency of electrons in the electroscope (the excess electrons having passed to ground), leaving the electroscope with a positive charge. This causes the leaves to spread. In general, the larger the charge, the greater will be the separation of the gold leaves.

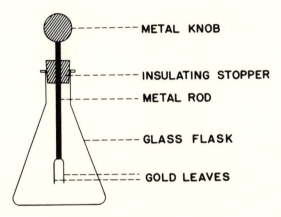

Figure 5.8. Gold leaf electroscope. When an electric charge is imparted to the electroscope, the leaves repel each other and spread apart.

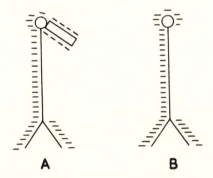

Figure 5.9. Charging an electroscope by contact.

An electroscope can be used not only to detect electric charges, but also to determine their sign, that is, whether they are positive or negative. When a positively charged body is brought near the knob of a negatively charged electroscope, some of the electrons will be attracted from the leaves up to the knob, and the leaves will come together. But when a negatively charged body is brought near the knob of a negatively charged electroscope, some of the electrons from the knob will be repelled down to the already negative leaves and they will diverge even further. We can therefore deduce that a charged body brought near a similarly charged electroscope causes the leaves to diverge farther. But a charged

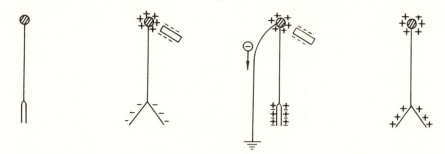

Figure 5.10. Charging an electroscope by induction. The successive steps are indicated in the diagrams from left to right.

body brought near an electroscope that has an opposite charge causes the leaves to come together.

A charged electroscope can be used to detect the presence of x rays because of the ionization they produce in air. The resulting ions neutralize the charge on the electroscope, causing the leaves to collapse.

Static Discharge

Suppose we have two oppositely charged bodies separated by an air space or insulator. The surplus electrons on the negatively charged body naturally tend to move over to the positively charged body, but this may be prevented by the air space or insulator between the bodies. However, if the electrons continue to be piled up on the negatively charged body, the electric field becomes correspondingly stronger. Eventually, continued buildup of electric charge causes breakdown of the insulator as the electrons jump from the more negative to the less negative body, producing a *spark.*

There is a difference of electrical potential energy between two oppositely charged bodies, or between two bodies having like charges of different sizes. The point where electrons are in excess is at a higher negative potential than the point at which they are deficient. Therefore, the electrons move downhill, so to speak, from the point of higher to the point of lower negative potential. This is analogous to the difference in mechanical potential energy in the example cited on pages 34 to 36.

The above discussion can best be brought home by considering the familiar **lightning discharge.** A thunder cloud may have a positive charge on its upper side and a negative charge on its lower side, or *vice versa,* built up by its motion through the air. The earth beneath develops an opposite charge by electrostatic induction. When the difference of electrical potential between a negatively charged cloud and the earth exceeds a certain critical value, electrons rush between the two in the form of a bolt of lightning. Since this usually involves a sudden transfer of excess electrons to the earth, they may rush back toward the cloud, then back to the earth, repeating the process of oscillation many times within a few millionths of a second.

QUESTIONS

1. Discuss electrification. By what three methods can it be accomplished?
2. What happens if two similarly charged bodies are brought close together? Oppositely charged bodies?
3. How many kinds of electricity are there? Which one is capable of drifting in a solid conductor?
4. If a negatively charged object is touched to a neutral object, what kind of charge does the latter acquire?
5. State the five laws of electrostatics.
6. Of what value is the electroscope? How does it work?
7. What is meant by an electric field?
8. Describe electrification by contact; by induction.
9. Explain static discharge and give an example.
10. Define conductor; insulator; semiconductor. Give examples.

Chapter 6

ELECTRODYNAMICS—ELECTRIC CURRENT

Definition

IN THE PRECEDING chapter we described the characteristics of stationary electric charges. Under certain conditions, electric charges can be made to drift through a suitable material known as a *conductor. The science of electric charges in motion is known as electrodynamics, or current electricity.*

The Nature of an Electric Current

Fundamentally, *an electric current consists of a flow of charged particles.* This may occur under the following conditions:

1. **In a Vacuum**—electrons may jump a gap between two oppositely charged electrodes in a vacuum tube. This will be discussed in detail later in connection with x-ray and valve tubes.

2. **In a Gas**—two oppositely charged electrodes placed in a gas will cause positive ions, usually present in the gas, to drift toward the negative electrode (cathode) and negative ions to drift toward the positive electrode (anode). This occurs in an ordinary neon tube.

3. **In an Ionic Solution**—as already described on page 54, upon being dissolved in water a salt separates into its component ions. If electrodes are then immersed in the solution and connected to a source of direct current, the positive ions move toward the cathode while the negative ions move toward the anode. Such migration of ions in opposite directions characterizes the flow of an electric current in an ionic solution (see Figure 4.14).

4. **In a Metallic Conductor**—this will be our main concern in the present chapter, and in later ones as well. *An electric current in a solid conductor consists of a flow of electrons only.* In the atoms of metallic conductors, the valence (outermost) electrons are "free" and, under certain conditions, these free electrons can be made to

drift through the conductor in the so-called conduction band (see pages 225–226). An electron does not move instantaneously from one end of a wire to the other, the individual electron drift being relatively slow — less than 1 mm per sec. However, the electrical impulse or *current* moves with tremendous speed, approaching that of light in a vacuum (3 × 10^8 meters per sec). To make electrons flow, the conductor must have an excess of electrons at one end, and a relative deficiency at the other. The net result is that electrons will then flow from the point of excess (higher negative potential) to the point of deficiency (lower negative potential), just as a body will fall from a point of higher potential energy to a point of lower potential energy (see pages 64–66 and Figure 6.1).

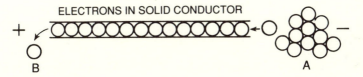

ELECTRONS IN SOLID CONDUCTOR

Figure 6.1. Diagrammatic representation of electron flow in a solid conductor. As an electron enters the conductor at *A* where there is an excess of electrons, another electron leaves at *B* where there is a deficiency of electrons.

An *electric circuit is defined as the path over which the current flows.* Electrons will not flow in a nonconductor such as glass or plastic because of the gap between the valence band and the conduction band (see pages 34–36), and, in semiconductors, they flow under certain conditions. Figure 6.2 shows diagrammatically the components of a simple electric circuit.

Sources of Electric Current

Various methods are available for piling up electrons at one point and simultaneously removing them at another point, thereby setting up a difference of electrical potential and causing an electric current to flow in a conductor. They all require the expenditure of energy as exemplified by the following two main types of devices:

1. *Cells* or *batteries* which convert *chemical energy* to electrical energy.

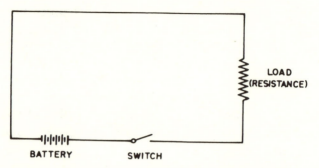

Figure 6.2. Simple resistance circuit. The load represents the total resistance of electrical devices in the circuit.

2. *Dynamo* or *generator* which converts *mechanical energy* to electrical energy.

Other sources of energy include solar (sunlight), atomic (nuclear), wind, and geothermal (underground natural steam). These are in various stages of development and use.

Factors in an Electric Circuit

Three main factors characterize a simple electric circuit, which is carrying a *steady direct current,* that is, a current of constant strength flowing always in the same direction. They include *potential difference, current,* and *resistance.* These factors can best be understood by comparing them with those governing the flow of water through a pipe. They need modification for alternating current (see pages 110–111).

1. **Potential Difference.** We may define potential difference as a *d ifference in electrical potential energy between two points in an electric circuit,* due to an excess of electrons at one point relative to the other. A piling up of electrons at one point and simultaneous removal at another point are brought about through the use of a battery or generator. The potential difference between two points actually represents the amount of *work* expended in moving a unit charge from one point to the other. This resembles the situation shown in Figure 6.3, in which there are two containers of water at different levels connected by a pipe closed by a valve. A difference of pressure exists between the water in both containers, proportional to the difference in height *d* of the water levels. As soon as

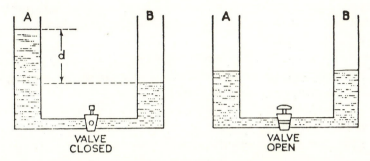

VALVE
CLOSED

VALVE
OPEN

Figure 6.3. Water pressure analogy of electrical potential difference or voltage. With the valve closed, there is a difference of pressure between the water in arms *A* and *B*, proportional to the difference in water levels, *d*. (This is analogous to the potential difference between the poles of a battery on open circuit with no current flowing.) When the valve is opened, water flows from arm *A* into arm *B* until the difference in pressure disappears. (In a closed electric circuit, current continues to flow as long as potential difference is applied.)

the valve is opened, the difference in pressure causes the water to flow from container *A* to container *B* until their levels become equalized and the pressure difference disappears.

Similarly, an excess of electrons at one end of a conductor causes a difference of electrical pressure or a *potential difference* to be set up, *even if the conductor is interrupted by an open switch* as in Figure 6.2. Closing the switch causes electrons to rush from the point of excess (higher negative potential) to the point of deficiency (lower negative potential). Thus, just as with the water analogy, it is *the difference of potential that drives the electrons.* If the difference of potential can be maintained by a battery or generator, electron current will persist as long as the switch is closed.

As the current flows, electrical potential gradually falls along the circuit. Similarly, with water flowing through a pipe, the water pressure decreases gradually along the pipe.

It would be well to review the section in Chapter 3 on the difference of potential energy between two points at different levels above ground, and the movement of a rock from a higher to a lower energy level. The analogy with the difference in water pressure and the difference of electrical potential is self-evident.

The term *electromotive force* (emf) applies to the maximum difference of potential between the terminals of a battery or generator. As noted above, this is not really a force but rather a quantity of work or energy needed to move a unit electric charge through the circuit.

The unit of potential difference (or emf) is the *volt, defined as that potential difference which will maintain a current of one ampere in a circuit whose resistance is one ohm.*

2. Current. The second factor in an electric circuit is *current, defined as the amount of electric charge flowing per second.* (In a water pipe, there is a definite amount of water flowing per second past a given point). Since an electric current consists of a flow of electrons, the more electrons flowing per second the "stronger" is the current. Similarly, the more water flowing per second in a pipe, the stronger is the current of water.

The unit of current is the *ampere, which may be defined as one coulomb quantity of electric charge flowing per second* (ie, 6.25 × 10^{18} free electrons per sec). A current of one ampere will deposit a standard weight of silver per second from a solution of silver nitrate by electrolysis (to be exact, 0.001118 g per sec).

3. Resistance. The third factor in an electric circuit is *resistance,* a property of the materials making up the circuit itself. *Electrical resistance is that property of the circuit which opposes or hinders the flow of an electric current.* As an analogy, there is resistance to the flow of water in a pipe because of the friction of the molecules of water against the molecules of the pipe material. Note that the very large resistance of a nonconductor prevents the flow of current.

The unit of electrical resistance is the *ohm* (symbol Ω, Greek *omega*), defined by the International Electrical Congress of 1893 as follows: *one ohm is the resistance of a standard volume of mercury under standard conditions* (to be exact, it is the resistance to a steady current by a column of mercury weighing 14.45 g, having a length of 106.3 cm and a uniform cross-sectional area, at 0°C).

On what does the resistance of a conductor depend? It depends on four things: material, length, cross-sectional area, and temperature.

a. *Material.* Certain materials, known as conductors, readily allow the flow of electrons. They are best exemplified by metals, such as copper and silver. Although copper is a poorer conductor than silver, it is much cheaper and therefore widely used in electric wires. Other materials, such as glass, wood, and plastic, being nonconductors, have virtually no free electrons and hence offer tremendous resistance to the flow of electricity. Such nonconducting materials are called *insulators* or *dielectrics.* Intermediate in resistance are the *semiconductors* (see pages 57, 225).

b. *Length.* A long conductor, such as a long wire, has more resistance than a short one. In fact, the *resistance is directly proportional to the length of the conductor.* Similarly, the resistance of a pipe to the flow of water increases with its length.

c. *Cross-sectional Area.* A conductor with a large cross-sectional area has a lower resistance than one with a small cross-sectional area. More specifically, *the resistance of a conductor is inversely proportional to the cross-sectional area.* Similarly, a water pipe having a large cross-sectional area offers less resistance than a small pipe.

d. *Temperature.* With metallic conductors, the *resistance becomes greater as the temperature rises.*

The following equation summarizes the determining factors in resistance:

$$R = \rho\frac{L}{A} \qquad (1)$$

where R is the resistance in ohms; ρ is the resistivity, a proportionality factor depending on the material of the conductor and the temperature; L is the length in meters; and A is the cross-sectional area in m^2.

Ohm's Law

Georg Simon Ohm (1787–1854), a German physicist, discovered that when a steady direct current is flowing in a resistance circuit (ie, one in which resistance is the only obstacle to current flow), there is a definite relation between the potential difference (volts), current (amperes), and resistance (ohms). This is expressed in *Ohm's Law,* which states that *the value of the current in a metallic circuit equals the potential difference divided by the resistance:*

$$I = \frac{V}{R} \qquad (2)$$

where I = current in amperes
V = potential difference in volts
R = resistance in ohms

If any two of these values are known, the third may be found by substituting in the equation and solving for the unknown. The letter E is often used instead of V to denote potential difference.

Two rules must be observed in applying this law:

1. When Ohm's Law is applied to a *portion of a circuit,* the current in that portion of the circuit equals the voltage across that portion of the circuit divided by the resistance of that portion of the circuit.

2. When Ohm's Law is applied to the *whole circuit,* the current in the circuit equals the total voltage across the circuit divided by the total resistance of the circuit.

Batteries or Cells

Let us turn now to some of the *chemical devices* that act as sources of electric current. There are two main types, the *dry cell* and the *wet cell.*

1. **Dry Cell.** An ordinary dry cell, such as that used in a flashlight, consists of a carbon rod surrounded by manganese dioxide and immersed in a paste composed of ammonium chloride, zinc chloride, cellulose, and a small amount of water. These ingredients are placed in a zinc can, the top of which is closed by a layer of asphalt varnish (see Figure 6.4).

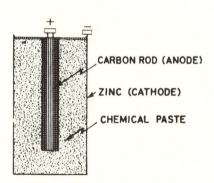

Figure 6.4. Diagram of an ordinary dry cell.

The *alkaline-type* dry cell consists of a manganese dioxide rod immersed in an alkaline potassium hydroxide paste contained in a zinc cylinder. The manganese dioxide rod forms the anode, and the zinc container the cathode. An alkaline cell has a useful life about ten times that of an ordinary dry cell.

Dry cells are actually moist. Note that the central rod nowhere comes in contact with the zinc can. A chemical reaction sets up a potential difference of 1.5 volts between the battery terminals or electrodes; that is, between the zinc and the central rod. The zinc becomes the negative electrode (cathode), while the central rod becomes the positive electrode (anode).

2. **Wet Cell.** An example of the wet cell is the lead storage cell used in automobiles. It consists of a hard rubber or plastic case containing sulfuric acid in which are immersed two electrodes, one lead and the other lead oxide. As the result of a chemical reaction with the sulfuric acid, the lead becomes negative while the lead oxide becomes positive. Each cell in a storage battery produces a potential difference of about 2 volts. Automobile batteries usually have three or six wet cells.

Of special interest is the sealed *rechargeable nickel-cadmium* or *nicad* cell. It has a nickel oxide anode and a cadmium cathode. The plates are each wrapped in a separator and immersed in potassium hydroxide.

Components of Elementary Electric Circuits

Certain fundamental principles having been presented, we can now examine the simplest type of electric circuit; that is, one supplied by a *battery* as the source of potential difference. The current is a *steady direct current* (DC) flowing in *one direction only.* Since the only hindrance to the flow of current in such a circuit is resistance, it is often termed a *resistance circuit.*

Essential Parts. The simplest electric circuit has three components: (1) battery, (2) conductor, and (3) load or resistance. The circuit may be represented by a diagram which shows, incidentally, some of the more common electrical symbols (see Figure 6.2).

Current flows only when the switch is *closed.* If the circuit is broken at any point, it is then called an *open circuit.* When completed it becomes a *closed circuit.* However, keep in mind that there is a potential difference even on open circuit.

Polarity of Circuits. The current in a circuit supplied by a battery has a definite direction, or *polarity.* Most physics books have again come to regard the direction of flow of the current to be from positive to negative. However, for the sake of simplicity, we shall assume the *direction of the current to be the same as that of the flow of electrons; that is, from the cathode (−), through the external circuit, to the anode (+)* (see Figure 6.5).

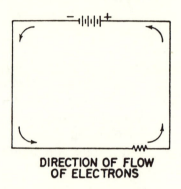

**DIRECTION OF FLOW
OF ELECTRONS**

Figure 6.5. The direction of *electron flow* determines the direction of an electric current. The *conventional direction* is opposite to this.

Connection of Meters. The polarity of DC circuits requires the proper connection of DC measuring devices or meters. Since these devices also have *polarity,* the two terminals on a meter are labeled (+) and (−) and, to avoid damage, must be so connected that the (+) terminal is on the (+) side of the circuit and the (−) terminal on the (−) side.

A *voltmeter* measures, in volts, the potential difference between any two points in a circuit. (Potential difference is often referred to as *voltage drop.*) Therefore, the voltmeter is always connected in *parallel;* that is, it is placed in a small circuit which is a branch of the main circuit, as in Figure 6.6. Obviously, a voltmeter connected successively between various pairs of points in a circuit where the resistances may differ, will give different voltage readings —they will be larger across the higher resistance and smaller across the lower resistance. Similarly, the drop in pressure between any pair of points in a water pipe will be different if the bore of the pipe is not uniform, there being a greater fall in pressure where the pipe is narrower, that is, where there is greater resistance to flow of water.

An *ammeter* (ampere + meter) measures, in amperes, the

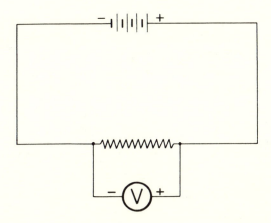

Figure 6.6. Correct connection of a voltmeter—in parallel.

quantity of electric charge flowing per second. It is connected directly into the circuit, that is, in *series* (see Figure 6.7). When so connected, *an ammeter will read the same no matter where it is placed in a series circuit.*

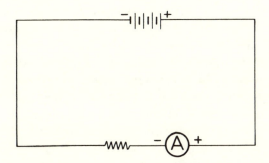

Figure 6.7. Correct connection of an ammeter—in series.

Ammeters and voltmeters look alike externally and consist of a box enclosing the mechanism, with a glass window in which a pointer moves over a calibrated scale, as shown in Figure 6.8. The instrument is always labeled "voltmeter" or "ammeter." The construction of these devices will be described later.

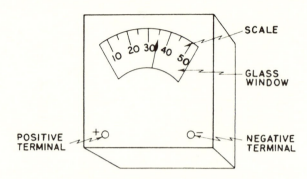

Figure 6.8. Voltmeter or ammeter.

Series and Parallel Circuits

Electric appliances (toasters, flatirons, fans, etc) and electric sources (batteries, generators) may be connected in a circuit in two principal ways. One is called *a series circuit, which may be defined as an electric circuit whose component parts are arranged end-to-end so that the current passes consecutively through each part* (see Figure 6.9). The other arrangement is called a *parallel circuit, wherein the component parts are connected as branches of the main circuit so that the current is divided among them* (see Figure 6.10).

1. Series Circuit. When *current-consuming devices* such as electric appliances are connected in series as shown in Figure 6.9, the total resistance is equal to the sum of the separate resistances. Thus, if R = total resistance of the whole circuit, and r_1, r_2, and r_3 represent the resistances of various devices connected in series, and disregarding the small internal resistance of the battery, then,

$$R = r_1 + r_2 + r_3 \qquad (3)$$

Suppose in this case, r_1 = 1 ohm, r_2 = 2 ohms, and r_3 = 3 ohms, and the ammeter reads 15 amp. (Remember that the current is the same everywhere in a series circuit.) What is the voltage across the *entire* circuit? First substitute the given values of resistance in equation (3),

$$R = 1 + 2 + 3 = 6 \text{ ohms}$$

Then apply Ohm's Law

$$I = V/R$$

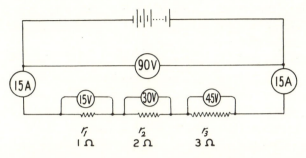

Figure 6.9. Resistors in series. Ohm's Law applies to the entire circuit, and to any portion of it. The total voltage drop, 90 V, is equal to the sum of the voltage drops across all the resistors, 15 V + 30 V + 45 V. The current, 15 amp, is the same everywhere in the circuit. The voltage drop across each resistor is proportional to the resistance, because $V = IR$ and I is constant in this circuit.

or

$$V = IR \qquad\qquad (4)$$

and substitute the values of I and R (obtained above)

$$V = 15 \text{ amp} \times 6 \text{ ohms} = 90 \text{ volts}$$

There is another approach to this problem — the application of Ohm's Law to **each part** of the circuit. Again, recall that the current I remains constant throughout the circuit. Since $V = IR$ the voltage across each resistor in Figure 6.9 is equal to the current times the resistance of each resistor:

$$v_1 = Ir_1 = 15 \times 1 = 15 \text{ volts}$$
$$v_2 = Ir_2 = 15 \times 2 = 30$$
$$v_3 = Ir_3 = 15 \times 3 = 45$$

The total voltage is then

$$V = v_1 + v_2 + v_3$$
$$V = 15 + 30 + 45 = 90 \text{ volts}$$

which is the same result obtained above by applying Ohm's Law to the entire circuit.

What is the real significance of the equation $V = IR$? It actually states that the **voltage across a resistor equals the current times the resistance.** But, besides this, it indicates that there is a fall or drop in potential across a resistor, and as the resistance increases so does the potential drop. Sometimes this is called a voltage drop or, what is the same thing, an *IR drop*. Keep in mind that the terms potential difference (in volts), voltage, voltage drop, and *IR* drop are used interchangeably.

In summary, then, voltage is a measure of the potential difference across a resistor. The voltage drop across the resistor increases as the resistance increases. It is obvious from the above examples that the voltage drop across an entire series circuit must equal the sum of the voltage drops across the resistors.

If *current-producing devices* such as electric cells are connected in series, *the total voltage equals the sum of the individual voltages*. In such a circuit, the positive pole of one cell is connected to the negative pole of the next, as in Figure 6.10. If V is the total voltage, and v_1, v_2, v_3, and v_4 are the voltages of the individual cells, then

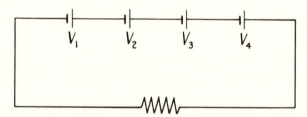

Figure 6.10. A circuit containing batteries in series.

$$V = v_1 + v_2 + v_3 + v_4 \tag{5}$$

The total current in such a circuit can be computed from the total voltage and the total resistance simply by applying Ohm's Law.

2. **Parallel Circuit.** Figure 6.11 shows a parallel circuit; careful comparison of A and B reveals that the circuits are essentially identical, although they have been drawn somewhat differently. The conductors themselves have so little resistance that it may be disregarded. In a parallel circuit the *voltage drop is the same across all branches, regardless of their resistance.* Thus, the voltage drop across r_1 is the same as that across r_2 and that across r_3, and the voltage drop in each case is therefore equal to V, the voltage supplied by the battery. However, application of Ohm's Law to each portion of the circuit will show that the current in each branch is *inversely* proportional to the resistance of that branch. In

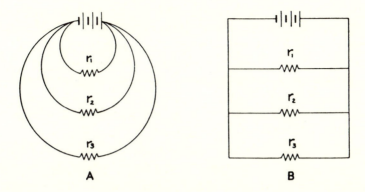

Figure 6.11. *A* and *B* are different methods of representing a parallel circuit. Note that they are fundamentally identical.

other words, the largest current will flow in the branch having the least resistance.

What is the resistance of a circuit with several parallel branches? It is obvious that in such a circuit, the current divides among the several branches or paths, so that the total resistance must be less than that in an individual path. Another way of looking at this is as follows: as more resistors are added in *parallel,* there is an increase in their total cross-sectional area so that the total resistance decreases.

The total resistance of a *parallel circuit* is expressed by the following equation:

$$1/R = 1/r_1 + 1/r_2 + 1/r_3 + \ldots 1/r_n \qquad (6)$$

where R is the total resistance and r_1, r_2, etc are the resistances of the separate branches.

Thus, the total resistance of a parallel circuit can be found by the rule based on equation (6): *the reciprocal of the whole resistance equals the sum of the reciprocals of the separate resistances in a parallel circuit.*

A typical example is presented in Figure 6.12 to clarify this rule. If the resistances in the branches of a parallel circuit are 1, 2, and 3 ohms respectively, what is the total resistance of the circuit? Substituting the values of r in equation (6), we find that the total resistance is less than any of the resistances in the individual branches.

$$1/R = 1/1 + 1/2 + 1/3$$

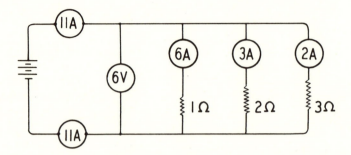

Figure 6.12. Resistors in parallel. The voltage drop, 6 V, is the same across the whole circuit and across each branch. The current differs in the branches, the smallest current flowing in the branch having the largest resistance. The total current, 11 amp, equals the sum of the branch currents: 6 amp + 3 amp + 2 amp. The total resistance is found by $1/R = 1/r_1 + 1/r_2 + 1/r_3$.

$$1/R = \frac{6 + 3 + 2}{6}$$

$$1/R = 11/6$$

Inverting both sides of the equation,

$$R = 6/11 = 0.55 \text{ ohm}$$

Suppose, in the above example, that the potential difference across the circuit is 6 volts. How do we find the current in each branch and in the main lines? Since this is a parallel circuit, the voltage drop across each branch must be the same—6 volts. Then, applying Ohm's Law to each branch,

$$i = v/r$$

$$i_1 = 6/1 = 6 \text{ amp}$$

$$i_2 = 6/2 = 3 \text{ amp}$$

$$i_3 = 6/3 = 2 \text{ amp}$$

The *current in the main lines of a parallel circuit equals the sum of the branch currents,*

$$I = i_1 + i_2 + i_3 \tag{7}$$

Substituting the values of the branch currents just obtained,

$$I = 6 + 3 + 2 = 11 \text{ amp}$$

To verify the answer, we apply Ohm's Law to the entire circuit.

Since the total current is 11 amp and the total resistance is 0.55 ohm,

$$V = IR \qquad\qquad V = 11 \times 0.55 = 6 \text{ volts}$$

As expected, the current divides among the branches so that the **smallest current flows in the branch having the greatest resistance.** Since V is equal in all branches, if r in a given branch is greater, i must be smaller, and *vice versa,* in accordance with $V = IR$.

With **current-producing devices** connected in parallel, the total voltage is the same as that of a single source, if all the sources are alike. Figure 6.13 shows this type of connection. The total current, that is, in the main line, equals the sum of the currents provided by all the sources.

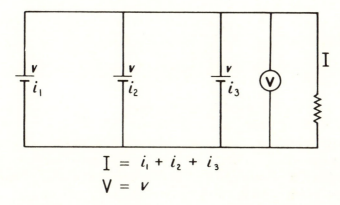

$$I = i_1 + i_2 + i_3$$
$$V = v$$

Figure 6.13. A circuit containing cells arranged in parallel. Total current, $I = i_1 + i_2 + i_3$. Voltage across entire circuit, $V = v$.

The parallel circuit is used in electrical wiring of homes, factories, and other buildings. With a series circuit, electrical failure of one appliance, such as a light bulb, causes all other lights in the circuit to go out. Besides, as more appliances are added in a series circuit, each appliance receives less voltage, since the total voltage drops must be divided among more appliances; thus, as more electric lights are added, all become dimmer.

The main advantage of the parallel circuit is that failure of one appliance does not prevent operation of the others, since they are on branches of the main circuit. As more branches are added, the total resistance decreases and the total amperage increases while the voltage remains unchanged. Thus, the more appliances added, the greater is the amperage in the *main circuit.* Adding too many appliances may cause the amperage in the main line to become excessive and *overload* the circuit. Under these conditions the wiring system could become hot enough to ruin one or more appliances, or start a fire. This danger can be avoided by connecting protective devices such as circuit breakers or fuses in the circuit. A *fuse* contains a wire which melts more easily than the wires of the circuit being protected (see Figure 14.3). When the amperage exceeds the safe maximum, the fuse melts, thereby opening the circuit and stopping the flow of current before the circuit can become overheated. Such a fuse, in which the protective wire has melted, is called a *blown fuse.* Contact between two non-insulated wires carrying a current results in a *short circuit;* there is a marked fall in resistance and a corresponding rise in amperage which is usually sufficient to blow a fuse. The *circuit breaker,* another type of protective device, will be discussed later.

Electric Capacitor (Condenser)

We have already noted that electrical energy can be obtained from a chemical reaction in a dry or wet cell; this may be regarded as *stored electricity.* There is another method of storing electrical energy that does not depend on chemical reaction. You should recall from the chapter on Electrostatics that a charge on a conducting body is distributed entirely on its surface. Such an electrically charged body *stores* electrical energy, provided it is insulated so that the charge does not "leak" away; it is an *electric capacitor* or *condenser.*

One of the simplest types is the *parallel-plate capacitor,* consisting of a pair of flat metallic plates arranged parallel to each other and *separated* by a small space containing air or special insulating material. Capacitor operation is shown in Figure 6.14. When the capacitor is connected to a source of direct current, such as a battery, electrons move from the negative terminal of the battery to the plate to which it is connected. At the same time, an equal number of electrons pass from the opposite plate to the positive terminal of the battery. This is indicated in Figure 6.14A. *No electrons pass from one plate to the other across the space between them,* unless there is a breakdown of the insulator due to application of excessively high voltage. When the capacitor acquires a full charge, the difference of potential (voltage) across its plates reaches a maximum that is equal and opposite in direction to that of the charging current. The current now ceases to flow in the wires.

If the capacitor is then disconnected from the battery, it retains its electrical charge until the plates are connected by a conductor as in Figure 6.14B, whereupon the capacitor discharges and once more becomes neutral. The direction of the discharging current is *opposite* to that of the original charging current, and stops flowing as soon as the capacitor is completely discharged. (With an alternating current the capacitor plates repeatedly change polarity and the current continues to flow.)

CAPACITOR

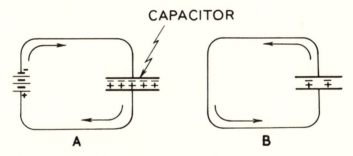

A   **B**

Figure 6.14. Parallel plate capacitor. In *A* the capacitor is connected to a battery and is being charged. In *B* the capacitor has been disconnected and its plates connected by a wire, whereupon it loses its charge. The arrows indicate the direction of electron flow in the wire.

How much electricity can a capacitor store? The quantity of electric charge in coulombs stored per volt applied to the capacitor is called its **capacitance,** the unit of capacitance being the *farad,* a very large unit; a more practical unit is the microfarad (millionth of a farad). The larger the area of the plates, and the smaller the space between them, the greater is the capacitance or electrical storing ability. The insulator or, as it is more often called, the *dielectric* between the plates also determines the capacitance for a given plate size and spacing; if the *dielectric constant* (insulating ability) of air is taken as 1, then wax paper has a dielectric constant of 2, and glass a constant of about 7. These materials will increase the capacitance of a given capacitor two times, and seven times, respectively, as compared with an air dielectric. In practice, a capacitor is usually made up of multiple parallel plates to increase the capacitance.

The Work and Power of a *Direct* Current

Electrical energy, just as any other form of energy, is capable of performing work. According to the Law of Conservation of Energy, a given amount of electrical energy is convertible to a definite amount of work and/or heat.

The **power** of a steady direct current, or the **electric energy expended per second,** is simply the product of the current and voltage:

$$P = IV \qquad (8)$$

power = amperage × voltage
(in **watts**)

The unit of power is the **watt,** 746 watts being equal to 1 horsepower.

It is well known that an electric current produces heat in a circuit. To find the power loss in terms of the heat produced per second, we must convert equation (8) to a form which contains the term resistance, since this remains constant regardless of the voltage or current. Applying Ohm's Law,

$$I = V/R$$
$$V = IR$$

Substituting *IR* for *V* in equation (8),

$$P = I \times IR$$

Power Loss = **I²R watts**   (9)

(heat production/sec)

where *I* is the current in amperes and *R* the resistance in ohms.

Two conclusions may be drawn from equation (9). First, the power loss due to the heating effect of the current is proportional to the resistance; thus, that portion of the circuit having the greatest resistance will sustain the greatest amount of heat production.

Second, the power loss in the form of heat increases very rapidly with an increase in the current. Since **the power loss is proportional to the square of the current,** doubling the current increases the power loss by a factor of 2^2, or four times. Tripling the current increases the power loss by a factor of 3^2, or nine times. The importance of this principle will be discussed later.

QUESTIONS AND PROBLEMS

1. How does an electric current flow along a wire? In a salt solution? Under what other conditions do electrons flow?
2. Describe the two main sources of electric current.
3. Define potential difference. What is the unit of measurement? State the difference between potential difference, electromotive force, and voltage?
4. What is meant by electrical resistance, and on what does it depend?
5. Define electric current strength and state its unit.
6. In which direction does an electric current flow? What is meant by the "conventional" direction?
7. State and explain Ohm's Law.
8. An electric circuit has, in series, appliances with these resistances: 4 ohms, 10 ohms, 6 ohms, 40 ohms. Find the value of the current when the applied potential difference is 120 volts?
9. What voltage is required to produce a current of 30 amperes in a circuit having a resistance of 5 ohms?
10. What type of electric circuit is used in ordinary house wiring? Why?
11. A circuit has the following resistances connected in *parallel:* 3 ohms, 6 ohms, 18 ohms. What is the total resistance of the circuit? With 12 volts applied across the circuit, what is the amperage in each branch *and* in the main lines?
12. State the power formula? Name the unit of power?
13. A circuit has a current of 2 amperes flowing under an impressed electromotive force of 12 volts. Find the power delivered to the circuit.
14. What is the function of an electric capacitor? Describe its construction and explain its operation with the aid of a diagram. Contrast its behavior in a direct current with that in an alternating current.

Chapter 7

MAGNETISM

Definition

M *agnetism may be defined as the ability of certain materials to attract iron, cobalt, or nickel.* In general, any material that attracts these materials is called a **magnet.** The various types of magnets will be described in the next section.

Classification of Magnets

There are three main types of magnets: *natural magnets, artificial permanent magnets,* and *electromagnets.*

1. **Natural magnets** include, first, the *earth* itself, a gigantic magnet which, in common with all others, can deflect the needle of a compass. Another example is the ore *lodestone,* consisting of iron oxide (magnetite) that has become magnetized by lying in the earth's magnetic field for an extraordinarily long period of time. In some parts of the world there are actually magnetic mountains consisting of such ores. The ancient Greeks recognized the magnetism of lodestone more than 2500 years ago.

2. **Artificial permanent magnets** usually consist of a piece of *hard steel* in the shape of a horseshoe or bar which has been *artificially* magnetized so that it is capable of attracting iron (see Figure 7.1). The *magnetic compass* is simply a small permanent magnetic needle swinging freely on a pivot at its center, used to detect the presence of magnetic materials and to locate the earth's magnetic poles. A widely used alloy of aluminum, nickel, and cobalt — *alnico* — can be made into an extremely powerful permanent magnet that has many times the lifting ability of a steel magnet. It is interesting to note, however, that aluminum itself cannot be magnetized.

3. **Electromagnets** are temporary magnets produced by means of an electric current, as will be described later.

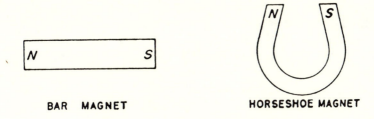

BAR MAGNET   HORSESHOE MAGNET

Figure 7.1. Permanent magnets.

Laws of Magnetism

The *three fundamental laws of magnetism* are as follows:

1. *Every magnet has two poles,* one at each end. This concept is based on the observation that when a magnet is inserted into a container of iron filings and then removed, iron particles will cling to its ends or *poles,* as illustrated in Figure 7.2. When allowed to swing freely (as in a magnetic compass), one pole of the magnet points north and the other end south. The poles are therefore called, respectively, *north* and *south* poles.

IRON   IRON
FILINGS   FILINGS

Figure 7.2. Polarity of a magnet. This is demonstrated by immersing first one end, then the other, of a magnet into a box of iron filings which cling mainly at the ends or *poles.*

2. *Like magnetic poles repel* each other; *unlike poles attract* each other. Thus, if two bar magnets are brought together with their south poles facing each other as in Figure 7.3A, there is a force which tends to push them apart. If, on the other hand, the magnets are brought together with opposite poles facing each other, a force tends to pull them together, as in Figure 7.3B.

3. *The force of attraction (or repulsion) between two magnetic poles varies directly as the strength of the poles, and inversely as the square of the distance between them.* Thus, doubling the strength

of one pole doubles the force between the poles. Tripling the strength triples the force. On the other hand, doubling the distance between two poles reduces the force between them to $(1/2)^2$ or one-quarter. Tripling the distance reduces the force to $(1/3)^2$ or one-ninth.

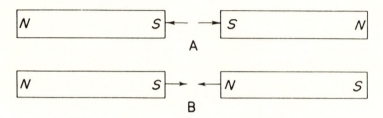

Figure 7.3. (*A*) Like poles repel. (*B*) Unlike poles attract.

Nature of Magnetism

Why are some materials magnetic (iron-attracting) whereas others are not? The most widely accepted explanation, an outgrowth of Weber's theory of magnetism first proposed in the early nineteenth century, is based on observations such as the following:

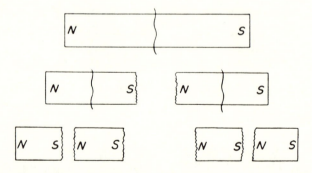

Figure 7.4. When a magnet is broken, each part becomes a whole magnet with a north and a south pole.

1. Breaking a magnet results in each fragment becoming a *whole magnet* with its own north and south poles, as in Figure 7.4.

2. Heating or hammering a piece of steel while it lies near a magnet causes the steel to become a magnet (i.e. magnetized).

3. Stroking an iron bar repeatedly with one end of a magnet causes the bar to become magnetized.

4. Gently jarring a test tube of iron filings near a magnet causes them to become magnetized. If the test tube is now shaken vigorously, it loses its magnetism.

These observations suggested that elements such as iron, nickel, and cobalt consist of extremely small, discrete magnets. But only with the advent of modern atomic science has the true nature of such elements been fully explained. According to atomic theory, the orbital electrons in magnetic elements spin predominantly in pairs, in the same direction, creating *atomic magnetic dipoles* (*magnetic moments*). These atomic magnets are arranged in groups called *magnetic domains.*

A *magnetic element* can exist either in the magnetized or nonmagnetized state. In a *nonmagnetized* piece of iron, the domains are in a state of random disorder as in Figure 7.5A. Therefore, the iron does not have magnetic poles. On the other hand, in *magnetized* iron the domains are aligned in orderly fashion, all their north poles pointing in one direction, and their south poles in the opposite direction, as shown in Figure 7.5B.

In *nonmagnetic elements* such as copper, just as many electrons spin in one direction as in the other, so their opposing magnetic effects cancel out. Therefore, in such elements the atoms are not magnetic, no domains are formed, and they *cannot* be magnetized.

A. NON-MAGNETIZED IRON

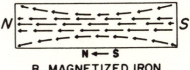

B. MAGNETIZED IRON (MAGNET)

Figure 7.5. Magnetic domains in a magnetic substance. In *A*, when a ferromagnetic substance such as iron is not magnetized, the domains are arranged haphazardly and there is no net polarity. In *B* the iron has been magnetized and the domains are arranged in orderly fashion to produce opposite poles.

Magnetic Fields

What is there about a magnet that causes it to attract iron? Surrounding a magnet there is a zone of influence called a *magnetic field,* representing a summation of the fields of all the domains. The magnetic field can be demonstrated by placing a piece of cardboard over a magnet and sprinkling iron filings on the cardboard. If the latter is gently tapped, the filings arrange themselves into a pattern resembling the one in Figure 7.6. The lines along which the filings arrange themselves are called *lines of force* or *magnetic flux.* In other words, the magnetic field consists of lines of force, and the more closely spaced these lines, the stronger is the field. In fact, the **strength of a magnetic field is proportional to the number of lines per square centimeter.**

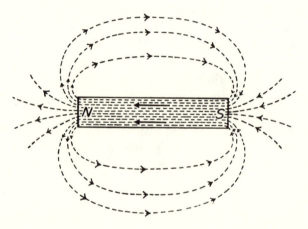

Figure 7.6. The magnetic lines of force about and in a magnet. The lines have been mapped by a small magnetic compass held successively in the indicated positions outside the magnet. They can also be shown by means of iron filings. (The direction of the lines within the magnet are not mapped, but are assumed.)

The lines of force can be mapped out by means of a small magnetic compass. Starting at the north pole of a bar magnet, one simply moves the compass around it toward its south pole and observes the behavior of the needle which acts as though it were following a line through space from the north to the south pole of the magnet, as shown in Figure 7.6.

Characteristics of Lines of Force

Magnetic lines of force may be pictured as follows:

1. They are directed from the north pole to the south pole of a magnet as curved lines in the surrounding space. Within the magnet they are directed from the south pole to the north pole (see Figure 7.6).

2. Lines of force seem to repel each other when they are in the same direction, and attract each other when they are in opposite directions (see Figure 7.7).

3. The field is distorted by magnetic materials, but is not affected by nonmagnetic materials (see Figure 7.8).

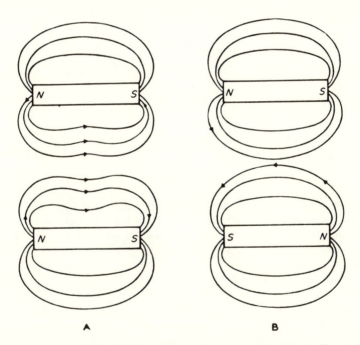

A   B

Figure 7.7. Magnetic lines of force act upon one another, depending on their relative directions. In *A* the lines of force are in the same direction and the resultant force tends to separate the magnets. In *B* the lines of force are in opposite directions and the magnets attract each other.

Magnetic Induction (Magnetization)

In an ordinary piece of iron the magnetic domains are jumbled helter-skelter by molecular motion and so the iron does not behave as a magnet (see Figure 7.9A). When such a piece of nonmagnetized iron is brought near one pole of a magnet, the end of the pole nearest the iron assumes the *opposite polarity* as a result of *magnetic induction;* the magnetic field of the pole exerts a force on the magnetic domains of the iron, making them line up in an orderly arrangement as in Figure 7.9B. The iron thus becomes temporarily magnetized and behaves as a magnet while lying in the field. When the iron is removed from the field, its magnetic domains return to their disorderly arrangement and the iron no longer is magnetized (ie, it no longer behaves as a magnet).

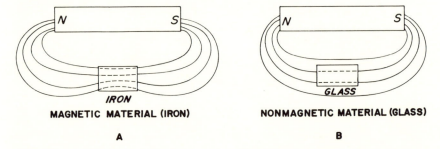

MAGNETIC MATERIAL (IRON)   **NONMAGNETIC MATERIAL (GLASS)**

A   B

Figure 7.8. The effect of magnetic and nonmagnetic materials on a magnetic field. In *A,* the iron "concentrates" the lines of force because of its magnetic permeability: this results from the summation of the lines of force associated with induced magnetism in the iron, and the lines of force of the magnetic field already present. In *B,* glass, a nonmagnetic material cannot become magnetized and so it has no effect on the magnetic field.

In summary, then, whenever a magnetic material (iron, cobalt, or nickel) is brought near a magnet, it first becomes magnetized (magnetic induction), its end nearer the magnetic pole acquiring the opposite polarity. This results in a force of attraction between the two.

Magnetic Permeability and Retentivity

Some materials are more susceptible than others to magnetic induction; that is, their magnetic domains can be readily lined up

A

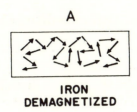

**IRON
DEMAGNETIZED**

B

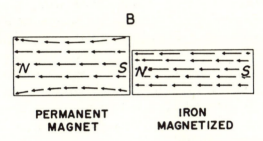

**PERMANENT   IRON
MAGNET   MAGNETIZED**

Figure 7.9. Magnetic induction. In *A*, the magnetic domains in the iron are normally jumbled, so the iron does not behave as a magnet. In *B*, the iron has been brought into the magnetic field of a magnet; the magnetic domains in the iron line up in an orderly arrangement and the iron is now magnetized. Note that the end of the iron near one pole of the magnet assumes the opposite polarity. When the iron is removed from the magnetic field of the permanent magnet, its magnetic domains return to their disorderly arrangement and the iron no longer behaves as a magnet.

when placed in a magnetic field. The ease with which a given material can be magnetized in this way is designated as its *magnetic permeability.*

Some materials such as steel do not readily become demagnetized after they have been removed from the magnetizing field; that is, once the magnetic domains have been aligned they tend to remain so. The ability of a magnet to resist demagnetization is called *magnetic retentivity.*

Examples of the above include:

1. "Soft" iron—high permeability, low retentivity.
2. "Hard" steel—low permeability, high retentivity.

Thus, a metal which is easily magnetized is also easily demagnetized (ordinary iron). On the other hand, a metal which is difficult to magnetize is also difficult to demagnetize (hard steel).

Magnetic Classification of Matter

All matter can be classified on the basis of its behavior when brought into a magnetic field.

1. **Ferromagnetic materials** (ordinarily called magnetic materials) include *iron, cobalt,* and *nickel.* They are strongly attracted by a magnet because of their extremely high permeability, that is, their great susceptibility to magnetic induction. Ferromagnetic substances become *permanently* magnetized under certain conditions. For example, steel is a form of iron that has been hardened by special treatment, as a result of which its magnetic permeability decreases while its retentivity increases; it is therefore susceptible to permanent magnetization. Another example is the alloy *alnico,* composed of aluminum, nickel, and cobalt, which, when magnetized, can lift many times its own weight. It has been used to remove magnetic foreign bodies such as nails and bobby pins from the stomach, under fluoroscopic control.

2. **Paramagnetic materials** are feebly attracted by a magnet. Platinum, an example of such a material, has extremely low magnetic permeability.

3. **Nonmagnetic materials** are not attracted by a magnet because they are not susceptible to magnetic induction. In such materials, just as many electrons spin in one direction as in the other so that the atoms are nonmagnetic and therefore do not form magnetic domains. Examples of nonmagnetic materials include wood, glass, and plastic.

4. **Diamagnetic materials** are actually repelled by a magnet, although this action is feeble. Only a few elements exhibit this property, among them beryllium and bismuth.

Detection of Magnetism

We can detect magnetism most easily by using a *magnetic compass.* This consists of a horizontal *magnetized steel needle* which swings freely about a vertical axis at its center. When brought near a magnet the needle, since it is a magnet, is deflected according to the laws of magnetism. Thus, the south pole of the compass needle turns toward a magnetic north pole, while the north pole of the compass needle turns toward a magnetic south pole. Notice that the north pole of a compass (or any other magnet) is really a geographic "north-seeking" pole, since it points north when allowed to swing freely. Hence the earth's geographic "north" pole is actually a south magnetic pole, whereas the earth's geographic "south" pole is a north magnetic pole.

QUESTIONS

1. Define magnetism in your own terms.
2. Name and discuss briefly the three main types of magnets.
3. What is the present concept of the intimate structure of a magnet? Show by diagram. State three facts which support this theory.
4. State the laws of magnetism.
5. Explain magnetic induction. How does a magnet attract a piece of iron?
6. Compare ordinary iron and hard steel with regard to their magnetization and demagnetization.
7. Define magnetic field; magnetic flux.
8. What happens to a magnetic field when a piece of ordinary iron is placed in it?
9. Indicate the direction of magnetic lines of force in space? Within the magnet?
10. How is magnetism detected?
11. State the magnetic classification of matter.
12. Define the following terms pertaining to magnetism:

    a. Domain
    b. Permeability
    c. Retentivity
    d. Flux
    e. Field
    f. Poles
    g. Compass
    h. Paramagnetism
    i. Ferromagnetism
    j. Diamagnetism

Chapter 8

ELECTROMAGNETISM

Definition

EVERY ELECTRIC CURRENT is accompanied by magnetic effects. *Electromagnetism is defined as the branch of physics that deals with the relationship between electricity and magnetism.*

Electromagnetic Phenomena

In 1820, Oersted, a Danish scientist, discovered that a magnetic compass needle turns when placed near copper wire carrying a direct current but returns to its normal position when the current is discontinued. Since this indicated the presence of a magnetic field, Oersted concluded that *a magnetic field always surrounds a conductor in which an electric current is flowing.* Actually, a magnetic field surrounds any stream of charged particles, whether in a metallic conductor or in space.

What is the direction of this magnetic field? It is specified by the *left thumb rule: if the wire is grasped in the left hand with the thumb pointing in the direction of the electron current [(−) to (+)], then the fingers encircling the wire will indicate the direction of the magnetic lines around the current* (see Figure 8.1).

These lines of force can be demonstrated by Davy's Experiment. A wire is pushed through a hole in a piece of cardboard and connected to a source of direct current. If iron filings are sprinkled on the cardboard, and the cardboard is then gently tapped, the iron particles arrange themselves in concentric circles around the wire. This is illustrated in Figure 8.2, in which the concentric circles represent the magnetic lines of force, or *magnetic flux,* surrounding the electric current in the wire.

Note carefully this very important fact: *a magnetic field exists around a wire only while an electric current is flowing.* The mag-

96

netic field disappears on open circuit because this terminates the current.

Figure 8.1. Left hand thumb rule. The thumb points in the direction of the electron current. The fingers designate the direction of the magnetic field around the conductor.

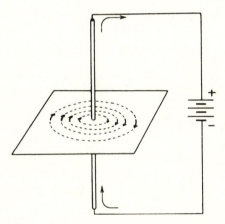

Figure 8.2. The magnetic lines of force around a conductor carrying a direct current. The dotted lines represent the magnetic flux when the electron current is in the indicated direction.

A magnetic field is present also around an alternating current, but because of the rapid reversal of field direction, it cannot be demonstrated by a compass or by Davy's Experiment.

The Electromagnet

While a current flows in a wire *helix* (coil), one end of the helix behaves as though it were the north pole and the opposite end the south pole of a magnet. By applying the thumb rule to each turn of the wire in the coil, we can predict not only the direction of the magnetic flux inside the coil, but also which end of the coil will be the north pole. Such a helix carrying an electric current is known as a *solenoid.* In modern x-ray equipment, solenoids are used to lock various movable parts, such as tube stand or Bucky tray, in almost any desired position.

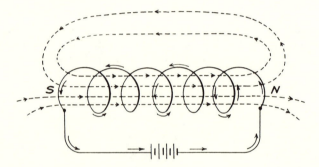

Figure 8.3. A helix carrying an electric current is called a *solenoid.* This shows the magnetic field associated with a solenoid.

As noted earlier (page 93), certain metals, such as iron, possess a high degree of magnetic permeability. Placing an iron rod inside a solenoid increases the strength of the associated magnetic flux because the iron becomes magnetized by the magnetic field of the solenoid. *Such a solenoid with an iron core is called an electromagnet.* Figure 8.4 shows the main features of a simple electromagnet.

The electromagnet has many practical applications, the best known being the electric bell and the telegraph. In x-ray equipment, electromagnets are used in remote control switches and relays of various kinds. These will be discussed later under the appropriate headings.

Electromagnetic Induction

You have just learned that electrons moving in a conductor are surrounded by a magnetic field. Under suitable conditions the

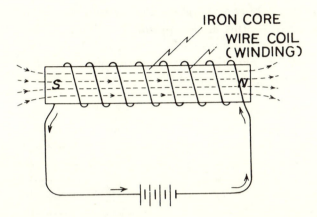

IRON CORE
WIRE COIL
(WINDING)

Figure 8.4. Electromagnet. Note that it is a solenoid into which has been inserted an iron core to increase the strength of the magnetic field. If the current were reversed in the illustration, the magnetic poles would be reversed. If the current were interrupted, the electromagnet would lose its magnetism because of the low retentivity of the soft iron core.

reverse is also true; *moving magnetic fields can produce electric currents.* In 1831, Michael Faraday discovered how to convert mechanical energy to electrical energy. On rapidly passing a closed wire between the poles of a horseshoe magnet, he found that an electric current was generated in the wire. This is the principle of *electromagnetic induction,* stated more precisely as follows: *whenever a conductor cuts across magnetic flux, or is cut by magnetic flux, an electromotive force (emf) or potential difference is induced in the conductor.* The practical unit of emf is the volt, which is also the unit of potential difference. Interaction between a line of force and a single loop conductor is called a *flux linkage.*

The magnitude or size of the induced emf depends on the number of lines of force cut per second, or the number of flux linkages per second. This is determined by four factors:

1. The *speed* with which a conductor cuts the magnetic lines of force or magnetic flux; *the more flux lines cut per second, the higher the induced emf.*

2. The *strength* of the magnetic field, or the degree of crowding of the magnetic flux lines — *the stronger the magnetic field the higher the induced emf.* A moment's thought will reveal that this is essentially the same as statement (1); the closer the spacing of the magnetic lines, the more flux cut per second.

3. The **angle** between the conductor and the direction of the magnetic field. As this angle approaches 90° the spaces between the lines of force become smaller, progressively more flux lines are cut per second, and therefore a correspondingly larger emf is induced, as shown in Figure 8.5.

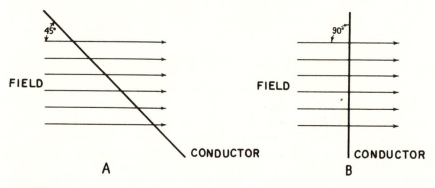

Figure 8.5. In *A*, where the conductor cuts the magnetic field at a 45° angle (down through this page), the induced emf is less than in *B*, where the wire cuts the field at 90°.

4. The number of *"turns"* in the conductor if it is wound into a coil, because the lines cut by each turn of the coil are additive; that is, there are more flux linkages per second. The induced emf is directly proportional to the number of turns in the coil.

Bear in mind that *an emf is induced only while the wire is moving across the field.* If the motion of the wire ceases, the induced emf drops to zero. But the motion of the wire need only be relative; thus, *if the wire is stationary and the magnetic flux passes across it, an emf will be induced in it* just as well, and the above four principles apply. Finally, *if the wire is stationary and the magnetic flux changes (varies in strength) an emf will be induced in the wire.*

An emf is induced in a conductor moving relative to a magnetic field *even if the circuit is open* because emf represents the potential difference between the ends of the conductor. However, no current flows unless the circuit is closed.

In summary, we may state that there are three ways to induce an emf or voltage in a wire by electromagnetic induction:

1. The wire may move across a stationary magnetic flux, or
2. The magnetic flux may move across a stationary wire, or

3. The magnetic flux may vary in strength while a stationary wire lies in it. This principle is extremely important in the electric transformer, to be described later.

We may conclude that *whenever a conductor and a magnetic flux move with relation to each other, an emf is induced in the conductor, the magnitude of the induced emf being directly proportional to the number of lines of force per second crossing the conductor.*

Direction of Induced Electric Current

An induced electric current resulting from an induced emf has a definite direction depending on the relationship between the motion of the conductor and the magnetic field. This is stated in the *left hand rule* or *dynamo rule: first, hold the thumb and first two fingers of the left hand so that each is at right angles to the others, as in Figure 8.6. Next, let the index finger point in the direction of the magnetic flux, and the thumb point in the direction the conductor is moving. The middle finger will then point in the direction of the induced electron current.*

The relationships among the direction of the magnetic flux, the motion of the conductor (wire or coil), and the induced electron current are summarized in Figure 8.7. This principle underlies the electric generator, which will be discussed in detail in the next chapter.

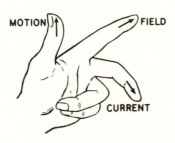

Figure 8.6. Left hand dynamo rule. This shows the direction of *electron* current flow.

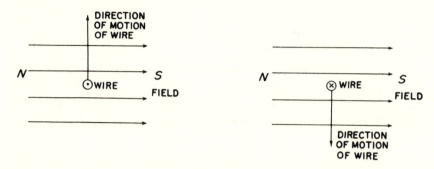

Figure 8.7. The direction of the electron current induced in a wire. The wire is shown in cross section. ⊗ indicates that the current is moving into this page. ⊙ indicates that it is moving out of this page, toward the reader.

Self-induction

Suppose we have a coil in a *direct current* circuit whose switch is open. Now suppose we close the switch; at this instant a magnetic field is set up about the coil, building up rapidly to a maximum, where it remains *while the primary current continues to flow.* During this *brief interval* in which the magnetic field "grows up" around the coil, immediately after closure of the switch, the magnetic flux is really expanding, thereby *cutting the coil itself.* By the rules of electromagnetic induction, an emf is induced in the coil in a direction *opposite* to the *applied emf,* a phenomenon known as *self-induction.* Thus, at the instant of switch closure, the self-induced current tends to *oppose* the flow of primary current in the coil (see Figure 8.8).

While the switch is closed and the primary current is flowing steadily in one direction, the surrounding magnetic field is also constant and therefore does not move relative to the coil. Consequently, *the self-induced current falls to zero.*

If the switch is then opened, the magnetic field collapses, sweeping across the coil in a reverse direction from that occurring when the switch was closed. As a result, the self-induced emf reverses and tends to maintain the current in the coil momentarily after the switch has been opened.

In summary, then, closing the switch results in momentary induction of an emf opposing the applied emf, while opening the

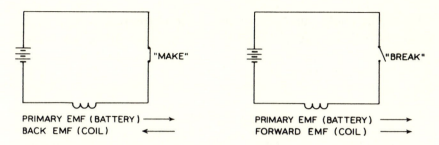

PRIMARY EMF (BATTERY) ———→
BACK EMF (COIL) ←———

PRIMARY EMF (BATTERY) ———→
FORWARD EMF (COIL) ———→

Figure 8.8. Self-induction. On the "make" (instant of switch closure) there is a *momentary* self-induced emf bucking or opposing the primary battery emf. On the "break" (instant of switch opening), the self-induced emf is reversed momentarily while the battery emf is unchanged.

switch causes momentary reversal of the self-induced emf which is now in the same direction as the applied emf. Figure 8.8 describes self-induction, and Figure 8.9 shows graphically the relation of self-induced emf to switch opening and closing.

An iron core inserted into the coil produces an increase in the self-induced emf which opposes or **bucks** the applied emf. The size of the bucking effect depends on the depth of insertion of the core.

Although we have explained the principle of self-inductance by means of a direct current, in actual practice self-induction devices

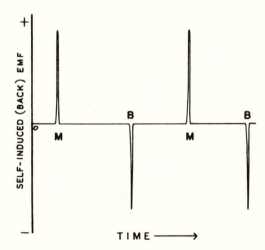

Figure 8.9. Self-induction of emf occurs only at the instant of "make" or "break" of a direct current. While a constant current flows in one direction, no emf is induced. M = make. B = break.

such as the choke coil operate only with **alternating current** (AC) as described on page 133. The rapid reversal in direction and magnitude of AC produces an effect similar to the manual opening and closing of a switch with DC.

QUESTIONS

1. Define electromagnetism in your own words. How can you easily determine its presence?
2. What is the left thumb rule? How is it applied?
3. Describe an electromagnet. In what two ways can you determine its polarity?
4. Define and discuss electromagnetic induction?
5. Discuss the four factors that determine the magnitude (size) of an induced emf? Name two terms used interchangeably with emf.
6. In Figure 8.10, what is the direction of the electromagnetically induced current?
7. Two coils of wire are lying parallel to each other. If one is connected to the terminals of a battery, what will happen in the second coil? What happens when the switch is opened? Closed?

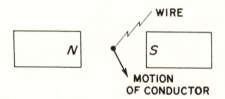

Figure 8.10.

8. Describe the difference between a helix, a solenoid, and an electromagnet.
9. Compare the self-inductance of AC and DC circuits.

Chapter 9

ELECTRIC GENERATORS AND MOTORS

ELECTRIC GENERATOR

Definition

An electric generator or dynamo is a device that converts mechani-cal energy to electrical energy by electromagnetic induction.

Essential Features

As already described in the last chapter, electromagnetic induc-tion is the setting up of an emf in a conductor as it cuts, or is cut by, magnetic flux. In actual practice, an electric generator has two components to bring about the continual cutting of magnetic flux by a conductor:

1. *A powerful electromagnet* to set up the necessary magnetic field. This is called the *field magnet.*

2. *An armature,* consisting of a coil of wire that is rotated mechanically in the magnetic field. The mechanical energy needed to turn the armature is usually supplied by a steam turbine which is heated by oil, gas, coal, or a nuclear reactor; or by a waterfall, as in Figure 9.1.

Simple Electric Generator

An *elementary form* of electric generator may be represented by a single loop of wire (armature) rotated mechanically between the poles of a magnet (see Figure 9.2). What are the size and direction of the current produced by such a generator? Examination of the relationship between the armature and the magnetic field in which it is rotated should give the answer. In Figure 9.2 the armature is represented by the wire loop *AB* whose ends are **connected separately** to the metal *slip rings* *E* and *F* which rotate with the armature in the direction indicated by *C*. Carbon *"brushes"* rest firmly against each slip ring, at the same time letting the rings rotate. In this way the circuit is completed through the external circuit containing *R* (resistance, representing load or current-consuming devices). Magnetic flux passes from the *N* pole to the *S* pole.

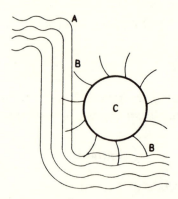

Figure 9.1. The waterfall at *A*, dropping on the blades *B* of a turbine, rotates it counterclockwise about the axis *C*. An armature connected to axis *C* will rotate with the turbine.

When the armature is rotating clockwise, as in Figure 9.2A, with *A* passing up through the magnetic field, an emf is induced in it in the direction indicated by the straight arrow (according) to the left hand rule). At the same time, *B* is passing down through the field, so that its emf is in the direction of the straight arrow near *B*. The net effect is that the current leaves *A* through slip ring *E*, passes through the external circuit *R*, and enters slip ring *F*.

As the armature coil rotates, *A* and *B* exchange positions, as shown in Figure 9.2B. Now *A* is moving down through the field while *B* is moving up, and the current *leaves slip ring F* and *enters slip ring E*. Note that the current is now *reversed in both the coil and the external circuit.* This reversal takes place repeatedly as the armature continues to rotate. A current which periodically reverses its direction is called an *alternating current,* usually abbreviated AC.

Let us now examine the magnitude of the induced voltage *from instant to instant* as the coil turns in the field. In Figure 9.3, the generator is simply shown end-on, *A* and *B* representing the cross

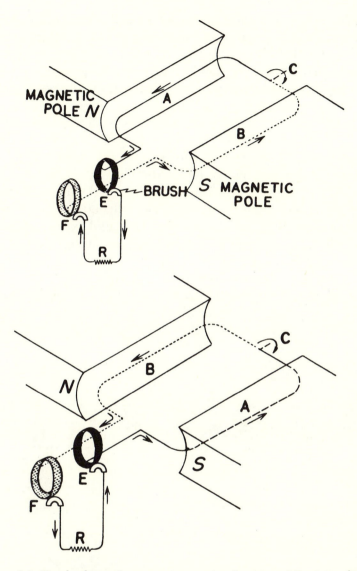

Figure 9.2. Simple alternating current generator. Armature *AB* rotates clockwise in the magnetic flux (which passes from *N* to *S*). In the upper diagram, side *A* moves up through the field while side *B* moves down. Arrows show the direction of the induced electron current which leaves the generator through slip ring *E*, passes through external circuit *R*, and completes the circuit through slip ring *F*. As the armature continues to rotate, sides *A* and *B* reverse positions as in the lower diagram, and the current is now reversed in the armature and in the external circuit *R*.

section of the wire armature, rotating in the direction of the arrows around axis *C*. At the instant the armature is in position (1) it is moving parallel to the magnetic field and is therefore not cutting magnetic flux. Because no emf is induced at this instant, the graph shows armature voltage to be zero. As the armature continues to move toward position (2) it cuts progressively more magnetic flux lines per second until, at position (2), it is moving perpendicular (90°) to the flux lines. Now the induced emf in the armature is at maximum, as represented by the peak of the curve. At (3) the armature is again moving parallel to the flux and the induced emf is zero. At (4), with the armature again cutting the flux at an angle of 90° but in the opposite direction, the induced emf is reversed, and the peak of the curve is below the horizontal axis of the graph. Finally, the cycle is completed when the armature returns to its original position.

The alternating current curve embodies certain useful information. Such a curve is presented in Figure 9.3 to show the relationship of the AC to the successive positions of the generator armature as it

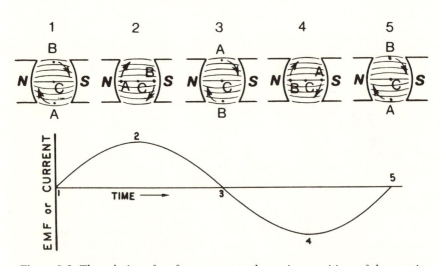

Figure 9.3. The relation of emf or current to the various positions of the rotating coil (armature) of an *AC* generator. *A* and *B* represent the single-loop coil in cross-section (see Figure 9.2). The induced emf depends on the angle at which the coil cuts the magnetic flux at any particular instant. *C* is the axis of rotation of the armature coil.

rotates between the magnetic poles. This is a *sine curve* because it depends on the sine of the angle, from instant to instant, between the plane of the armature and the plane perpendicular to the magnetic field.

In Figure 9.4 the AC curve is shown again in its usual form. The YY' axis (vertical) may represent either the induced emf or the current while the X axis (horizontal) represents time. The distance between two successive *corresponding points* on the curve, for example AB, represents *one cycle* of the AC and indicates that the armature has made one complete revolution. In common use in the United States is 60-cycle (60-Hz) AC, consisting of sixty complete cycles such as AB in each second, and corresponding to sixty revolutions of the armature per second. *The number of cycles per second is called the frequency of an alternating current.* Note that *each cycle consists of two alternations;* that is, the voltage starts at zero, rises gradually to a maximum in one direction (shown above the X axis), gradually returns to zero, increases to a maximum in the opposite direction (shown below the X axis), and finally returns to zero. Thus, there are 120 alternations per second with a 60-cycle current, which means that the *current flows back and forth in the conductor 120 times in each second.* The single sine curve represents a *single phase AC.*

Since the AC voltage and amperage vary from instant to instant,

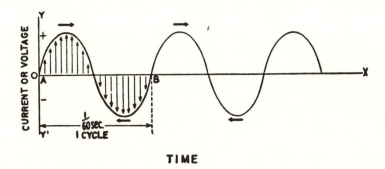

TIME

Figure 9.4. Sine curve representing a 60-cycle (60-Hz) alternating current (AC). The thick horizontal arrows show the direction of the *current* during different parts of the AC cycle. Note that the current reverses its direction every 1/120 sec. The thin vertical arrows indicate the variation in voltage or amperage from instant to instant.

they cannot be expressed in the same way as in a direct current which flows steadily in one direction (see Figure 9.5). Instead, we must use the concept of *effective* or *root mean square (rms) value of an alternating current,* defined as *that direct current which has the same heating effect in a resistor as the alternating current in question.* For example, if a steady direct current of 1 ampere produces the same heating effect in a 1-ohm resistor as does a given AC, then the AC may be said to have an effective value of 1 amp. If we have a pure sine wave as in Figure 9.4, then the relationship between the peak and the *effective* or *root mean square value* of the current is as follows:

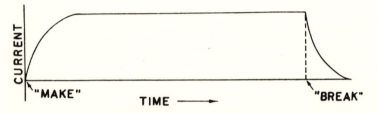

Figure 9.5. Curve representing a steady direct current with a coil in the circuit. The circuit is closed at "make." The current rises rapidly to a maximum and maintains that value at a constant level until the circuit is opened at "break," when it rapidly drops to zero.

effective or rms current = 0.707 × maximum current

or

maximum current = 1.41 × effective or rms current

The *effective* or *rms voltage* has the same relationship to the *maximum voltage:*

effective or rms voltage = 0.707 × maximum voltage

or

maximum voltage = 1.41 × effective or rms voltage

Properties of Alternating Current Circuits

In Chapter 6 we found that a circuit carrying a *steady direct current* has a property called resistance, which tends to oppose the flow of current. A *coil* of wire connected in series in such a circuit holds back the flow of current at the instant of switch closure, over and above its electrical resistance. This results from the *inductance* of the coil, which sets up a *back emf* (see pages 133–134). Therefore, the direct current (DC) rapidly rises to a plateau and remains constant (see Figure 9.5), since the associated magnetic field becomes constant and there is no longer a back emf.

On the other hand, with an *alternating current* the rapid change in magnitude and direction of the current causes a corresponding fluctuation in the strength and direction of the magnetic field set up around a coil in the circuit. Because of the coil's inductance, the persistently changing magnetic field around the coil supplied by an AC sets up a *back emf* in a direction opposite the applied alternating voltage, first in one direction and then in the other. This "bucking" tendency of inductance is called *inductive reactance,* and is measured in ohms because it opposes the flow of current.

A *capacitor* in a DC circuit allows current to flow only until it has reached full charge, whereupon the current ceases to flow. However, if a capacitor is added to an AC circuit, it periodically reverses polarity as the current reverses. This offers a certain amount of hindrance to the flow of current, known as *capacitive reactance.*

The apparent total resistance of an AC circuit is called its *impedance,* which depends on the inductive reactance, capacitive reactance, and true electrical resistance. The calculation of the impedance from these three factors is *not* that of simple addition, and will not be pursued further.

It is interesting to note that Ohm's Law applies to AC if the above discussed properties are taken into consideration,

$$\textbf{effective current} = \frac{\textbf{effective voltage}}{\textbf{impedance}}$$

$$I = \frac{V}{Z} \tag{1}$$

where I is the effective current measured in amperes, V is the effective voltage measured in volts, and Z is the impedance measured in ohms.

Direct Current Generator

The principle of the AC generator also applies to the DC generator, but in the latter a *commutator* instead of slip rings feeds the generated current into the external circuit. Whenever the A and B segments of the armature exchange positions as shown in Figure 9.6, the corresponding half of the commutator always supplies current to the external circuit in the same direction even though there is an AC in the armature itself. The commutator actually consists of a metal ring split in two, with the halves separated by an insulator, each half being permanently fixed to its corresponding end of the wire.

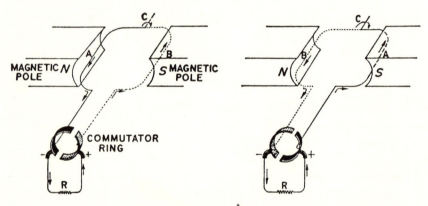

Figure 9.6. Simple direct current (DC) generator. Note that a commutator ring is used, with the result that the current is always supplied to the external circuit in the same direction. (Compare with Figure 9.2, which illustrates the AC generator.)

The wave form of the current obtained from such a generator is represented in Figure 9.7. Note that the part of the sine wave which with AC would be below the horizontal line lies above the line. This is a *pulsating direct current.* In actual practice, a DC generator consists of numerous coils and a commutator which is divided into a corresponding number of segments. Since the coils are so arranged that some are always cutting magnetic flux during rotation of the armature, emf is always being induced. At no time does the voltage fall to zero, as in the simple one-coil DC generator illustrated above. Consequently, the voltage is steadier than that generated by a single-coil armature. Figure 9.8 shows the DC curve obtained with a multiple coil generator.

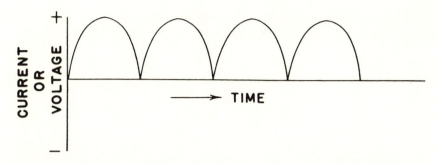

Figure 9.7. Wave form of a direct current curve obtained from a simple DC generator such as that shown in Figure 9.6. Note that the curve differs from a sine curve in that the portions of the curve that lie below the horizontal axis are turned up to lie above the axis. This is a *pulsating direct current,* in contrast to the steady DC supplied by a battery.

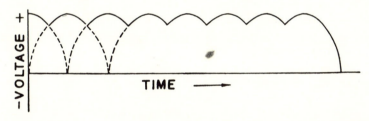

Figure 9.8. Direct current wave form produced by a multiple-coil generator. The solid line curve represents the voltage variation from instant to instant. This fluctuation is much less than that produced by any one of the coils alone (shown by the broken line).

Advantages of Alternating Current

For at least two reasons, AC is employed much more frequently than DC.

1. AC is needed to operate transformers which are indispensable in radiographic equipment as well as in many areas of industry. A transformer will not operate adequately on voltage with the wave form shown in Figure 9.8.

2. The electric current generated at the power plant is often transmitted over a great distance. It has been shown earlier that

with *direct current* the power loss (in the form of heat) is related to the current and resistance as follows:

$$P = I^2R \tag{2}$$

where P = loss of power in the form of heat,
                measured in watts
         I = current in amperes
         R = resistance in ohms

According to this equation, the power loss is proportional to the *square of the amperage.* It is obvious, therefore, that if a current is transmitted at a very *low* amperage there will be a much smaller loss of electric power than if transmitted at high amperage. An example will illustrate this point.

Suppose we wish to transmit 50,000 watts of electric power over a distance of ten miles (a total of twenty miles of wire), the wire having a resistance of 0.15 ohm per mile. The total *R* is 0.15 × 20 = 3 ohms. If we use *DC* at 500 volts and 100 amps (500 V × 100 A = 50,000 watts), the power loss is obtained by the power equation:

$$\begin{aligned} \textit{Power loss} &= I^2R \\ &= 100 \times 100 \times 3 \\ &= 30{,}000 \text{ watts} \end{aligned}$$

Thus, three-fifths of the power is wasted as a result of the heating effect of the current in the transmission wires!

If, instead, we used *AC* at 5000 volts and 10 amperes (again 50,000 watts), the power loss would be much less. The above equation for power consumption cannot be applied accurately to AC without a correction known as the *power factor,* but it does give a rough approximation. Neglecting the power factor,

$$\begin{aligned} \textit{Power loss} &= I^2R \\ &= 10 \times 10 \times 3 \\ &= 300 \text{ watts} \end{aligned}$$

as contrasted with a power loss of 30,000 watts with DC transmitted at lower voltage and higher amperage.

Upon reaching its destination the high voltage (often called high tension) must be reduced to useful values, ordinarily 120 volts for lighting and 120 or 240 volts for most x-ray equipment. This is accomplished by means of a *transformer*, an electromagnetic device which will be described in detail in the next chapter.

ELECTRIC MOTOR

Definition

An electric motor is a device that converts electrical energy to mechanical energy. In other words, an electric current can be made to do useful work by means of a motor.

Principle

Whenever a conductor carrying an electric current is placed in a magnetic field, there is a force or side-thrust on the conductor. It should be emphasized that two conditions must prevail: *the conductor must carry a current, and it must be located in a magnetic field.* The conductor experiences a force that is directly proportional to its length or number of turns, the strength of the magnetic field, and the size of the current. The force is increased when the conductor is in the form of a coil.

The motor effect results from the interaction between the magnetic field around the electron current in the conductor, and the other magnetic field. Therefore, a magnetic field acts on a current-carrying wire as shown in Figure 9.9, where it is represented in cross section and the following conditions prevail:

1. The electric current in the wire is directed away from the reader in this particular case.

2. Magnetic flux surrounds the wire as shown by the small arrows (associated with the current in the wire).

3. Magnetic flux provided by the *field* magnets is directed to the right between the poles of the magnets.

Now look at the upper side of the wire; there the magnetic flux surrounding the wire is directed opposite to that of the flux between the field magnets. Such oppositely directed flux lines "attract" each other. At the same time, note that on the lower side of the wire the surrounding magnetic flux is in the same direction as the flux between the field magnets. Flux lines in the same direction "repel" each other. The net effect is for the wire to be thrust upward by the interaction of the two magnetic fields, as shown by the vertical arrow.

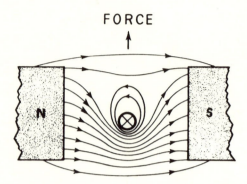

Figure 9.9. The dark circle represents a wire in cross section carrying a current "into the page." As explained in the text, the wire is forced upward by the weaker magnetic field above (flux lines in opposite directions) and by the stronger magnetic field below (flux lines in the same direction). Adapted from *The New Physics and Chemistry Dictionary and Handbook* by Robert W. Marks. Copyright © 1967 by Bantam Books, Inc.

This principle is embodied in the *right hand or motor rule* which mirrors the left hand rule. *The right hand or motor rule states that if the thumb and first two fingers of the right hand are held at right angles to each other; and if the index finger indicates the direction of the magnetic field, and the middle finger the direction of the electron current in the conductor, then the thumb will point in the direction that the conductor will move.* Applying this rule to Figure 9.9, we find that the same result is obtained as before: the wire is thrust upward. Note that according to this rule a conductor experiences a force that is perpendicular to both the magnetic field and the electron current.

Simple Electric Motor

The diagrams shown in the section on the electric generator apply also to the electric motor, except that current is supplied at the brushes instead of being withdrawn. An AC motor has *slip rings,* whereas a DC motor has a *commutator.*

In its elementary form, an electric motor may be represented, as in Figure 9.10, by a cross-sectional diagram of a single coil armature carrying an electric current and lying in a magnetic field. Applying the **right hand rule**, we find that as *A* is thrust upward *B* is thrust downward. The net result is that the armature turns in a clockwise direction about axis *C*.

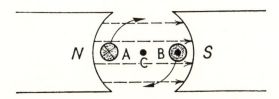

Figure 9.10. Cross section of an elementary motor. *A* and *B* are the coil wire seen end-on. *C* is the axis of rotation. The curved arrows show the direction in which the coil armature rotates when the proper current is fed to the armature.

At this point we should mention the **back emf** of a motor. When the coil of a motor rotates in a magnetic field, an emf is induced in the coil by electromagnetic induction. This self-induced emf is in a direction **opposite** to the direction of the emf already applied to the coil and is therefore called **back emf**. Thus, an electric motor acts at the same time as a generator, sending some current back to the main lines. This is one of the reasons why a motor consumes less current than a heating device.

Types of Electric Motors

There are two main types of electric motors:

1. **Direct current motor.** This is the reverse of a direct current generator.

2. **Alternating current motors.** There are two kinds of alternating current motors, **synchronous** and **induction**.

The **synchronous motor**, supplied by AC, is built like a single-phase AC generator. However, current is fed to the armature, causing it to rotate in the magnetic field. Certain conditions are necessary for the operation of a synchronous motor. Its armature must rotate at the same speed (revolutions per minute) as the armature of the AC generator supplying the current, or at some fixed multiple of that speed. In other words, the speed of this type of motor **must be synchronous with the speed of the generator.**

Special devices are required to bring the speed of the motor armature up to the required number of revolutions per minute. The synchronous motor is used in electric clocks and synchronous timers. The main *disadvantage* of the synchronous motor lies in the limited speeds with which it will operate, that is, it must be in step with the generator. At the same time, a fixed speed may be of *advantage* in certain types of equipment, such as electric clocks.

The *induction motor* has a *stator* which has an even number of stationary electromagnets distributed around the periphery. In the center is a rotating part or *rotor,* consisting of bars of copper arranged around a cylindrical iron core, and resembling a squirrel cage. There are no slip rings, commutators, or brushes because this type of motor works on a different principle from other types of motors and generators. The stator is supplied by a *multiphase* current, which means that the AC supplying one opposite pair of coils is out of phase (out of step) with that supplying the next pair of coils. Thus, successive pairs of coils set up magnetic fields which successively reach their peak strength, acting as though they were rotating, and dragging along the rotor. The drag on the rotor is caused by eddy currents induced in its copper bars by the "moving" magnetic field. Each copper bar, now carrying an induced current while lying in a magnetic field, experiences a push or force according to the motor principle. This force causes rotation of the rotor.

Although an induction motor is usually designed for modified single-phase or three-phase operation, its principle can be more simply explained with a now obsolete two-phase system. Figure 9.11 describes such a motor. Its power source would have to be a two-phase generator with an armature consisting of two coils perpendicular to each other. The current produced by such a generator may be represented by two curves, one of which is one-fourth cycle ahead of the other, as shown in the upper part of Figure 9.11. This two-phase current is applied to the stator to induce a rotating magnetic field. The rotor is shown in cross section at the center.

A modified single-phase induction motor is used to rotate the anode in an x-ray tube.

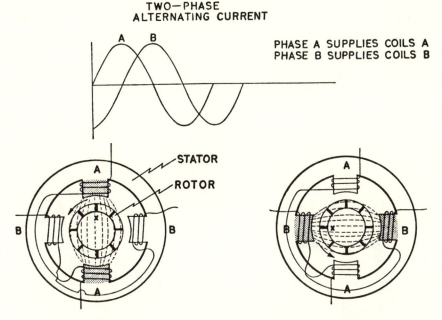

Figure 9.11. Principle of the induction motor. In phase *A* (left diagram) the magnetic flux is directed downward as shown. In the next ¼ cycle, phase *B* activates its stators and the magnetic flux is now directed from right to left (the flux in the preceding ¼ cycle having disappeared). In the next phase, the magnetic flux is directed upward, and finally from left to right. Eddy currents are set up in the rotor by electromagnetic induction. This process continues as long as current is supplied. In effect, then, we have a magnetic field that rotates clockwise, dragging the rotor with it and making it rotate. (*The two-phase system is now obsolete but aids in understanding the principle of the rotating field.*)

Current-measuring Devices

An interesting and important application of the *motor principle* is in the construction of instruments for the measurement of current and voltage. The basic device employing this principle is the *moving coil galvanometer,* which consists essentially of a coil of fine copper wire suspended between the poles of a horseshoe field magnet. One end of the coil is fixed and the other is attached to a hairspring helix. When a *direct current* is passed through the coil as it lies in the magnetic field of the permanent magnet, the coil moves according to the motor principle. In this case, the construction of the instrument, as shown in Figure 9.12, is such that the coil must rotate through an arc that is usually less than one-half circle (180°). Rotation takes place against the spring, so that when the circuit is opened the coil returns to the zero position. Since the degree of rotation of the coil is proportional to the current, the attached pointer moving over a *calibrated scale* measures the current as shown in Figure 9.12.

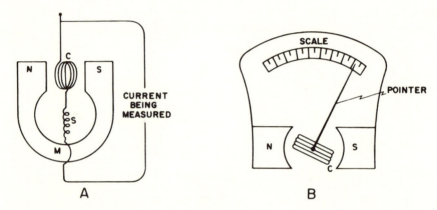

Figure 9.12. Moving coil galvanometer. In *A*, side view. In *B*, adapted as direct reading ammeter or voltmeter.

A galvanometer with a low resistance in parallel is capable of measuring *current* when connected in *series* in a DC circuit; if it has been previously calibrated to read amperes, it is an *ammeter* (see Figure 6.8).

On the other hand, a galvanometer protected by a high resistance in series measures *potential difference* when connected in *parallel* with the circuit being measured; if it has been previously calibrated to indicate volts, it is a voltmeter (see Figure 6.8).

Thus, modifications of the same basic instrument—the galvanometer—allow measurement of amperage or voltage in a DC circuit, provided the instrument has been calibrated against known values of amperage or voltage, respectively.

Further modification is necessary to measure the voltage and amperage of an *alternating current,* because the coil would "freeze" due to the rapid alternation of the current in the coil. (Recall that a 60-cycle AC sustains 120 reversals of direction per second.) Instead of using a horseshoe magnet, one must provide a magnetic field which reverses its direction in step with the current entering the coil. To do this, the AC being measured is, at the same time, used to supply the coils of a pair of field electromagnets. Since the current alternates simultaneously in the rotating coil and in the field coils, the rotation of the coil and hence the deflection of the pointer are unaffected by the alternations of the current, but are still affected by the size of the current. This device, called an *electrodynamometer,* can be designed to read AC voltage or amperage. The same type of instrument can be modified to measure electric power in a circuit, reading the values directly in watts.

QUESTIONS

1. What is an electric generator? The generator rule?
2. List some of the commonly used sources of energy in a generator.
3. Describe, with the aid of a diagram, a simple electric generator. Show how it produces an alternating current and relate it to the AC sine curve. Label the curve in detail.
4. How does an AC generator differ from a DC generator?
5. Compare the current curves of single phase AC, single phase DC, and an electric battery.
6. Why is AC generally preferred to DC? Prove by means of an example.
7. Define the effective (RMS) value of alternating current or voltage. How is it derived from peak values?
8. Find the RMS value of 80 kVp.
9. The effective value of an alternating current is 30 A. What is the peak amperage?
10. Define an electric motor. What is the motor principle?
11. Discuss the right hand or motor rule.
12. Under what conditions will a conductor, placed in a magnetic field, tend to move up through the field?
13. What are the two main types of motors?
14. Describe the principle and construction of an induction motor.
15. What is a synchronous motor? How does it differ from an induction motor?

16. Discuss the importance of the induction motor in radiographic equipment.
17. State the underlying principle of the galvanometer. Show with the aid of a simple diagram the construction of a DC voltmeter.

Chapter 10

PRODUCTION AND CONTROL
OF HIGH VOLTAGE—REGULATION OF CURRENT

TRANSFORMER

Electric power to the consumer is usually supplied at 120 to 240 volts, but much higher voltage is needed to give the electrons in an x-ray tube enough energy to generate x rays. Such high voltages are produced by the *transformer*, also referred to as an *x-ray generator*. It serves as a major component of an x-ray machine. In this chapter we shall discuss only the single-phase circuit carrying 60-cycle (60-Hz) alternating current (AC).

Principle

A transformer is an electromagnetic device which changes an *alternating current* from low voltage to high voltage, or from high voltage to low voltage, without loss of an appreciable amount of electrical energy (usually less than 5 percent). The transformer transfers electrical energy from one circuit to another *without the use of moving parts or any electrical contact between the two circuits,* employing the principle of *electromagnetic mutual induction.*

The simplest type of transformer—the *air core* transformer— consists of two highly insulated coils of wire lying side by side. One of these, the *primary coil,* is supplied with an alternating current (AC). The other or *secondary coil* develops AC by mutual induction (see Figure 10.1). The primary coil is the *input side* of the transformer, whereas the secondary coil is the *output side.*

How does a transformer function? An AC in the *primary* coil sets up, in and around it, a magnetic field that varies rapidly in direction and strength, just as does the AC itself. This changing magnetic flux *cuts* or *links with* the secondary coil, inducing in it an alternating emf by the process of electromagnetic induction. But the resulting AC in the secondary coil sets up an induced emf

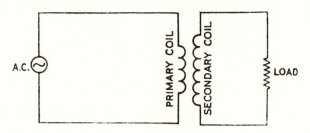

Figure 10.1. Simplest type of transformer. An electromotive force is set up in the secondary coil by electromagnetic mutual induction.

in the primary coil (opposite in direction to the primary current). Since emf is induced in each coil by the changing magnetic flux in the other, we call this process *mutual induction,* which is obviously a special case of electromagnetic induction.

Note here, again, the important principle of electromagnetic induction: the emf induced in any coil is directly proportional to the number of turns in the coil that cuts, is cut by, or links with, a given magnetic flux. This also applies to the transformer. Suppose the primary and secondary coils each have one turn. The AC in the primary coil sets up a magnetic field that cuts the one turn in the secondary coil; therefore, the emf is the same in both coils (neglecting any loss of energy during the process). But suppose the primary coil has one turn and the secondary has two turns. Now the *same* magnetic flux set up by the AC in the single-turn primary coil cuts the double-turn secondary coil, *doubling the emf in the latter.* Similarly, if the primary coil has two turns and the secondary one turn, the secondary emf will be half the primary emf.

Because the *induced* emf depends directly on the total number of turns (for a particular changing magnetic flux), we may assume that each turn makes the same contribution toward the total induced emf; that is, their individual induced emfs are additive. Therefore the relationship may be simplified in terms of *emf per turn,* or V/N, where V is the emf in volts, and N is the number of turns in the coil. For example, if the primary emf is 100 V and there are ten turns, then

$$100 \text{ V}/10 = 10 \text{ volts per turn in the primary coil}$$

Now, if the secondary coil has ten turns (note that they must link with the changing magnetic flux associated with the current in the primary coil), the induced emf in the secondary coil will be

$$10 \text{ turns} \times 10 \text{ volts/turn} = 100 \text{ V}$$

or the same as the primary voltage. But if the secondary coil has twenty turns, the induced emf in it will be

$$20 \text{ turns} \times 10 \text{ V/turn} = 200 \text{ V}$$

or twice the primary voltage.

The preceding discussion is embodied in the ***transformer law:*** ***the emf induced in the secondary coil is to the emf in the primary coil, as the number of turns in the secondary coil is to the number of turns in the primary coil,*** expressed simply in practical units in the following equation:

$$\frac{V_S}{V_P} = \frac{N_S}{N_P} \tag{1}$$

where V_S = voltage in secondary coil
V_P = voltage in primary coil
N_S = no. of turns in the secondary coil
N_P = no. of turns in the primary coil

Stated in words, this means that if the number of turns in the secondary coil is twice the number in the primary coil, then the voltage in the secondary will be twice the voltage in the primary. If the number of turns in the secondary coil is three times that in the primary coil, then the voltage induced in the secondary will be three times as great as in the primary, etc. Such a transformer, having more turns in the secondary coil than in the primary, puts out a ***higher*** voltage than is supplied to it and is therefore a ***step-up transformer.***

If the secondary coil has fewer turns than the primary coil, the output voltage will be ***less*** than the input voltage, and the transformer is a ***step-down transformer.***

What happens to the value of the ***current*** in a transformer? According to the Law of Conservation of Energy, there can be no more energy coming out of the transformer than is put in, and similarly, the power output (energy per unit time) can be no greater than the power input. Since the power in an electric circuit equals voltage multiplied by amperage, then (neglecting the power factor),

$$I_S V_S = I_P V_P \tag{2}$$

where I_S = current in amperes in the secondary coil
I_P = current in the primary coil
V_S = voltage in the secondary coil
V_P = voltage in the primary coil

Rearranging this equation, we get the proportion:

$$\frac{I_S}{I_P} = \frac{V_P}{V_S} \tag{3}$$

In other words, if the voltage is increased, as in a step-up transformer, the amperage is decreased; and if the voltage is decreased, as in a step-down transformer, the amperage is increased. Thus, a *step-up transformer increases the voltage, but decreases the amperage in an inverse ratio.* The power output of an x-ray transformer is rated in kVA (kilovolt-amperes or kilowatts). Thus, an x-ray generator that can deliver a tube current of 200 mA at 140 kVp is rated at 28 kVA (obtained from 140 kVp × 0.2 A).

In roentgen *diagnostic* equipment the step-up transformer takes 120 or 240 volts and multiplies this voltage to 30,000 to 150,000 volts (30 to 150 kilovolts), thereby providing the high voltage necessary to drive the electrons through an x-ray tube with sufficient speed. At the same time it decreases the current to thousandths of an ampere (milliamperes).

Construction of Transformers

Four main types of transformers will now be described.

1. **Air Core Transformer.** This consists simply of two insulated coils lying side by side as shown in Figure 10.1.

2. **Open Core Transformer.** An iron core inserted into a coil of wire carrying an electric current causes a marked intensification of the magnetic flux within the coil because of the magnetization of the core. Therefore, a transformer becomes more efficient if each *insulated* coil has an iron core, as shown in Figure 10.2. This is known as an *open core transformer.* Although more efficient than an air core transformer, the open core type is still subject to an appreciable waste of power due to loss of magnetic flux at the ends of the cores, a condition termed *leakage flux.*

3. **Closed Core Transformer.** Here the heavily insulated coils,

IRON CORES

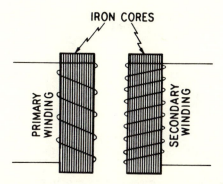

Figure 10.2. Open core transformer. There is some waste of power due to leakage of magnetic flux in the air at the ends of the iron cores.

often called *windings,* are wound around a square or circular iron ring or core as in Figure 10.3. With this type of transformer, the closed core provides a continuous path for the magnetic flux, so that only a small fraction of the magnetic energy is lost by leakage. An AC in the primary winding sets up an alternating magnetic flux in the iron core. Because this same flux links with (or cuts) both the primary and the secondary windings, it induces the same *voltage per turn in both windings* (coils). Therefore, the induced emf in *each* winding is proportional to the number of turns in that winding. Thus, the transformer law applies, namely, $V_s/V_p = N_s/N_p$ (see also pages 124–125).

The core is *laminated,* that is, it is made of layers of metal plates. Eddy currents (see page 129) are set up in the core by electromagnetic induction during transformer operation. Such eddy currents take energy from the current in the transformer coils, causing a waste of power known as *transformer loss.* Lamination of the core hinders the formation of these eddy currents and thereby increases the efficiency of the transformer; in other words, there is less power wasted. Silicon steel is often used in laminated cores because its high electrical resistance further reduces eddy current power loss. Since the closed core transformer is much more efficient than the open core, it is the *one most commonly used in x-ray-generating equipment.* The highly insulated coils are submerged in a metal box containing a special type of oil for additional insulation. Furthermore, the oil helps to keep the transformer cool by improving the dissipation of heat produced during operation.

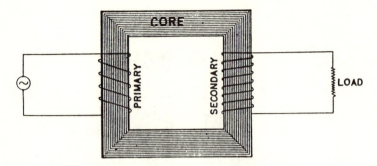

Figure 10.3. Closed core transformer. This type of iron core provides a continuous path for the magnetic flux, thereby minimizing flux leakage.

4. Shell-type Transformer. This, the most advanced type, is used as a commercial or power transformer. Here, too, a laminated core is used, consisting of a pile of sheets of silicon steel, each having two rectangular holes, as illustrated in Figure 10.4. The primary and secondary coils are both wound around the central section of the core for maximum efficiency, as shown in Figure 10.4. These windings must be highly insulated from each other by a special coating. In addition, the entire transformer is submerged in a tank filled with a special type of oil for maximum insulation and cooling.

Transformer Efficiency and Power Losses

The efficiency of a transformer is the ratio of the power output to the power input:

$$\% \text{ efficiency} = \frac{\text{power output}}{\text{power input}} \times 100$$

Ideally, the power input and power output would be equal, with an efficiency of 100 percent. However, in practice the efficiency is more likely to be about 95 percent or more; that is, there is up to about 5 percent less power output than power input, the lost energy appearing as heat. In contrast to other types of electrical and mechanical equipment, this represents a small waste of energy, and the transformer is therefore a highly efficient device.

We shall now summarize the various kinds of ***losses of electrical***

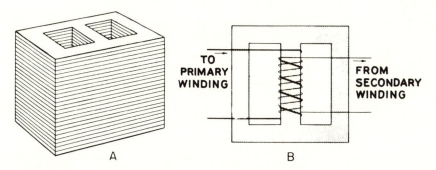

Figure 10.4. *A*, core of shell-type transformer. *B*, shell-type transformer, top view. Note the windings of the primary and secondary coils around the *same* central section of the core to provide high efficiency. Very heavy insulation is required.

power in a transformer and the methods of minimizing them to improve efficiency.

1. **Copper Losses.** These include mainly the loss of electrical power due to the *resistance* of the coils. Such loss of power can be reduced by using copper wire of adequate diameter. In a step-up transformer the primary winding carries a large current, so its wire must be thicker than the wire of the secondary coil which carries a small current (milliamperes). Recall that a thick wire, having less resistance than a thin one, can carry more current with less waste of energy in the form of heat (*Power Loss* = I^2R).

2. **Eddy Current Losses.** The alternating magnetic flux set up in the transformer *core* by the AC in the windings induces electrical *eddy currents* (swirling currents) in the core itself by electromagnetic induction. Eddy currents, in turn, produce heat in the core because electrical power loss due to heat = I^2R. The wasted electric power must come from that supplied to the transformer, thereby contributing to the loss of efficiency. Eddy currents can be minimized by the use of *laminated (layered) silicon steel plates,* highly insulated from each other by a special coating. Lamination and high-resistance silicon steel increase the electrical resistance of the core, thereby decreasing the size of the eddy currents according to Ohm's Law (see Figure 10.5).

3. **Hysteresis Losses.** Since the transformer operates on and puts out AC, the tiny magnetic domains in the *core* are repeatedly rearranging themselves as the core is magnetized first in one direction and then the other by the AC in the windings. This

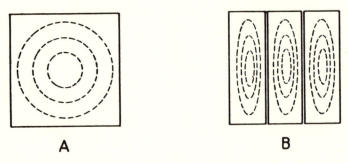

A   **B**

Figure 10.5. *A.* Dashed lines suggest the paths taken by eddy currents in a solid iron core. *B.* With the same volume of iron having three laminations (layers), the total paths for the eddy currents have been significantly increased; this means increased resistance and therefore smaller eddy current effect.

rearrangement of the domains produces heat in the core, thereby wasting electrical power. Such a loss of power, called *hysteresis loss*, is reduced by the laminated silicon steel core.

REGULATION OF HIGH VOLTAGE

Properly designed x-ray equipment must provide some means of regulating the input voltage to permit the choice of a variety of kilovoltages to be applied to the x-ray tube. This allows the technologist to obtain x rays of appropriate penetrating power for a particular technic. Without flexible voltage regulation, modern radiographic procedures would be seriously hampered. The device for changing *high voltage* is the *autotransformer.*

Autotransformer

According to the description of the step-up transformer, you should realize that there is a *fixed ratio* of voltage output to voltage input. Furthermore, the voltage input is fixed at 120 or 240 volts, depending on the power supply and type of equipment. Without any further modification, we would have available only a single kilovoltage, a situation that would seriously limit the range of radiography, since various kilovoltages must be applied to the x-ray tube to obtain x-ray beams with a variety of penetrating abilities.

How can we modify the basic equipment so as to obtain the required range of kilovoltages? This is accomplished simply by *varying the voltage input to the transformer primary.* Thus, if the ratio of the step-up transformer is 500 to 1 and the input to the primary coil is 100 volts, then the transformer will put out 500 × 100 = 50,000 volts (50 kV). If the input is 180 volts, the output will be 500 × 180 = 90,000 volts (90 kV). To obtain the required variety of input voltages we connect a device known as an *autotransformer* between the source of AC and the primary side of the transformer. As you will see, the autotransformer is a *variable transformer* or *Variac.**

Construction. The autotransformer is made up of a single coil of insulated wire wound around a *large iron core.* At regular intervals along the core, insulation has been omitted and the bare points connected or tapped off to metal buttons, as shown in Figure 10.6. A movable contactor, *C,* varies the number of turns included in the secondary circuit of the autotransformer, thereby varying its output voltage. Thus, the autotransformer serves as a *kV selector.*

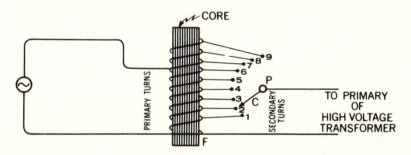

Figure 10.6. Diagram of an autotransformer (kV selector), which has only one winding or coil. The *autotransformer* secondary voltage stands in the same ratio to the primary voltage, as the ratio of the number of tapped turns (in this case, 2) is to the number of primary turns (7); that is, $^2/_7$, so the voltage is stepped down. If C is turned above 7, the ratio exceeds 1 and the voltage is stepped up.

Principle. The autotransformer is an electromagnetic device which operates on the principle of *self-induction. A single coil or winding serves as both the primary and secondary, the number of turns being adjustable.* (This is in sharp contrast to the transformer, wherein the number of turns cannot be changed.) By turning the contactor to include more or fewer turns on the secondary side of the autotransformer, we can vary the ratio of the number of

*General Radio Corporation

secondary turns to the number of primary turns, thereby varying the ratio of voltage output to voltage input. This relationship, similar to that in the transformer, is embodied in the *autotransformer law:*

$$\frac{\text{autotransformer secondary voltage}}{\text{autotransformer primary voltage}} = \frac{\text{number of tapped turns}}{\text{number of primary turns}}$$

Thus, in Figure 10.6, if the primary voltage is 240,

$$\frac{\text{autotransformer secondary voltage}}{240} = \frac{2}{7}$$

$$\text{autotransformer secondary voltage} = 68.6 \text{ volts}$$

The secondary voltage of the autotransformer is applied to the primary side of the main step-up transformer (see Figures 10.6 and 14.12).

For those desiring a more detailed description of the operation of an autotransformer, a wiring diagram is shown in Figure 10.7. We may regard the voltage drop across the coil as being related to the number of turns, that is, *volts per turn.* Suppose that the applied potential is 240 volts and there are seven primary turns. We would then have 240/7 = 34 volts per turn. Now, looking at the secondary side of the autotransformer, we see that there are two tapped turns; therefore, the voltage drop across this side is 2 turns × 34 volts per turn = 68 volts. Accordingly, the ratio of the voltage output to input of the autotransformer is 68/240 = 2/7, the same result as that obtained by means of the autotransformer law.

Note that an autotransformer can be used only where there is a relatively small difference between its input and output voltage. Furthermore, *the position of the contactor should not be changed while the exposure switch is closed because sparking may occur between the metal buttons, thereby damaging them.*

CONTROL OF FILAMENT CURRENT AND TUBE CURRENT

We have just learned that control of high voltage by means of an autotransformer allows convenient kV selection. Of equal importance is the capability of x-ray equipment to vary tube current or milliamperage (mA).

The construction and operation of an x-ray tube will be explained

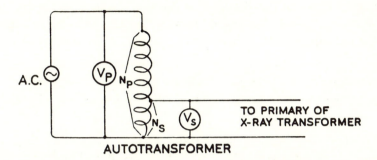

AUTOTRANSFORMER

Figure 10.7. Further simplification of wiring diagram of autotransformer. If there are 7 primary turns N_P, and the applied potential is 240 volts, then there would be $240/7 = 34$ volts per turn. On the secondary side there are 2 tapped turns N_S; therefore the voltage across the secondary side $V_S = 2$ turns $\times$ 34 volts per turn $= 68$ volts. So the voltage output to input $= 68/240 = 2/7$, the same result as that obtained with the autotransformer law.

in detail later. At present it is enough to say that an x-ray tube has two circuits: (1) a filament circuit carrying the current needed to heat the filament and (2) the tube circuit itself carrying the tube current which passes between the electrodes of the x-ray tube and produces x rays. Of utmost significance is the fact that *a small change in filament current produces a large change in tube current.* Therefore, by regulating the filament current we can control the x-ray output in terms of exposure per unit of time, or exposure rate.

Four options allow control of filament current: choke coil, rheostat, saturable reactor, and high-frequency circuit.

Choke Coil

Principle. The choke coil, an electromagnetic device operating on the principle of *self-inductance,* requires *alternating current.*

In Chapter 8 we learned that an electric current in a coil sets up a magnetic field so that one end of the coil becomes a north pole and the opposite end a south pole. If the magnetic field varies in direction and strength, as when it is generated by an *alternating current,* the magnetic flux links with or cuts the coil itself and induces a voltage which *opposes* the voltage already applied across the coil. This is the *back emf of self-induction.*

The introduction of an *iron core* into the coil intensifies the magnetic field and increases the inductive reactance of the coil (analogous to, but not identical with, resistance) producing the following effects:

1. A larger voltage drop across the choke coil.
2. A smaller remaining voltage for the rest of the circuit.
3. A *decrease in current* due to the larger inductance of the circuit.

These effects become more pronounced with deeper insertion of the core into the coil. The principle of the choke coil is explained further in Figure 10.8.

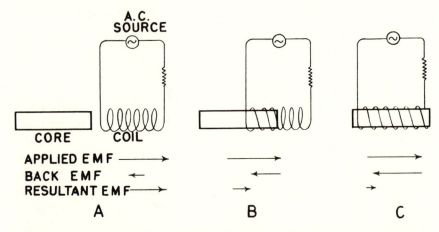

Figure 10.8. Operation of a choke coil. Note that AC is required. Arrow length indicates the emf in that direction. In *A*, with the core completely out, back emf is minimal and voltage drop across the coil is small, so the resultant emf is only slightly less than the applied emf. In *B*, partial insertion of the core increases back emf and voltage drop across the coil. In *C*, with the core completely inserted, back emf of self-induction is maximal, producing a large voltage drop across the coil with a small resultant emf available for the rest of the circuit; *at the same time, the current is decreased.* The adjustable position of the core permits selection of the desired emf and current.

The choke coil was previously used to regulate filament current in the x-ray tube, to permit control of tube current (mA). However, it has been largely replaced by a rheostat, saturable reactor, or high-frequency control.

Rheostat

A rheostat is a *variable resistor*, designed to regulate *filament current* in radiographic equipment, as shown in Figure 10.9. A variable resistor controls not only the voltage but also the current, according to Ohm's Law, $I = V/R$. If the rheostat resistance is increased, the current in the filament circuit decreases, so there will be less heating of the filament, a smaller electron emission, and a much smaller tube current. But at the same time, the increased rheostat resistance causes an increased voltage drop across the rheostat, thereby leaving a smaller voltage for the filament, provided the total voltage drop across the entire filament circuit is kept constant.

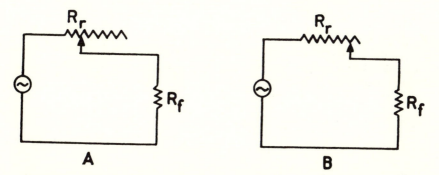

Figure 10.9. Diagram of a filament circuit where R_r is the resistance setting of the rheostat, and R_f is the resistance of the tube filament. In *A*, with a smaller value of rheostat resistance, a large current flows in the circuit including the filament, ie, filament circuit. In *B*, with a larger value of rheostat resistance, there is a small current in the circuit. In each instance there is a corresponding change in the filament voltage.

Saturable Reactor

The saturable reactor is a more advanced device than the rheostat for regulating filament current. It depends on the electromagnetic principle, not discussed heretofore, that if an iron core within or near a coil is saturated with magnetic flux by an independent source of direct current, the inductance of the coil increases as the degree of saturation of the core increases. This variable degree of core saturation is accomplished by varying the size of the applied

DC. As usual, an increase in the inductance of a circuit causes a decrease in the current, *I*. Figure 10.10 shows a simplified version of a saturable reactor.

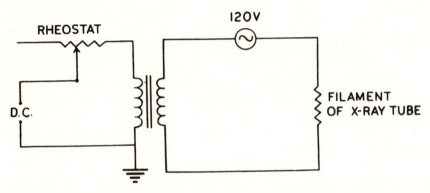

Figure 10.10. Current control by a saturable reactor. The DC voltage in the left coil saturates the iron core with a steady magnetic flux whose strength can be changed by changing the voltage by means of a rheostat. The higher the saturation of the core, the greater will be the inductance of the right coil and the lower will be the current in the filament circuit.

High-Frequency Filament Control

The most modern radiographic-fluoroscopic units, designed especially for the larger hospital departments, have filament voltage controlled by a *high-frequency circuit* instead of a rheostat or a saturable reactor. Several methods of high-frequency control serve this purpose.

Shown in Figure 10.11 is a block diagram of a typical circuit. It starts with a fixed 120-V DC supply, obtained by conversion from an ordinary 120-VAC° outlet. The DC voltage is fed, through a *transistor filament driver* to the *center* of the primary winding of the filament stepdown transformer. A 120-Hz "chopper" power supply to the two *ends* of the transformer primary produces an AC at this frequency by alternately opening and closing the upper and lower switches in the figure. Now comes the interesting part: *during an x-ray exposure*, a *pulse width modulator (PWM)*, operating at a constant frequency of 5 kHz, receives feedback signals from

---

°VAC = volts alternating current

two sources—the *filament* and the *mA selector.* If the filament feedback signal is *lower* than the mA selector signal, the PWM instantly lengthens (ie, widens) the squared voltage pulse (see Figure 10.11A); this increases the filament voltage and current just enough to restore the selected mA (tube current).

On the other hand, if the filament feedback signal is higher than that of the mA selector, the PWM shortens (ie, narrows) the squared voltage pulse so as to restore the selected mA.

In some systems, filament control is achieved by feedback to a frequency modulator, which alters AC frequency to adjust filament current.

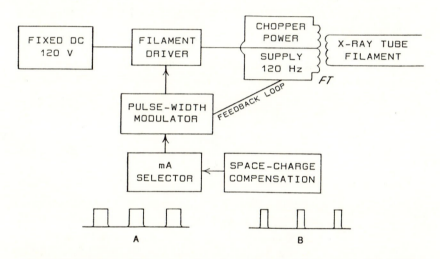

Figure 10.11. Block diagram of high-frequency control of filament voltage. *FT* is the filament transformer. *A* and *B* show wider and narrower squared voltage pulses, respectively.

Thus, by either system, mA is fine-tuned to a constant value during tube operation. In Figure 13.2 you will find that a small change in filament current entails a large change in tube current (mA).

QUESTIONS AND PROBLEMS

1. Why is a high voltage needed in x-ray equipment?
2. What device is used to change low voltage to high voltage?
3. Discuss the transformer law. A transformer is so constructed

that there are 100 turns in the primary coil and 100,000 turns in the secondary. If the input is 110 volts, what is the output voltage?

4. Show by diagram a closed core transformer, first as a step-up transformer, second as a step-down transformer.

5. How are eddy currents reduced in a transformer? Why should they be kept to a minimum?

6. Why is a core used in a transformer?

7. What two types of insulation are used in a transformer and why?

8. Discuss the purpose and principle of an autotransformer. How does it operate? What is the autotransformer law?

9. Why is the autotransformer preferred in the control of high voltage in an x-ray machine?

10. The main electrical supply to an x-ray machine is 240 volts. If the x-ray transformer has a step-up ratio of 1000 and we wish to obtain 100 kV, what will be the step-down ratio of the autotransformer?

11. Describe a rheostat. In what type of x-ray equipment is it most widely used today? How does it operate?

12. Explain the principle of the choke coil. By what devices has it been replaced in radiographic equipment?

13. Which type of current (AC or DC) is needed for the operation of choke coil; autotransformer; rheostat?

14. Why is it necessary to regulate the filament current of an x-ray tube? How is tube current related to filament current.

15. What purpose is served by the saturable reactor? How does it operate?

16. Describe the operation of a high-frequency filament circuit. What are its advantages over the rheostat and saturable reactor?

Chapter 11

RECTIFICATION

Definition

A S WE HAVE pointed out before, an alternating current (AC) periodically reverses direction and varies in strength. While an x-ray tube can operate on *any* high voltage AC, it *operates most efficiently when supplied by high voltage direct current* (DC) — a current that is *unidirectional*, flowing always in the same direction. *Rectification may be defined as the process of changing alternating current to direct current.*

There are two possible ways of rectifying the high voltage AC supplied by the secondary side of the step-up transformer: (1) by suppressing that half of the AC cycle represented by the portion of the curve that lies below the line (ie, the negative half cycle) or (2) by changing the negative half cycle to a positive one. The resulting curves are shown in Figure 11.1.

As an aid to understanding the purpose and process of rectification, we shall summarize the structure and operation of x-ray tubes, although Chapter 13 contains a detailed account of this important subject.

An *x-ray tube* consists of a glass bulb from which the air has been evacuated as completely as possible. Sealed into the ends of the bulb — or *tube,* as it is better known — are two terminals or electrodes: the *cathode* and the *anode.* The cathode consists of a thin tungsten wire (filament) which, when heated to incandescence (white hot) by a *low voltage current,* gives off electrons. These form a cloud or *space charge* near the filament. If, now, a *high voltage* from the secondary of the transformer is applied between the cathode and anode so as to produce a large negative charge on the cathode and a large positive charge on the anode, the resulting strong electric field drives the space charge electrons toward the anode at tremendous speeds. When the electrons strike the anode, their *kinetic energy* is changed to heat and x rays.

139

During ordinary operation, this electron current can pass in *only one direction, from cathode to anode.* If the applied AC has first been rectified so that the resulting current is always directed from cathode to anode, the tube can withstand greater energy loading (ie, exposure factors) than if the current has not been rectified.

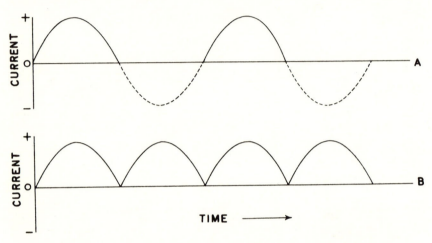

Figure 11.1. Two fundamental types of rectified circuits. In *A,* with half-wave rectification, the negative half lying below the line is *suppressed.* In *B* the negative half of the alternating current now lies above the line and becomes positive. In either case, a *pulsating* direct current is produced.

Methods of Rectifying an Alternating Current

There are two main systems of rectification: (1) self-rectification and (2) vacuum tube or solid state diode rectification. Vacuum tube diodes, also called *valve tubes* were used for many years in radiographic equipment. However, solid state diodes (see pages 225–233) have supplanted valve tubes in modern diagnostic units, both single-phase and three-phase. *All rectifying systems are connected between the secondary side of the x-ray transformer and the x-ray tube.*

1. **Self-rectification.** In this, the simplest type, the high voltage is applied *directly* to the terminals of the x-ray tube. Under ordinary conditions, an x-ray tube allows passage of electrons only from the cathode to the anode during the positive half cycle of the

AC curve, when the anode is positively charged. This half of the voltage cycle is the *useful voltage* or *forward bias*.

During the negative half cycle the "anode" is negative and the "cathode" is positive, but despite the presence of a high voltage across the tube (however, in the "wrong" direction) no current will *normally* flow because there is no space charge near the anode. At the same time, the reverse voltage — also known as the *inverse voltage* or *reverse bias* — is actually higher than the useful voltage. This is caused by *transformer regulation*, a condition associated with power loss in the transformer core and characterized by a slight fall in kV accompanying the flow of tube current (mA) during positive or forward bias. Such a fall in kV does not occur during negative or reverse bias because there is no current in the tube.

The principal disadvantage of self-rectification is the low heat-loading capacity (ie, limitation to low exposure factors) of an x-ray tube in such a circuit. In a self-rectified circuit the anode must never be heated to the point of electron emission because then, during the negative half cycle, the inverse voltage would drive the electrons in the wrong direction — toward the filament — causing the filament to melt and ruining the tube. Therefore, with self-rectification, exposure factors (kV and mAs) must be limited to lower values (lower tube rating) than with full-wave rectification.

An important point to remember is that the milliammeter indicates average mA. In a self-rectified circuit, because only one-half the cycle is used, the *peak* mA and consequent anode heat must be greater, for the same milliammeter reading than in a full-wave rectified circuit in which both halves of the cycle are used.

Figure 11.2 describes the current through the tube in a self-rectified circuit. This type of rectification is known as *self-half-wave* or single-pulse rectification (one voltage peak per cycle).

Self-rectification was formerly used in mobile x-ray units, but it has been replaced by more sophisticated methods of rectification, to be described next.

2. **Diode Rectification.** A diode normally passes current in one direction only, that is, *electrons flow from cathode to anode.* Although it is customary in solid state circuitry to show the current passing from anode to cathode (flow of "holes," see Figure 13.26), we shall consistently trace the *flow of electrons* in x-ray circuits in this course since it is conceptually easier.

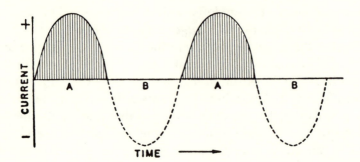

Figure 11.2. Curve of tube current in a self-rectified circuit (single-pulse rectification). During the first half cycle, *A,* the current passes through the tube; the applied voltage is the *effectual* or *useful voltage.* During the second half, *B,* no current flows from anode to cathode and now the applied voltage is *ineffectual* or *inverse voltage.* Note that the useful voltage is less than the inverse voltage.

Either one or two diodes can provide half-wave rectification, also known as *one-pulse* rectification (one voltage peak per cycle). Figure 11.3 shows a circuit containing *one rectifier diode.* During the *positive half* of the AC cycle, the polarity of the secondary coil of the transformer is that shown in the figure, the direction of the electron current being from cathode to anode. The current flows through the rectifier and the x-ray tube as indicated by the arrows. During the negative half cycle, when the polarity of the transformer secondary is reversed, the rectifier diode is nonconductive; in fact, it behaves as an open switch, thereby preventing application of high voltage to the x-ray tube. Thus, the rectifier *suppresses the inverse voltage,* diminishing the possibility of reverse flow of electrons in the x-ray tube, should the anode become hot enough during operation to be a source of electron emission. Such a reverse flow of electrons could easily destroy the filament. Therefore, the diode allows greater loading of the x-ray tube than is possible with self-rectification, although we obtain half-wave or one-pulse rectification in either case (see Figure 11.2).

With *two rectifier diodes,* as shown in Figure 11.4, the high voltage during the negative half cycle is divided among the diodes and the x-ray tube, thereby increasing the efficiency of the system and improving the heat-loading capacity of the x-ray tube. The wave form of two-diode rectified current resembles that of self-rectified current (see Figure 11.2); in other words, it, too, is single- or one-pulse rectification.

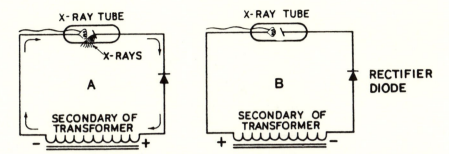

Figure 11.3. Half-wave rectification with a single rectifier diode. In *A*, electrons flow readily in both the x-ray tube and rectifier diode; this is *forward bias*. In *B*, during the next half cycle, the polarity of the transformer has reversed and electrons cannot flow from anode to cathode in the rectifier diode, thereby protecting the x-ray tube from the inverse voltage, also called *reverse bias*.

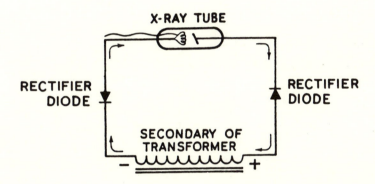

Figure 11.4. Half-wave rectification with two rectifier diodes. When the polarity of the transformer reverses, no current flows in the circuit.

Four rectifier diodes can be arranged to provide *full-wave rectification* as shown in Figure 11.5; this is *two-pulse rectification* since there are two voltage peaks per cycle. In *A*, the current flows from the negative pole of the transformer through the successively numbered portions of the circuit to the cathode of the x-ray tube, eventually returning to the positive end of the transformer. (Remember that each rectifier diode permits the current to flow only in one direction, from cathode to anode.) In *B*, the polarity of the transformer reverses due to alternation of the current. The current again follows the successively numbered portions of the circuit to the cathode of the x-ray tube, finally reaching the positive end of

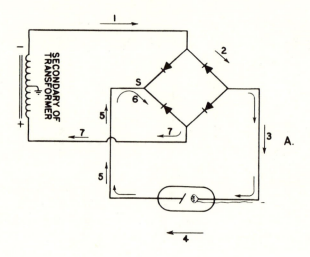

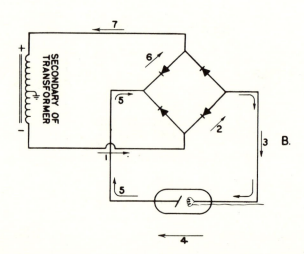

Figure 11.5. Full-wave rectification with four rectifier diodes. By following the numbered segments of the circuits in *A* and *B*, you can see that regardless of the polarity of the transformer engendered by the AC, the electron current always reaches the x-ray tube in the same direction; that is, both half cycles are used. This is called two-pulse rectification because of the two voltage peaks per cycle.

the transformer. Note that at some time during its course through the circuit, the current arrives at the junction of two rectifier cathodes, as at point *S* in *A;* the current passes through one of the diodes because passage through the other diode would lead it back to a point of higher potential and this is physically impossible (analogy: a ball will not, of itself, roll uphill to a point of higher potential energy).

You should encounter no difficulty in learning the proper connection of the rectifier diodes once you have mastered the following simple rules:

(1) First arrange a circuit without indicating the diodes, as shown in Figure 11.6.

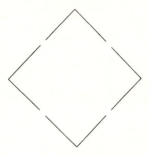

Figure 11.6.

(2) Then insert two cathodes joined as in Figure 11.7.

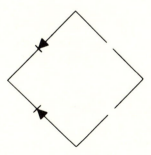

Figure 11.7.

(3) Now connect these cathodes to the anode of the x-ray tube, as in Figure 11.8.

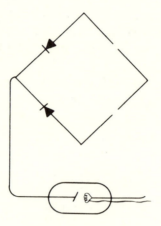

Figure 11.8.

(4) Next, insert two diodes on the opposite side of the circuit, with their anodes joined, and then connect them with the cathode of the x-ray tube, as in Figure 11.9.

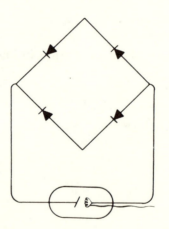

Figure 11.9.

(5) Finally, connect the remaining terminals to the terminals of the transformer, as in Figure 11.10.

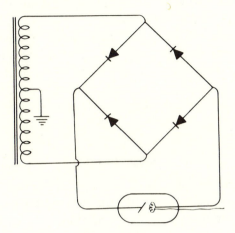

Figure 11.10.

The above rules are necessary because there is only one correct way to connect the rectifier diodes to each other and to the x-ray tube and transformer. If this relationship is altered in any way, the system will fail to rectify. Notice that in any particular half cycle, one parallel pair of rectifier diodes is conductive (see Figure 11.5). The wave form of the full-wave rectified current is shown in Figure 11.1.

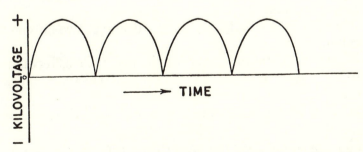

Figure 11.11. Wave form of a current rectified by four diode rectifiers. Note that all the values of kV are represented, from zero to the maximum value, or kVp. This is two-pulse rectification since there are two voltage peaks per full cycle.

There is one disadvantage in four-diode rectification. Using the entire AC wave results in the production of a high percentage of low energy x rays, that is, low kV range. But these can be minimized either by using appropriate filters or by boosting the lower kV in the rectified wave form to higher kV by means of capacitors, as in constant-potential equipment.

Full-wave or two-pulse rectification permits a large increase in the tube rating or heat-loading capacity (relative to half-wave or single-pulse rectification), which means that larger exposures can be used without damaging the tube. But we must emphasize that a tube must be used in accordance with the rating furnished by the manufacturer for that particular tube.

QUESTIONS

1. Define rectification.
2. Why is it desirable to rectify the current for x-ray generation?
3. Name two main types of rectification.
4. How are x rays produced?
5. Show with the aid of a diagram: (a) single-diode rectification and (b) four-diode rectification. Draw the shapes of the current waves produced by each.
6. What is meant by a constant potential unit? Under what conditions, and why, is it desirable?
7. Draw and label the current curves in a self-rectified circuit; a single-diode rectified circuit; a four-diode rectified circuit. Which are half-wave rectified? Full-wave rectified?
8. Define inverse voltage; reverse bias; useful voltage. What is their importance?
9. What is meant by single-pulse rectification? Two-pulse rectification?
10. What property of diodes makes them suitable as rectifiers?
11. In what type of equipment might you find self-rectification used?
12. Why are diodes not conductive during the inverse half of the AC cycle?
13. Show by aid of a diagram a rectifier using four diodes. Draw the wave form of the rectified current. Why is it called full-wave or two-pulse?

Chapter 12

X RAYS (ROENTGEN RAYS)

How X Rays Were Discovered

FOR A NUMBER of years before the discovery of x rays physicists had been observing high voltage discharges in vacuum tubes. In 1895 Wilhelm Konrad Roentgen, a German physicist, was studying these phenomena in a Crookes tube operated at high voltage in a darkened room. Suddenly, he noticed the *fluorescence* (glowing) of a barium platinocyanide screen lying several feet from the end of the tube. He soon realized that the fluorescence was caused by a hitherto unknown type of invisible radiation. Moreover, he found that this radiation could pass completely through solid materials such as paper, cardboard, and wood, since they did not prevent fluorescence when placed between the tube and the barium platinocyanide. However, the rays could be stopped by denser materials such as lead.

Thus, by the sheerest accident, Roentgen had discovered a new type of radiation which, upon leaving the Crookes tube, was capable of passing through certain solids that are opaque to ordinary light.

Roentgen quickly discovered the potentiality of the new radiation in medicine when he placed a hand (presumably his wife's) between the tube and a piece of cardboard coated with barium platinocyanide. Imagine his excitement on seeing the bones depicted on this fluoroscopic screen for the first time!

Roentgen gave the name *x rays* to his newly found, invisible, penetrating radiation because the letter x represents the unknown in mathematics, although within the next few months he had ascertained most of the properties of x rays.

The discovery of x rays changed the course of medical history, to say nothing of its impact on science and industry. Roentgen's name is often directly linked with x rays in that they are also known as *roentgen rays*. *Roentgenology* is that branch of medicine dealing

149

with the use of x or roentgen rays in diagnosis and treatment. The term *radiology* includes not only x rays but also natural and artificial radionuclides, computerized tomography, magnetic resonance imaging, and ultrasonography. *Radiography* deals with the art and science of recording x-ray images on film, paper, magnetic tape, and magnetic disc. *Fluoroscopy* includes the observation of x-ray images on a screen coated with fluorescent material, either directly or with the aid of an intensifying system.

Nature of X Rays

An x-ray beam consists of a group of rays that are related to white light, ultraviolet, infrared, and other similar types of radiant energy. They are all classified as *electromagnetic radiation* — wavelike fluctuations of electric and magnetic fields set up in space by oscillating (vibrating) electrons. A mechanical analogy would be, for example, the waves produced on the surface of water by a dropped stone.

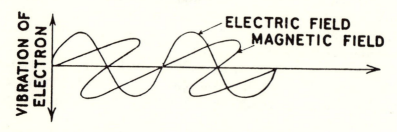

Figure 12.1. The electrical and magnetic components of an electromagnetic wave oscillate (vibrate) in mutually perpendicular planes. From J. Selman: *The Basic Physics of Radiation Therapy*, 3rd Ed. Springfield, Thomas, 1990.

The electric and magnetic fields fluctuate perpendicular to their direction as well as to each other, as shown diagrammatically in Figure 12.1. As they are identical in form, we usually depict only one of them as in Figure 12.2.

All electromagnetic waves such as radio, heat, light, x rays and gamma rays have the same general form and travel with the same *constant speed* as light — 3×10^8 meters, 3×10^{10} cm, or 186,000 miles per sec in a vacuum or air. However, they differ in *wavelength* — the distance between two successive crests in the wave, such as

A to *B* in Figure 12.2. The number of crests or cycles per second is the *frequency* of the wave, the unit of frequency being the *hertz* defined as one cycle per sec. You can see, by comparing the upper and lower waves in Figure 12.2, that if the wavelength is decreased, the frequency must increase correspondingly. This can also be shown in another way. The speed with which the wave travels is equal to the frequency multiplied by the wavelength:

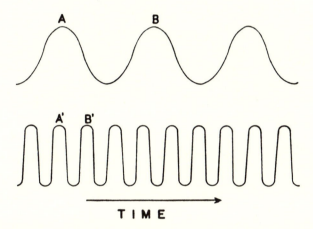

T I M E

Figure 12.2. Electromagnetic waves. The upper and lower wave trains differ in wavelength, that is, the distance between two crests such as *A* to *B* and *A'* to *B'*. The lower wave thus has a shorter wavelength and therefore more crests or cycles in a given period of time, than does the upper wave, because their speeds are identical. Thus, the wave with the shorter wavelength has greater frequency or number of cycles per sec (hertz or Hz).

$$c = \nu\lambda \qquad (1)$$

where c = speed of x rays or light in vacuum or air
$(3 \times 10^8$ m per sec$)$
ν (Greek *nu*) = frequency in hertz or Hz (cycles per sec)
λ (Greek *lambda*) = wavelength in m (meter)

Since c, the speed of all electromagnetic waves, is constant in a given material, an increase in frequency (ν) must always be accompanied by a corresponding decrease in wavelength (λ) and, conversely, a decrease in frequency by an increase in wavelength. **Frequency is inversely proportional to wavelength.**

The wavelengths of x rays are extremely short; for instance, in **ordinary radiography,** their useful range extends from about 0.1 to

0.5 Å (recall that 1 Å = 10^{-10} or one ten-billionth m = 0.1 nm). The ultrashort wavelengths are associated with enormous frequencies—3 × 10^{19} to 6 × 10^{18} hertz (Hz).

The various kinds of electromagnetic radiation, ranging from radio waves with wavelengths measured in several thousands of meters, down to high energy x and gamma rays with wavelengths measured in nanometers (10^{-9} meter), as shown in Figure 12.3.

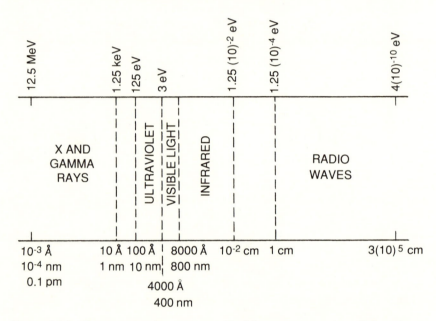

Figure 12.3. Electromagnetic spectrum. Useful energy range in radiology is as follows:

Radiography	40 to 120 kVp
Cobalt 60 gamma rays	1.25 MeV
Linear accelerator	4 to 25 MV

As we shall explain in more detail later (see pages 166–168), each electromagnetic wave also behaves as a bit of energy that depends on the frequency of the wave. Such an energy bit is called a ***photon*** or ***quantum***. It is preferable to designate x and gamma rays according to their photon energy rather than their wavelength.

Source of X Rays in Radiology
The X-ray Tube

Whenever a stream of fast moving electrons suddenly undergoes deceleration (ie, reduction of speed), x rays are produced. While these conditions may prevail in nature, they can also be brought about artificially by means of an *x-ray tube*. Gamma rays, which are identical to x rays, are emitted by the *nuclei* of certain radionuclides.

At this point we shall outline the main features of a typical x-ray tube, reserving a more detailed discussion of the subject for a later section.

Historically, the early x-ray tubes were cold cathode gas tubes. Such a tube consisted of a glass bulb in which a partial vacuum had been produced, leaving a small amount of gas. In it were sealed two electrodes, one negative (cathode) and the other positive (anode). The cathode terminal was *not* heated. Application of a high voltage across the terminals caused ionization of the gas in the tube with release of a stream of electrons which produced x rays upon striking the positive terminal (anode). The gas tube is now obsolete not only because of its inefficiency but also because tube mA could not be changed independently of kV, making it difficult to control x-ray quantity and quality.

In 1913 W. D. Coolidge, at the General Electric Company Laboratories, invented a new type of x-ray tube based on a radically different principle (Edison effect); this was the *hot cathode diode tube,* which revolutionized radiographic technic because it made possible the independent control of mA and kV. The principles of construction and operation are surprisingly simple and must be clearly understood by all those concerned with the use of x rays.

An x-ray tube has the following components (see Figure 12.4):

1. A *glass envelope* or *tube* from which the air has been evacuated as completely as possible. Air must be removed not only from the interior of the tube, but from the glass and metal parts as well, by prolonged baking before the tube is sealed. This process is called *degassing.* A high vacuum is necessary for two reasons:

   a. To prevent collision of the high speed electrons with gas molecules, which would cause significant slowing of the electrons (see below).

b. To prevent oxidation and burning-out of the filament.

2. A *hot filament* supplied with a separate low-voltage heating current. The filament serves as the *cathode*, or negative electrode, when the high voltage is correctly applied.

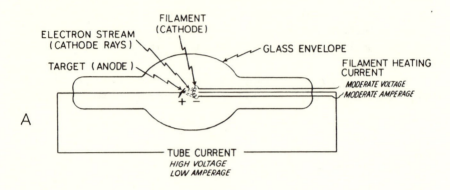

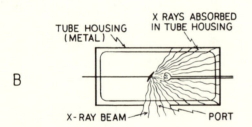

Figure 12.4. Model of an x-ray tube. *A* shows the two circuits and the general features. The inset at *B* indicates that x rays are emitted from the target in all directions, but only those approaching the port (window) pass out as the *useful beam,* the remaining x rays being absorbed in the protective housing.

3. A *target* which serves as the *anode*, or positive electrode, when high voltage is correctly applied.

4. A *high voltage* applied across the electrodes, charging the filament negatively and the target positively.

Notice that there are *two circuits in the x-ray tube:* one, a low voltage heating circuit through the cathode itself; and the other, a high-voltage circuit between the cathode and anode to drive the electrons (see Figure 12.4). However, at the cathode end, the high voltage conductor is connected to one side of the filament circuit.

Conditions Necessary for the Production of X Rays

Whenever (1) *fast-moving electrons undergo rapid deceleration* or (2) *electrons drop from an outer shell to a hole in an inner atomic shell, x rays are produced.* The x-ray tube is a device for obtaining free electrons, then speeding them up, and finally stopping them. In addition to these basic processes, the electrons must also be concentrated on a small area of the anode known as the *focus.* These four conditions required for the production of x rays in a hot cathode tube will now be described in some detail because of their extreme importance.

1. **Separation of Electrons.** In common with all other atoms, the tungsten atoms in the filament have orbital electrons circulating around a central nucleus. How can these electrons be liberated from their atoms? The *filament current* supplying the filament causes it to become glowing hot or *incandescent,* with resulting separation of some of its outer orbital electrons. We may regard such electrons as escaping from the filament to form a cloud or *space charge* nearby. The electrons liberated in this manner are called *thermions,* and the process of their liberation through the heating of a conductor by an electric current is called *thermionic emission.*

2. **Production of High Speed Electrons.** If, now, a high potential difference, ie, kV, is applied across the tube, giving the filament a very high *negative charge* (cathode) and the target an equally high *positive charge* (anode), the resulting strong electric field causes the space charge electrons to rush at an *extremely high speed through the tube from cathode to anode,* this electron stream constituting the *cathode rays* or *tube current. The speed of these electrons approaches one-half the speed of light,* and even more in some modern equipment.

3. **Focusing of Electrons.** The electron stream in the tube is confined to a very narrow beam and is concentrated on a small spot on the anode face—*the focal spot*—by a negatively charged metal focusing cup in which is suspended the filament; the narrower the electron beam the smaller the focus and the sharper the x-ray images.

4. **Stopping of High Speed Electrons in Target.** When the fast electrons enter the target of the x-ray tube, their *kinetic energy* changes to other forms of energy. The efficiency of ordinary

radiographic equipment is such that *at 80 kV$_p$ only about 0.6 percent of this energy is converted to x rays while the remaining 99.4 percent appears as heat.* (See page 159 for a discussion of efficiency of x-ray production.) In fact, only about one part in a thousand of the kinetic energy of the electrons eventually results in radiologically useful x rays! As shown in Figure 12.4B, x rays are emitted in *all* directions, but only those leaving the window of the x-ray tube comprise the *useful beam.*

You must avoid confusing the electrons flowing in the tube, with the x rays emerging from the tube. Imagine a boy throwing stones at the side of a barn. The stones are analogous to the electron beam in the x-ray tube, whereas the emerging sound waves are analogous to the x rays that come from the target when it is struck by electrons.

Electron Interactions with Target Atoms X-ray Production

When the fast electron stream enters the tube target, the electrons interact with target atoms producing x rays by the following two processes:

1. **Brems Radiation.** Upon approaching the strongly positive *nuclear field* of a target atom, the negatively charged high-speed electron is deviated from its initial path because of the attraction between these opposite charges (see Figure 12.5). As a result, the electron is slowed down or *decelerated,* thereby losing some of its kinetic energy; the *lost kinetic energy is radiated as an x ray of equivalent energy.* The term *bremsstrahlung* or *braking radiation* has been applied to this process. Brems radiation is *heterogeneous* or *polyenergetic,* that is, nonuniform in energy and wavelength because the amount of braking or deceleration varies among electrons according to their speed and how closely they approach the nucleus. With each different deceleration, a corresponding amount of kinetic energy is converted to x rays of equivalent energy; just as the deceleration varies, so does the energy of the x rays. An electron that happens to approach the nucleus head-on is completely stopped by the nuclear electrostatic field. In this special case of the brems effect, all of the kinetic energy of the electron is converted to an x ray of equivalent energy. The deceleration of electrons depends also on the atomic number of the target. Thus, targets of

higher atomic number are more efficient producers of brems radiation (eg, rhenium with atomic number 75 as opposed to tungsten with atomic number 74). It should be pointed out that electrons also interact with, and are repelled by, the electrons swarming around the target nuclei. In this process heat is transferred to the target atoms; such heat production far exceeds x-ray production at ordinary values of tube potential (kV).

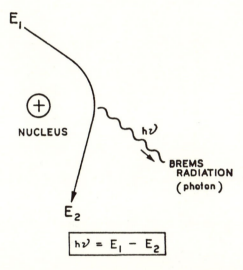

Figure 12.5. Production of bremsstrahlung (brems radiation). An electron with energy *E*, on approaching an atomic nucleus, experiences an attractive electrostatic force which causes it to change direction along a curved path. In changing direction the electron radiates energy *h* which must come from the kinetic energy of the electron. Therefore the electron moves away from the nucleus with energy E_2. The radiated brems photon energy $h\nu = E_1 - E_2$.

2. **Characteristic Radiation.** This is the other process of x-ray production in an x-ray tube. An electron with a sufficient minimum kinetic energy may interact with an *inner orbital electron* (for example, in *K* or *L* shell) of a target atom, ejecting it from its orbit (see Figure 12.6). To free this electron, work has to be done to overcome the attractive force of the nucleus. The atom is now unstable, being *ionized* (electron missing from atom) and in an *excited state* (electron vacancy in a shell). Immediately, the space or "hole" vacated by the electron is filled by electron transition from one of the outer shells. Since, in the first place, energy was

put into the atom to free the electron, a like amount of energy must be given off when an electron from a higher energy level enters to fill the hole in the shell (satisfying the Law of Conservation of Energy). This energy is emitted as a *characteristic x ray* because its energy is characteristic of the target element and the involved shells. (In fact, there are series of characteristic rays because, for example, the replacing electron leaves a hole which must, in turn, be filled by an electron from a still higher energy level or shell.)

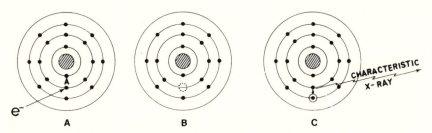

Figure 12.6. Production of characteristic radiation. In *A*, the incoming *electron* collides with an inner shell electron. In *B*, the atom is in an "excited state," an electron now being displaced from its shell. In *C*, an electron jumps from an outer shell to replace the ejected electron; this is accompanied by characteristic radiation, as the atom returns to its normal state.

We may now summarize the nature of the *primary radiation,* that is, the x rays emerging from the target:

a. *General Radiation*—"white," polyenergetic, or heterogeneous radiation with a continuous range of energies (and wavelengths) due to deceleration of electrons by strongly positive electric fields of nuclei in target atoms. This is *brems radiation,* constituting about 90 percent of emitted x rays when 80 to 100 kV is applied to the tube.

b. *Characteristic Radiation*—consists of limited, discrete energies (and wavelengths), constituting about 10 percent of emitted x rays when the tube potential is in the range of 80 to 100 kV. No characteristic radiation is produced at tube potentials less than 69 kV with a tungsten target.

Target Material

The target metal must be selected with two aims in mind. First, it must have a very **high melting point** to withstand the extremely high temperature to which it is subjected. Second, it must have a **high atomic number** because (1) the resulting **characteristic** radiation is of high energy and therefore more penetrating and (2) there is increased production of brems radiation. Tungsten, a metal with a melting point of 3370 C and an atomic number of 74, generally satisfies the above requirements. Rhenium (atomic number 75) and tungsten, coated on a molybdenum-tungsten disc, forms the target in rotating anode radiographic tubes for more efficient x-ray production. To facilitate the dissipation of heat from the anode in **stationary** anode tubes, a tungsten button is imbedded in a block of copper, which is a better conductor of heat than tungsten. The electron beam in the tube is focused on a small spot on this tungsten target. A molybdenum target is used in mammographic units to produce low-energy x rays (see pages 345–346).

Efficiency of X-ray Production

As already indicated, only a small fraction of the energy of the cathode-ray electrons undergoes conversion to x rays in the target. Most of the electrons' energy is degraded to heat as the result of collisions with outer shell electrons of the tungsten atoms. X rays are produced by the previously described process of brems radiation and, if the applied potential (kV) is high enough, by the additional process of characteristic radiation.

The efficiency of x-ray production may be defined as the percentage of the kinetic energy of the electrons that is converted to x rays. The efficiency is directly proportional to the atomic number of the target and the applied potential, as expressed in the following equation:

$$E = K \times Z \times kV_p \qquad (2)$$

where E is the efficiency in percent, K is a constant = 1×10^{-4}, Z is the atomic number of the target, and kV_p is the peak kilovoltage. For example, with a tungsten target (Z = 74) and 80 kV_p.

$$E = 1 \times 10^{-4} \times 74 \times 80$$
$$= 0.6\%$$

This means that of the total kinetic energy of the electron stream, only about 0.6 percent appears as x rays and the remaining 99.4 percent as heat in the anode when the applied potential is 80 kV_p.

Properties of X Rays

We shall now list some of the more important properties of x rays:

1. *Highly penetrating, invisible rays,* which belong to the general category of electromagnetic radiation and behave both as *waves and particles.*

2. *Electrically neutral,* so they cannot be deflected by electric or magnetic fields.

3. *Polyenergetic* or *heterogeneous,* having a wide span of energies and wavelengths; useful energy range in radiography is about 25 to 120 kVp (maximum photon energy 25 to 120 keV).

4. Liberate *minute* amounts of heat on passing through matter.

5. Travel ordinarily in *straight lines.*

6. Travel only at the *same speed as light,* 3×10^8 m per sec or 186,000 miles per sec in a vacuum.

7. Ionize gases *indirectly* because of their ability to remove orbital electrons from atoms.

8. Cause *fluorescence* of certain crystals, making possible their use in fluoroscopy, and in radiographic intensifying screens.

9. *Cannot be focused* by a lens.

10. *Affect photographic film,* producing a latent image which can be developed chemically.

11. Produce *chemical and biologic changes* by ionization and excitation.

12. Produce *secondary and scattered radiation.*

SPECIFICATIONS OF THE PHYSICAL CHARACTERISTICS OF AN X-RAY BEAM

One may describe a given x-ray beam on the basis of two properties—*intensity* and *quality.* Strictly speaking, the intensity of the radiation at a given point in the beam refers to the quantity of radiant energy flowing per second through a unit area of the surface perpendicular to the direction of the beam at the designated point. However, because of its simplicity and adequacy, the concept of *exposure rate* has been adopted in medical radiology. Furthermore, with the main interest in radiotherapy being centered on the interaction of radiation with the patient's tissues, another unit of quantity has had to be devised—the *absorbed dose.*

The *quality* of an x-ray beam refers to its *penetrating ability.* We shall now discuss the concepts of exposure and quality, leaving absorbed dose for a later section.

X-ray Exposure (Quantity)

Under certain controllable conditions, the degree of ionization (see pages 53–54) produced by radiation in matter is proportional to the quantity of radiation of a particular type. Therefore, ionization would be expected to serve as an appropriate basis for measuring x- and gamma-ray quantity in terms of exposure. Indeed, in 1928 the Second International Congress of Radiology established the *roentgen,* now symbolized by R, as the international unit of exposure *based on ionization in air.* Special instruments known as *ionization chambers* are used for the measurement of exposure (see page 163).

In 1938 the roentgen was redefined as *"that exposure of x or gamma radiation such that the associated corpuscular emission per 0.001293 gram of air, produces in air, ions carrying one electrostatic unit of quantity of electricity of either sign."* This may be simply explained as follows: a roentgen is a definite radiation *exposure* which, in a standard quantity of air (chosen as 0.001293 g of air, which is really the weight of 1 cc of air at a temperature of 0 C and a pressure of 760 mm mercury), produces ionization consisting of a certain number of ions and electrons (called corpuscular emission in the definition). Eventually, all the usable energy of the corpuscles produced by one roentgen is spent in releasing pairs of oppositely charged ions. If the ions released by an exposure of 1 R are collected and measured, they will be found to carry one electrostatic unit (esu) of positive or negative charge. In other words, *1 R of x or gamma rays causes ionization in 1 cc of air under standard conditions, such that the ions of only one sign carry 1 esu of electric charge.* It should be pointed out that the roentgen is a valid unit only with x or gamma rays *up to 3 million volts.*

A revised definition of the roentgen as the unit of exposure was adopted by the International Commission on Radiation Units and Measurements (ICRU) in 1962, as expressed in the following equation:

$$X = \Delta Q / \Delta m \tag{3}$$

where X is the exposure in R, and ΔQ is the sum of all the electric charges on all the ions of *one sign* produced in air, when all the electrons released by photons (x and gamma rays) in a mass of air, Δm, are completely stopped in air.

Note that the most recent definition of exposure (equation [3]) implies two steps in the process: (1) photons first release electrons by interaction with atoms in the air, and (2) these primary electrons go on to produce ions whose total charge of one sign (positive or negative) is a measure of exposure in air. As specified in the definition, all the electrons released in (2) must give up all their kinetic energy in the formation of ions in air. Thus, the conditions laid down in the definition of 1938 were stated more precisely in 1962, so the value of the R is really unchanged in the newer definition. On this basis, the roentgen may be expressed as an amount of charge released per unit mass of air, specifically,

$$1\ R = 2.58 \times 10^{-4}\ \text{coulomb/kg air}$$

In the modern International System of Units — SI — the *exposure unit* replaces the roentgen, and is an exposure of one coulomb per kilogram; this unit is still not widely used.

The radiation exposure per unit time, for example R/min, at a given location is the *exposure rate:*

$$\underset{(R/min)}{\text{exposure rate}} = \frac{\text{exposure in R}}{\text{time in min}}$$

The *total exposure* in R is obtained by multiplying the exposure rate by the exposure time:

$$\underset{(R)}{\text{exposure}} = \underset{(R/min)}{\text{exposure rate}} \times \underset{(min)}{\text{time}}$$

In measuring radiation exposure, we must be careful that our instrument collects and measures all the produced ions; and that more ions are not accidentally lost to the surroundings than enter from the surroundings. The measuring instrument, designed to have a high degree of precision, is the **standard free air ionization chamber.**

For practical calibration of the roentgen output of an x-ray therapy unit, we may use a **thimble type ionization chamber** which has been calibrated initially against the more precise type of instrument. The thimble chamber was developed by H. Fricke and O. Glasser, and incorporated in the Victoreen R-meter (see Figure 12.7). Not only is this device convenient and simple to use, but it is sufficiently accurate for medical radiology if sent to the manufacturer or an approved standardization laboratory for calibration at regular intervals. However, its accuracy holds only for a limited span of x-ray or gamma-ray energies, so that outside this range, special caps must be placed over the thimble.

Other types of measuring devices have been introduced, such as the Baldwin-Farmer Sub-standard Dosemeter, as well as those using solid state circuits.

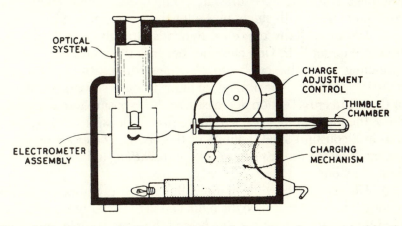

Figure 12.7. Diagram of a Victoreen-R-meter®, based on data furnished by the Victoreen Instrument Corporation. From J. Selman: *The Basic Physics of Radiation Therapy*, 3rd Ed. Springfield, Thomas, 1990.

Radiation exposure as measured in roentgens is not the same as the total quantity of radiant energy in a beam. The former is of more practical value in radiology because the exposure in R is based on the ionizing ability which, after all, is responsible for biologic and radiographic effects.

The **exposure rate** of an x-ray beam may be altered by varying four factors: (1) tube current, (2) tube potential, (3) distance, (4) filtration, and (5) target metal (discussed on pages 208, 211).

1. *Tube Current.* An increase in the filament current liberates more electrons *per second* at the filament. As a result, the tube current (mA) also increases and more electrons strike the target per sec, increasing the amount of x radiation emitted per sec. Thus, *the exposure rate is directly proportional to the mA*, if the other factors remain constant. It should be emphasized that a change in mA does not affect the quality of the radiation.

2. *Tube Potential.* A rise in *kilovoltage* (kV) increases exposure rate because the electrons in the tube are speeded up and produce more x-ray photons per sec at the target. The resulting x-ray beam also contains photons of higher energy than those produced at lower kV. Note that an increase in tube potential increases both the exposure rate and penetrating ability of an x-ray beam, but that this relationship is not directly proportional. In fact, exposure rate is approximately proportional to kV^2. As will be shown later (see page 255), the exposure rate is higher with three-phase x-ray generators than with single-phase for the same applied kV.

3. *Distance.* X-ray exposure depends strongly on the distance from the target to the ion chamber because the x rays proceed in a spreading, cone-shaped beam. As the distance from the target increases, the number of x-ray photons in each square centimeter of the beam cross-section decreases, so the exposure rate decreases. Since the ion chamber has a definite size, it measures the exposure rate in a constant small volume of the beam. *The exposure rate is governed by the inverse square law of radiation*, to be explained in detail in Chapter 18.

4. *Filtration.* The thicker the filter and the higher its atomic number, the greater will be the reduction in the exposure rate beyond the filter. At the same time, *the beam is hardened due to the greater removal of soft (low energy) than of hard (high energy) rays by the filter*, except at the absorption edge (see page 198). Preferential removal of lower energy, poorly penetrating photons by a suitable filter incidentally reduces patient exposure (see pages 526–529).

Quality of X Rays

Definition. The quality of an x-ray beam refers to its ability to penetrate matter. In radiology, we may define penetrating ability as the fraction of a particular x-ray beam that can pass through a given part of the body and onto an x-ray film; or that can enter the body and deposit energy in a tumor-containing volume at a particular depth.

However, we need a more precise concept of *penetrating ability* than that just stated. As shown in Figure 12.8A, for a diagnostic x-ray beam, suppose the measured exposure at the point where the beam enters the body were 1 R and the point where the beam leaves the body (at the film) were 0.2 R, then the transmission or penetrating ability would be $0.2/1 \times 100 = 20$ percent. With a more penetrating beam giving an exposure of 1 R at the entrance point as in 12.8B, the exposure at the film might be 0.4 R, giving a transmission or penetrating ability of $0.4/1 \times 100 = 40$ percent. Thus, the beam in 12.8B is more penetrating than that in 12.8A.

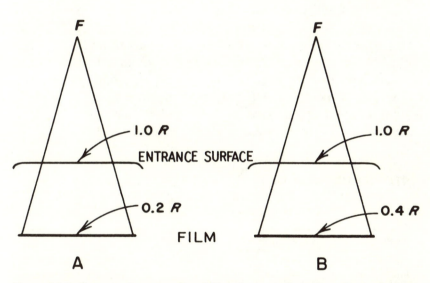

Figure 12.8. Concept of penetrating ability of an x-ray beam in radiography. In *A*, for an assumed exposure of 1.0 R at the entrance surface of the patient, the exposure at the film might be 0.2 R because of beam attenuation (absorption, scattering, distance) in the patient; this represents a transmission or *penetrating ability* of $0.2/1 \times 100 = 20\%$. In *B*, for the same entrance exposure of 1.0 R, the exposure at the film might be 0.4 R, representing a transmission or penetrating ability of 40%. Thus, beam *B* is more penetrating than beam *A*.

X-ray Energy and Quality. To understand the relationship between x-ray energy and quality (ie, penetrating ability) we must first look into the meaning of x-ray energy. An individual x ray may be regarded as a bit of energy called a *photon* or *quantum,* represented by the equation

$$E = h\nu \tag{4}$$

where E is the energy of the photon in ergs, h is a constant (Planck's constant), and ν (Greek *nu*) is the frequency of the x ray in hertz (cycles per second). This equation tells us that the *energy of a particular photon is directly proportional to the frequency of its associated wave;* therefore, x rays of high frequency have higher energy than x rays of low frequency. But since the frequency times the wavelength of an x ray equals a constant (speed of light), as shown in equation (1) on page 151, a wave of high frequency must have a short wavelength; and, conversely, a wave of low frequency must have a long wavelength.

Since the quality of x rays may be defined as their penetrating ability which, in turn, is determined by their energy, we may summarize the statements in the preceding paragraph as follows: a highly penetrating x-ray beam consists of photons which, on the average, have high energy, high frequency, and short wavelength. Conversely, a poorly penetrating x-ray beam consists of photons which, on the average, have low energy, low frequency, and long wavelength. This relationship is shown in Table 12.1.

TABLE 12.1
INTERRELATIONSHIP OF FACTORS IN RADIATION QUALITY SPEED
OF RADIATION IS CONSTANT
$c = frequency \times wavelength$

Frequency	Wavelength	Energy	Quality
high	short	high	good penetrating ability
low	long	low	poor penetrating ability

What influences the *energy* of x rays? We learned earlier that x rays consist of brems and characteristic radiation resulting from the interactions of high speed electrons with target atoms. The larger the kinetic energy of the electrons in the x-ray tube, the higher will be the energy of the resulting x radiation. Since the kinetic energy of the electrons depends on their speed which in turn depends on the potential applied to the tube, it follows that an increase in kV will ultimately produce x rays of higher energy.

This discussion may be summarized as follows: the higher the kV, the greater the speed of the electrons, the greater their kinetic energy, the greater the energy of the x rays produced, and the greater the penetrating ability of the x-ray beam. We may therefore conclude that the *penetrating ability of x rays is increased by increasing the kilovoltage, and decreased by decreasing the kilovoltage.* However, this relationship is *not* a direct proportion.

How is the energy of an x-ray beam stated? Since the energy of x rays depends on the speed of the electrons in the tube, which depends on the applied potential, we customarily use the *peak* (maximum) voltage (kVp) for the energy of the beam. But in this book we shall simply use the term kV, except where needed for emphasis. However, we must bear in mind that the beam is actually *heterogeneous* or *polyenergetic,* consisting of photons whose energies range from a minimum to the peak value, in this case 100 kV (often stated as kVp).

On the other hand, since the energy of monoenergetic radiation such as characteristic x rays and gamma rays is uniform for any given beam, it may be expressed in units of electron volts. One *electron volt* is the energy acquired by an electron when it is accelerated through a potential difference of 1 volt. Thus, the energy of monoenergetic radiation or of a particular photon is stated in electron volts (eV) or in some multiple thereof, such as kiloelectron volts (keV), million electron volts (MeV), or billion electron volts (BeV); for example, the gamma rays emitted by technetium 99m all have the same energy — 140 thousand electron volts, usually stated as 140 keV.

Insofar as equivalence is concerned, an x-ray beam having a particular kVp has a quality resembling a monoenergetic x-ray beam of about one-third to one-half the peak energy. Thus, a polyenergetic 120-kV x-ray beam is roughly equivalent to 50-keV monoenergetic x rays.

Causes of the Polyenergetic Nature of an X-ray Beam. As we have already noted, an ordinary x-ray beam is *polyenergetic,* consisting of a tremendous number of photons that vary in energy and penetrating ability. Since photon energy is directly proportional to frequency and inversely proportional to wavelength, we may say that an ordinary x-ray beam is heterogeneous, consisting of *innumerable rays of different energies and wavelengths.* Analogy with white light may help clarify this. White light comprises a mixture of various colors. When passed through a glass prism, white light separates into its components which are identical to the colors of the rainbow, or the *spectrum*. These colors differ in frequency and wavelength. For instance, the wavelength of "red" is about 7000 Å (700 nm°), whereas the wavelength of "violet" is about 4000 AÅ (400 nm). A light beam made up of a single pure color is called *monochromatic*. Similarly, an x-ray beam consists of different x-ray "colors"—energies or wavelengths—although they are invisible. X rays can be refracted by certain crystals and the photons separated according to their energy or wavelength.

We shall now explain the four major reasons for the polyenergetic nature of an x-ray beam:

1. *Fluctuating Kilovoltage.* The applied kV varies according to the wave form of the pulsating current (see Figure 9.4). Since changes in kV are manifested by changes in the energy of the resulting x rays, there must necessarily be a range of photon energies corresponding to the fluctuation of kV.

2. *Processes in X-ray Production.* The x-ray beam leaving the target consists of *general radiation* (brems radiation resulting from deceleration of electrons by target nuclear fields), and *characteristic radiation* (arising from electron transfers within the atoms of the target). The general and characteristic x rays constituting the primary beam, therefore, have a variety of energies.

3. *Multiple Electron Interactions with Target Atoms.* Electrons traveling from cathode to anode in the x-ray tube undergo varying numbers of encounters with target atoms before being completely stopped. In these encounters x rays are produced.

4. *Off-focus Radiation.* Some electrons may strike anode metal other than the focus (target), causing the emission of x rays. This *off-focus radiation* has been reduced by special design of modern tubes. Further decrease in off-focus radiation that happens to enter the x-ray beam can be achieved by special beam-limiting devices (collimators), which fit close to the tube aperture, thereby improving recorded detail (image sharpness).

X-ray Spectra. A particular x-ray beam can be precisely characterized by sorting out its photons according to their energy. This concept may be clarified by an analogy. Suppose we wish to classify a population of high school senior boys on the basis of

---

°1 nm = 10^{-9} m, or 1 billionth m

height. We would first measure all the boys in the designated population and group them according to height. These data would then be set down as in Table 12.2 and plotted graphically as in Figure 12.9, the number of boys in each group being the dependent variable, and the height the independent variable.

Similarly, we can sort out the rays or photons in a given x-ray

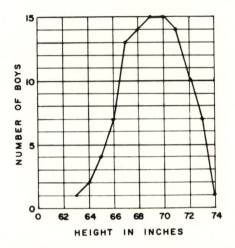

Figure 12.9. Distribution curve of a class of high school senior boys according to their height. Note that the peak incidence is at 68 to 70 inches.

TABLE 12.2

HEIGHT DISTRIBUTION OF A CLASS OF SENIOR HIGH SCHOOL BOYS

Height	Number of Boys
in.	
63	1
64	2
65	4
66	7
67	13
68	14
69	15
70	15
71	14
72	10
73	7
74	1

beam by means of a special instrument known as an x-ray spectrometer, determining the *spectrum,* or the range of intensities or relative numbers, of x rays of various energies. When these data are plotted as in Figure 12.10, a *spectral distribution curve* is obtained. In the figure, we have shown the spectral distribution curves for x-ray beams produced in a tungsten target by three different applied peak tube potentials — 30, 40, and 80 kV, respectively.

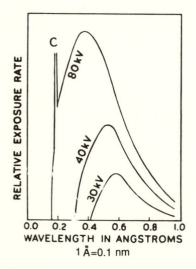
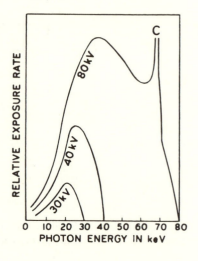

Figure 12.10. Spectral distribution curves of x rays generated by peak tube potentials of 30, 40, and 80 kV, respectively. Those on the left show relative exposure rates as a function of wavelength, and those on the right as a function of photon energy. The latter is preferred. *Characteristic radiation* appears as a peak at *C* (actually, a group of closely spaced peaks), requiring an applied potential of at least 69 kV. The remaining curves (indicated by 30 kV, 40 kV, and 50 kV) represent *general radiation.* Note that in each case the maximum photon energy in the beam is the same as the applied peak kV (figure on the right).

You can see that in each case the curve crosses the horizontal axis at a point corresponding to the peak or maximum applied kV. Thus, *the energy of the most energetic and most penetrating photons in a particular x-ray beam is governed solely by the applied peak kV.*

A comparison between these curves shows that an increase in kV causes not only an increase in the output of higher energy photons, but also an increase in output of photons of all other energies in the beam, as indicated by the areas under the respective curves. Thus, the higher the kV, the higher will be the exposure rate, and conversely.

If a sufficiently high kV (at least 69 kV) is applied across the tube, the curve is modified by the appearance of a sharp *peak* at 69 kV representing the *characteristic radiation* emitted by the tungsten target. This is shown in the 80-kV curve.

Although photons are preferably specified according to their energy rather than their wavelength, the following equation shows how they are related:

$$photon\ energy = \frac{1.24}{\lambda}\ keV \tag{5}$$

where λ is the wavelength in nanometers (nm). Thus, the energy of a photon $h\nu$ whose associated wavelength is 0.02 nm, is obtained from equation (5):

$$h\nu = \frac{1.24}{0.02} = 62\ keV$$

The minimum wavelength λ_{min} of the radiation in an x-ray beam can be derived from equation (5):

$$\lambda_{min} = \frac{1.24}{h\nu_{max}}\ nm \tag{6}$$

In Figure 12.11 we see that if the kV is kept constant, an increase in mA simply increases the exposure rate. But, as shown in Figure 12.11, there is no shift in the curves—the maximum photon energy remains the same because it depends only on the applied peak kV. This is also evident from equation (5).

Specification of X-ray Quality—Half-value Layer and Kilovoltage. Because of the polyenergetic nature of the photons in an x-ray beam, the spectral distribution curves described in the preceding section depict most completely the distribution of photon energies in any given beam. However, this degree of precision is not required for our purpose in medical radiology. Instead, we use two methods of specifying beam quality, which include (1) the *half-value layer* and (2) the *accelerating potential* (applied voltage).

1. *Half-value Layer. The half-value layer (HVL) of an x-ray beam is defined as that thickness of a specified material (usually a metal) which reduces the exposure rate to one-half its initial value.* The material is used in a thin sheet called a *filter.* First we measure the exposure rate of the x-ray beam whose quality is to be specified. We then make a series of measurements with increasing thicknesses of filter in the beam. That thickness of the selected metal which reduces the exposure rate 50 percent is the half-value layer. For example, if the exposure rate is initially 90 R/min and it requires 1.5 mm Cu (copper) to decrease the exposure rate by just one-half (ie, to 45 R/min), then the HVL = 1.5 mm Cu.

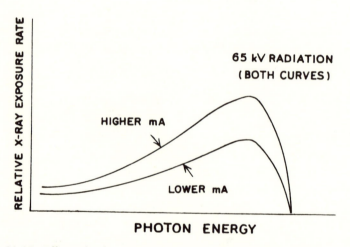

Figure 12.11. Effect of tube mA on relative exposure rate. With the kV held constant, a higher mA produces a larger exposure rate for all photon energies in the beam. However, the mA has no effect on the maximum photon energy, which is the same for both curves.

How is the *half-value layer of a beam measured?* A brief outline of the procedure will now be presented:

a. Select the x-ray technical factors of kV, mA, and filtration for the *initial beam whose half-value layer is being determined.* Collimate the beam to 10 cm × 10 cm.

b. Place a thimble ionization chamber in the center of the beam at a *fixed distance* from the focus of the x-ray tube. The thimble should be at least several feet from any possible scattering objects such as table, floors, or walls.

c. Determine the exposure rate in R/min (roentgens per minute) without any additional filtration of the beam (other than that which may already be present). Tabulate this exposure rate under *initial beam.*

d. Repeat the procedure after adding a 0.25 mm copper filter, if *orthovoltage* therapy is used, that is, 200 to 300 kV. All other exposure factors remain constant. (Different filters are needed for various photon energy ranges.)

e. Repeat the procedure for each of several further additions of copper filters. Assemble the results as in Table 12.3.

TABLE 12.3
EXPOSURE RATES WITH ADDED COPPER FILTERS
OF VARIOUS THICKNESSES. kV, mA,
AND DISTANCE CONSTANT THROUGHOUT

Added Copper Filtration	Exposure Rate
mm	R/min
0	100
0.25	50
0.50	37
1.0	25
1.5	19
2.0	15

f. Plot the data **graphically** with exposure rate in R/min as the dependent variable and the thickness of added filter as the independent variable, as in Figure 12.12.

g. Locate the exposure rate corresponding to one-half the initial value on the vertical axis. Draw a horizontal line from this point to the curve, and then a vertical line from the point of intersection on the curve, to the horizontal axis, where the vertical line locates the thickness of copper responsible for the halving of the initial exposure rate. This halving thickness is the half-value layer. For an actual example, refer to Table 12.3 and Figure 12.12.

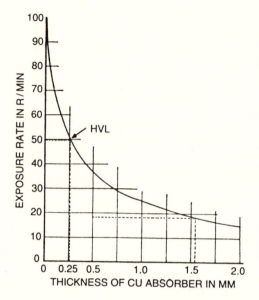

Figure 12.12. Attenuation curve of 220-kVp x rays in copper filters of increasing thickness based on data in Table 12.3. Half-value layers can be obtained from the curve. If, with no added filter, the exposure rate is 100 R/min, we must add a filter of 0.25 mm Cu to decrease the exposure rate to one-half its initial value, or 50 R/min. The half value layer (HVL) of this beam is therefore 0.25 mm Cu.

If the initial beam is then filtered by 0.5 mm Cu, we move the vertical axis to the 0.5 mm line. The corresponding exposure rate is found to be 37 R/min; half of this is 18.5 R/min, corresponding to a reading on the horizontal axis of 1.6 mm Cu. But the initial filter was 0.5 mm Cu. Therefore, the HVL of this particular beam is 1.6 mm − 0.5 mm = 1.1 mm Cu.

As more filtration is added, the beam becomes more homogeneous, that is, the HVL increases and becomes constant.

Depending on the kV of the initial beam, we use other metals to specify HVL. Thus, *aluminum* is used for superficial therapy beams (50 to 150 kV$_p$) and *lead* for megavoltage therapy beams (1 MV and above).

2. *Applied Potential (Voltage)*. The International Commission on Radiation Units and Measurements (ICRU) recommends that under certain conditions, the voltage applied to the source of radiation be included in the specification of beam quality. The following data are based on the ICRU report of 1963:

a. *For radiography,* the peak kV is customarily used to designate beam quality, since the HVL does not routinely serve a useful purpose, *except that* HVL in aluminum for radiographic beams has an important application in patient protection (see pages 528–529).

b. *For x-ray therapy up to 2 million volts (MV), the kV or MV and the HVL should be stated* because in this range of radiation energy, filters cause significant changes in the quality of the beam.

c. *For x-ray therapy above 2 MV, the MV only need be stated* because in megavoltage therapy (2 MV and above) filters have very little effect on beam quality. With *gamma-ray* and *electron beams,* energy is stated in MeV.

Table 12.4 gives the usual classification of radiotherapy beams. In addition, some dermatologists apply extremely low energy x rays, known as *Grenz* rays, having an energy of about 10 to 15 kV_p, in treating extremely superficial, noncancerous skin lesions. The German word Grenz means boundary, and refers to the fact that these rays are so soft that they lie between ultraviolet rays and ordinary x rays in the electromagnetic spectrum.

TABLE 12.4
CLASSIFICATION OF RADIATION THERAPY BEAMS

Type of Therapy	Energy	Filter	Distance
			cm
Superficial X rays	50 to 120 kV	1 to 3 mm Al	5 to 25
Orthovoltage (deep) X rays	200 to 300 kV	1 to 2 mm Cu or 0.5 mm Sn	50
Megavoltage X rays	4 to 25 MV	—	100
^{60}Co gamma rays	1.25 MeV (av.)	—	80

Other methods, such as the equivalent constant potential and the photographic, can be used, but their application is limited and will not be described.

"Hard" and "Soft" X Rays

X rays may be roughly classified as *"hard"* or *"soft"* on the basis of their penetrating ability. This simply expresses the quality of a beam in nonscientific terms.

1. **Hard x rays** are higher energy photons with relatively high penetrating ability. Hard x-ray beams may be obtained with

a. Higher generating voltage.
b. Filters such as copper, which absorb relatively more of the less penetrating rays. The higher the atomic number of the filter and the greater its thickness, the greater is its hardening effect on the x-ray beam. This applies to orthovoltage beams.

2. **Soft x rays** are lower energy photons with relatively low penetrating ability. Soft x-ray beams may be obtained with

a. Lower generating voltage.
b. Filters of lower atomic number, such as aluminum.

THE INTERACTIONS OF
IONIZING RADIATION AND MATTER

On passing through a body of matter, an x-ray beam undergoes *attenuation;* that is, its exposure rate gradually decreases. Attenuation consists of two related processes: (1) *absorption* of part of the radiation by various kinds of interactions with the atoms of the body in the path of the beam (see pages 179–183) and (2) emission of radiations comprising *scattered* and *secondary* radiation resulting from interactions in (1). A third factor is the *inverse square law.*

Scattered radiation refers to those x-ray photons that have undergone a *change in direction* after interacting with atoms.

Secondary radiation comprises the characteristic radiation *emitted* by atoms after having absorbed x-ray photons.

It must be emphasized that many of the x-ray photons do not interact with atoms at all, but pass through the body unchanged. This results from two factors: (1) x rays are electrically neutral so there is no electric force between them and orbital electrons, and (2) atoms contain mainly empty space.

Dual Nature of X Rays. The various types of interactions between x rays and atoms make up one of the most fascinating chapters in x-ray physics. To simplify the discussion, as well as to make it more accurate scientifically, we must regard an x-ray beam as consisting of discrete bits of energy traveling with the speed of light. Such a tiny unit of energy is called a *photon* or *quantum* (pl., quanta). This may seem to contradict the electromagnetic wave theory of radiation discussed earlier, but the phenomena of radiation absorption and emission can be explained only by the *apparently contradictory theories* that a given x ray exists at the same time *both as a wave and as a particle of energy.*

This *dual nature* of electromagnetic radiation is now an accepted fact in physics. It tells us how characteristic radiation is produced in an x-ray tube when an electron drops from a higher to a lower energy level in the atom of the target metal (see Figure 12.6). Such a characteristic x ray is a quantum or photon of energy, and its energy and frequency are peculiar to the target element. Thus, an *atom of a given element can emit definite characteristic photons only,* and these are different from those of any other element. The energy of a characteristic photon represents the difference in the energy levels of the shells between which the electron has shifted.

On the basis of this concept, known as the *quantum theory,* an x-ray beam consists of showers of photons, or bits of energy, traveling with the speed of light and having no electric charge. Also known as the *corpuscular theory of radiation,* it applies equally to other forms of electromagnetic radiation such as light and gamma rays.

What determines the amount of energy in a given photon? Equation (4) cited above, formulated by the German physicist Max Planck, provides the answer:

photon energy = a constant × frequency

$$E = h\nu$$

In this equation, h is Planck's constant (does not vary) and ν (Greek "nu") represents the frequency of the associated electromagnetic wave in cycles per sec (hertz = Hz). This equation tells us that the *quantum* or *photon energy is directly proportional to the frequency,* in agreement with the observation that the penetrating ability of x rays, which depends on their energy, increases as the frequency increases. And since the speed of x rays is constant and equals frequency times wavelength, an increase in frequency is associated with a decrease in wavelength. Thus, *highly penetrating x-ray photons have high frequency and short wavelength,* while conversely, poorly penetrating x-rays have low frequency and long wavelength.

X-ray Interactions with Matter. As a basis for understanding the interactions of ionizing radiation and matter we must know something about the electronic *energy levels* or *shells* within the atom. These are analogous to the mechanical model of the rock on the cliff (see pages 35–36). Since the nucleus carries a positive charge, it exerts an electric force of attraction on the orbital electrons, the force being greater the nearer the electron shell is to the nucleus. Hence, more work would be required to remove an electron from the K shell and out of range of the nuclear electric field than would be required to remove an electron from one of the outer shells. Inasmuch as work must be done in moving an electron away from the nucleus in the direction of the outer shells, *the shells are at progressively higher energy levels the farther they are located from the nucleus.* Thus, the K shell represents the lowest energy level, while the Q shell represents the highest.

The energy required to remove an electron from a particular shell and beyond the range of the nuclear positive electrostatic field is called the *binding energy* of that shell. Therefore, the binding energy is largest for K-shell electrons because of their proximity to the positively charged nucleus, and decreases progressively for successive shells. The binding energy is characteristic of a given element and shell; thus, it is about 70,000 eV for the K shell of tungsten, but only about 500 eV for the K shell of the average atom in the soft tissues of the body.

Owing to the relatively large amount of energy needed to remove an electron from an inner shell, such an electron is said to be *bound.* On the other hand, almost no energy is needed to liberate the outermost shell electrons, and they are therefore called *free* or *valence* electrons.

Let us examine in detail the possible sequence of events following the penetration of a body of matter by x-ray photons. Any of the four types of interactions may occur between the photons and atoms lying in their path: (1) photoelectric interaction, (2) coherent or unmodified scattering, (3) Compton interaction with modified scattering, and (4) pair production. Remember that most photons pass completely through the body without experiencing any type of interaction.

1. **Photoelectric Interaction (Photoelectric Effect).** This type of interaction is most likely to occur when the energy—hν—of the incident (incoming) photon is *slightly* greater than the binding energy (see above) of the electrons in one of the inner shells such as the *K* or *L*. The incident photon gives up all of its energy to the atom; in other words, the photon is *truly absorbed* and disappears during the interaction. Immediately, the atom responds by ejecting an electron, usually from the *K* or *L* shell, leaving a *hole* in that shell (see Figure 12.13). Now the atom is *ionized positively* (why?) and in an *excited state.* Note that the energy of the incident photon ultimately went to (1) free the electron from its shell and (2) set it in motion as a *photoelectron.*

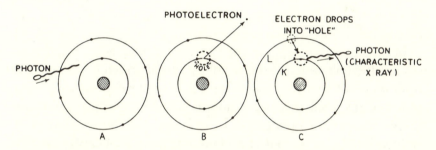

Figure 12.13. Photoelectric interaction with true absorption. *A.* Incident photon loses all its energy on entering an atom, being *absorbed* in the process. The atom responds by ejecting an *inner shell* electron which becomes a photoelectron. *B.* Atom is in an *excited state* (also ionized—electron missing). *C.* Electron from a higher energy level fills vacancy or *hole* in the *K*-shell, a *characteristic* x-ray photon being emitted.

The energy exchange during a photoelectric interaction is beautifully expressed in Einstein's equation:

$$h\nu_{photon} = W_K + \tfrac{1}{2}mv_e^2 \qquad (7)$$

energy of	binding energy	kinetic energy
incident	of	of
photon	K shell	K photoelectron

where hν is the energy of the incident (entering) photon, W_K is the binding energy or *work function* of the *K* shell of the element concerned, and $\tfrac{1}{2}mv_e^2$ is the kinetic energy of the photoelectron.

The same equation applies to photoelectric interactions in other shells. However, when photoelectric interaction occurs with electrons in *outer* shells, the incident photons are more likely to be in the ultraviolet or visible light region; as indicated above, inner shell photoelectric interactions are more likely to occur with x or gamma photons.

What happens next? Suppose that a K electron has been ejected. The hole in the K shell is *immediately* filled by transition of an electron from one of the outer shells, usually the L shell, because they are at higher energy levels than the K shell (see pages 178–179). During the transition of an L electron to the hole in the K shell the energy difference between these shells is radiated as a photon known as a *fluorescent* or *characteristic x ray* because its energy (also its frequency and wavelength) is typical of the shells and the element concerned. Thus, the characteristic x rays of copper are different from those of lead.

In pursuing the sequence of events, you may ask, "What happens to the hole in the L shell?" It is promptly filled by an electron from an outer shell, that is, from a still higher energy level. This continues as each successive hole is filled by electron transition from some higher energy level, until the atom loses its excited state, returning to *normal* or *ground state.* The total energy carried by the characteristic photons in a single photoelectric interaction equals the binding energy of the shell from which the electron was initially expelled—in this case, the K shell.

In summary, then, the energy of the incoming photon in the *photoelectric interaction* involving the K *shell* has the following fate:

a. Photon enters atom and completely disappears.
b. A K-shell electron is ejected, leaving a hole.
c. Atom has excess energy—excited state.
d. A part of photon's energy was used to liberate electron and the rest to give it kinetic energy; ejected electron is a photoelectron.
e. Hole in K shell is filled by electron transition from a shell farther out, accompanied by emission of a characteristic x-ray photon.
f. Holes in successive shells are filled by electron transitions

from shells still farther out, each transition being accompanied by a corresponding characteristic photon.

g. Sum of energies of all the characteristic photons equals binding energy of shell from which the photoelectron originated, in this case, the *K* shell.

Thus, the photoelectric effect gives rise to two kinds of secondary radiation — photoelectrons and characteristic x rays. Because of their negative charge, photoelectrons ionize atoms in their path. However, they usually have very low energy. For example, a 100-keV photon releases, in soft tissue, a 99.5-keV photoelectron which is absorbed in only about 1 mm of tissue. Still, this is one of the important ways in which x rays transfer energy to tissues and produce biologic changes in them. The characteristic x rays also have very low energy — about 0.5 keV — and are locally absorbed.

The photoelectric effect is very important in radiography because the chance of its occurrence varies *directly* with $\overline{Z}^3$ (atomic number^3) of the irradiated tissue, and *indirectly* with photon energy^3. Bone has a high atomic number relative to soft tissue and photon energy is low; hence, the large difference in photoelectric absorption by these structures, a major factor in *radiographic contrast* — the differential darkening of various areas of a radiograph. But differences in tissue thickness and density also contribute to contrast (see pages 338–341).

2. **Coherent or Unmodified Scattering.** If a *very low energy* x-ray photon interacts with a relatively bound orbital electron, it may set the electron into vibration. This produces an electromagnetic wave identical in energy to that of the incident photon, but *differing in direction* (see Figure 12.14). Thus, in effect, the entering photon has been *scattered without undergoing any change in wavelength, frequency, or energy.* Unmodified scattering occurs mainly with x rays whose energy is well below the range that is useful in clinical radiology.

3. **Compton Interaction with Modified Scattering.** If an incident (entering) photon of sufficient energy encounters a *loosely bound,* outer shell electron it may dislodge the electron and proceed in a different direction, as shown in Figure 12.15. The dislodged electron is called a *Compton* or *recoil electron.* It acquires a certain amount of kinetic energy which must be subtracted from the energy of the entering photon, in accordance with the Law of

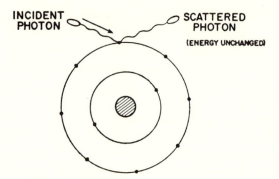

Figure 12.14. Unmodified scattering. The entering photon has undergone a change of direction only.

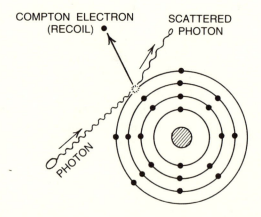

Figure 12.15. Compton interaction with modified scattering. *Part* of the photon's energy has been used up in removing a loosely bound orbital electron. Therefore, the emerging photon has *less energy,* and it has also undergone a *change in direction.*

Conservation of Energy. Consequently, the energy of the incident photon is decreased, so it comes out of the atom with less energy (also, longer wavelength and lower frequency). This phenomenon, discovered in 1922 by the renowned physicist, A. H. Compton, is called the ***Compton Effect.*** The emerging photon, having undergone a change of direction, is called a ***scattered photon.*** The characteristic radiation, arising from electron transitions between outer shells, may be disregarded in radiology because of its extremely low energy.

The Compton Effect may be expressed as follows:

$$h\nu_i \quad = \quad E_r \quad + \quad h\nu_s \tag{8}$$

| energy of | energy given | energy of |
| incident photon | to recoil electron | scattered photon |

Thus, the frequency and energy of the scattered photon are less than those of the incident photon. A scattered photon, in turn, behaves in the same manner as other x rays on interacting with atoms. The recoil electrons cause ionization of atoms in the same way as other high speed electrons.

As the energy of the incident photon *increases*, the probability of occurrence of the Compton interaction *decreases*. But, at the same time, the *scattered photons* (1) have more energy and (2) tend to be scattered more and more in a forward direction, increasing the likelihood of their passing completely through the body and reaching the film. This constitutes the *scattered radiation* which, in radiography, impairs image contrast by an overall fogging effect—a form of noise that degrades recorded detail.

Despite the decreased chance of occurrence of the Compton Effect at higher photon energies, the probability of the photoelectric effect decreases even more sharply, so as the energy of the incident photon increases above 70 keV (about 180-kV x rays) the Compton Effect becomes the predominant type of interaction.

4. **Pair Production.** A *megavoltage* photon with energy of at least 1.02 million electron volts (MeV), upon approaching a *nucleus*, may disappear and give birth to a *pair*—a negative electron or *negatron*, and a positive electron or *positron* (see Figure 12.16). These charged particles share unequally, as kinetic energy, the photon's energy in excess of 1.02 MeV. The positron, as it comes to rest, combines with any negative electron and disappears giving rise to two photons with an energy of 0.51 MeV, moving in opposite directions. This is the *annihilation reaction.* The two processes exemplify, first, the conversion of energy to matter, and then matter to energy according to Einstein's equation $E = mc^2$. Note that the "magic" number 0.51 MeV is simply the energy equivalent of the mass of an electron at rest. Pair production becomes significant at about 10 MeV, but not until about 24 MeV does it predominate over the Compton Effect insofar as energy absorption is concerned.

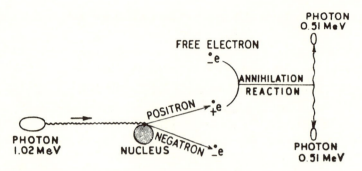

Figure 12.16. Pair production and annihilation reaction.

Secondary and Scattered Radiation Consists of:

1. Primary electrons
   a. Photoelectrons
   b. Compton or recoil electrons
   c. Positron-negatron pairs
2. Secondary and scattered x rays
   a. Characteristic
   b. Coherent or unmodified scattered
   c. Modified scattered (Compton Effect)
   d. Annihilation radiation

The secondary and scattered photons are less energetic (softer) than the incident photons, except for the relatively insignificant factor of unmodified scattering. The production of primary electrons accounts for the absorption of radiation by matter, since this removes energy from the primary beam. Hardening of an x-ray beam by a *filter* results mainly from photoelectric absorption of *relatively more low energy than high energy photons,* so that the emerging beam retains a relatively greater *percentage* of higher energy, more penetrating photons. A copper filter emits its own soft characteristic radiation which can be absorbed by an aluminum secondary filter. The aluminum also emits characteristic rays, but these are so soft that they are absorbed in a few centimeters of air.

Relative Importance of Various Types of Interaction in Radiology

The various types of interactions between x or gamma radiation and matter will now be discussed as they apply to diagnostic and therapeutic radiology, especially with regard to their relative importance.

1. **Diagnostic Radiology.** In the usual range of 30 to 140 kV, the photoelectric effect predominates insofar as energy absorption from the beam is concerned. Because photoelectric absorption is about four to six times greater in bone than in an equal mass of soft tissue in the radiographic kV range, this type of interaction is responsible for much of the *radiologic contrast* (manifested by differences in film darkness) between these tissues. Differences in density (grams/cc) of various tissues also contribute to radiologic contrast.

On the other hand, the Compton Effect makes its presence known mainly by the *scattered radiation* resulting from interaction with the atoms of the tissues, table top, etc. In fact, in the radiographic range, scattered radiation may contribute as much as 80 to 90 percent of the beam in soft tissues.

As will be shown later, the scattered radiation approaches the film from many directions and degrades radiographic quality, making it necessary to use grids and collimators. As kV is increased the scattered radiation becomes progressively more damaging to radiographic quality because (1) its energy increases and it is therefore more likely to pass completely through the body and reach the film and (2) it is scattered more and more forward toward the film. Note that the characteristic radiation arising in the photoelectric process is also scattered in many directions, but is so soft that it is absorbed locally in the tissues.

2. **Therapeutic Radiology.** When x or gamma photons traverse the body, their interaction with the atoms of the tissues results in the transfer of energy to orbital electrons, liberating them. The freed electrons then expend their energy in ionizing other atoms, thereby causing tissue damage and concomitant therapeutic effects. This will be discussed more fully below.

The photoelectric effect predominates below about 140 kV_p. Primary photons in the incident beam are completely absorbed upon interaction with the tissue atoms, setting in motion photoelectrons which ionize atoms of the tissue along their paths.

As the applied voltage is increased from about 150 kV to 3 MV, the probability of the Compton interaction decreases, but the probability of the photoelectric interaction decreases even more, so that the Compton Effect becomes predominant.

Although pair production begins at about 1.02 MeV, it does not contribute significantly to energy absorption until about 10 MeV. Only at 24 MeV and above does ionization by negatrons and positrons become the main process of energy absorption.

Detection of Ionizing Radiation

There are several ways in which the effects of ionizing radiation make its presence known:

1. **Photographic Effect** — causes changes in a photographic emulsion so that it can be developed chemically.

2. **Luminescent Effect** — causes certain materials to glow in the dark. These include zinc cadmium sulfide, calcium tungstate, lead barium sulfate, and the newer rare earth phosphors such as lanthanum and gadolinium.

3. **Ionizing Effect** — ionizes gases and discharges certain electrical instruments such as the electroscope.

4. **Thermoluminescent Effect** — undergoes conversion to light in certain crystals after they have been heated.

5. **Chemical Effects** — changes the color of certain dyes.

6. **Physiologic Effect** — reddens the skin, destroys tissues, and causes genetic damage.

RADIATION DOSIMETRY

The changes produced in tissues by ionizing radiation result from the transfer of energy to the atoms and molecules of the tissues by the processes of *ionization* and *excitation.*

We must emphasize here that x and gamma rays represent *indirectly ionizing* radiation. While it is true that they liberate electrons by one or more of the interaction processes, it is these primary electrons which play, by far, the major role in causing subsequent ionization and excitation of atoms. In other words, the ionizing effect of x and gamma photons occurs indirectly by means of the *primary electrons* they first release in the tissues — photons transfer energy to orbital electrons, which, in turn, transfer energy to atoms to produce pairs of oppositely charged ions, or *ion pairs.* In pair production, electrons arise by materialization of very high energy photons.

The amount of energy transferred by ionizing radiation per unit length of path is called *linear energy transfer—LET.* We saw in the sections on the interactions of radiation with matter that they cause ionization. With a given type of radiation, such as x rays, the degree of ionization and the resulting tissue effects are related to the amount of energy absorbed at a particular site. This *absorbed dose* of radiation depends on the quantity (exposure in R) and quality (energy) of the radiation, and on the nature of the tissue (muscle, bone, etc). We know that more hard than soft radiation is needed for a given tissue effect as, for example, skin reddening (erythema). Such dependence on beam hardness or quality—so-called *wavelength* or *energy dependence* —is due to differences in the number and distribution of ions liberated by radiation of various energies, that is, differences in LET.

Exposure—the Roentgen

As described earlier, we use the *roentgen* (R) as the unit of *exposure,* based on the ionizing ability of the radiation. Thus, exposure is a measure of the amount of radiation delivered to a particular site. In addition, we must state the *quality* of the beam, namely, the peak kV applied to the tube as well as the half-value layer in the orthovoltage region, or the energy of the radiation in MV or MeV in the megavoltage region. The mA, time, and distance should also be included where applicable, although they have no bearing on the quality of the x-ray beam. The total duration of a therapy course in terms of days or weeks has a distinct influence on the effect of treatment, since a given radiation exposure at one sitting has a much more pronounced effect than the same total exposure given in fractions over a period of days or weeks.

Absorbed Dose—the Rad

The preceding paragraph dealt with exposure in roentgens, no mention being made of the dose *absorbed* in the tissues. Since only the energy actually absorbed in the irradiated volume is responsible for the ensuing biologic changes, *we may define the absorbed dose of radiation as the amount of energy absorbed per gram of irradiated matter.* The unit of absorbed dose is the *rad (radiation absorbed dose),* and is defined as an *energy absorption of 100 ergs per gram.* This applies to any type of radiation (x rays, gamma rays, beta rays, etc) and to any kind of matter, including tissues. In the International System of units the rad has been replaced by the *gray* (Gy) = 1 joule/kg; thus, 1 Gy = 100 rad. On this basis, 1 centigray (cGy) = 1 rad.

The *roentgen,* it should be recalled, is limited to x rays and gamma rays up to 3 million volts (MV).

What is the relationship of the unit of absorbed dose—the rad—to the unit of exposure—the roentgen? The answer is not a simple one, because it depends on a number of factors. However, in the usual therapy range of 100 kV to 20 MV the *absorbed dose in soft tissues exposed to 1 R is approximately 1 cGy,* with an error of a few percent. In any case, this one-to-one conversion is *not* sufficiently accurate in radiotherapy.

But absorption in *bone* (higher average atomic number) depends *strongly* on the energy of the incident radiation. For example, with an x-ray beam of half-value layer 1 mm Cu the absorbed dose in bone is about twice that in soft tissues for the same exposure. Thus, 1 R would produce an absorbed dose of about 1 rad in soft tissue, and about 2 rads in cellular elements in compact bone. Higher-energy beams in the usual megavolt region, deliver the same dose per gram of soft tissue or bone. This is of practical importance in therapy when the beam has to traverse a bony structure; with the softer radiation, one must be careful not to exceed the tolerance of the bone cells while delivering a therapeutic dose to the soft tissues. With megavoltage radiation, the equal absorption per gram of bone and soft tissues reduces, but does not entirely prevent the possibility of serious bone damage.

Determination of Tissue Dosage

A *simplified method* of tissue or tumor dose determination, applicable to a cobalt 60 beam, will be covered under four headings: (1) in-air exposure, (2) backscatter, (3) given dose, and (4) tumor dose.

In-Air Exposure. When the exposure rate is measured at a given point in a beam, *with no scattering material in the vicinity of the measuring device* (other than air), it is called the *in-air exposure rate* and is usually expressed in R per min (abbreviated R/min). The total in-air exposure is obtained by multiplying the in-air exposure rate by the exposure time:

in-air exposure = in-air exposure rate × exposure time   (9)

$$R = R/min \times time$$

This is, in fact, one method of calibrating a radiotherapy machine, the measuring device being either a Victoreen condenser-R-meter with a thimble chamber or some other type of radiation-measuring instrument. Several factors influence the in-air exposure rate in the orthovoltage region: it increases with an increase in kV, an increase in mA, an increase in beam cross-sectional area, a decrease in filtration, and a decrease in the source-skin distance (according to the inverse square law—see pages 333–337). Therefore, an orthovoltage x-ray machine should be calibrated for every combination of treatment factors (kV, mA, filtration, and distance) that is to be used. With cobalt 60 gamma rays (1.25 MeV) we are concerned with source-skin distance and beam cross-sectional area in measuring in-air exposure rate. With megavoltage x rays, we must take into account beam energy, source-skin distance, and cross-sectional area.

Backscatter. If a radiation measuring device such as a thimble chamber is placed on the surface of the body (or some equivalent material such as hard plastic or water) the recorded exposure rate is larger than that in the absence of the body. In other words, the measured *entrance exposure rate* in R/min exceeds the in-air exposure rate. This results from *backscatter radiation* which is simply radiation scattered back from matter lying beneath the thimble chamber. Note that this increases the exposure to the skin and must be included in planning therapy. Another factor contributing to the in-air exposure rate is scattered radiation arising in the beam-limiting device (collimator).

We usually designate the degree of backscatter as the ratio of the measured entrance-R to in-air-R at the same point:

$$\text{backscatter factor} = \frac{\text{entrance-R}}{\text{in-air-R}} \qquad (10)$$

For example, if the entrance exposure is 125 R and, in the same time interval, the in-air exposure at the same point (in absence of body or phantom) is 100 R, then

backscatter factor = 125/100 = **1.25**

Obviously, the larger the backscatter factor, the larger will be the entrance exposure for a particular in-air exposure.

Published tables give the backscatter factors obtained experimentally under various conditions, but it is preferable to have an experienced radiation physicist obtain this information directly.

The factors affecting the backscatter factor are as follows:

1. *Area of Treatment Field.* As this is increased, the backscatter factor increases due to the larger volume of irradiated tissue and the correspondingly larger source of scattered radiation.

2. *Thickness of Irradiated Part.* As this is increased, the backscatter factor increases for the same reason as in (1). However, beyond a certain critical thickness, further increase in thickness does not influence the backscatter factor.

3. *Energy of Beam.* Above 1 mm Cu, an increase of HVL decreases the backscatter factor because an increasingly greater percentage of the radiation is scattered in a forward direction, and less in a lateral or backward direction. It should be recalled that HVL increases with an increase in kV and filtration. With *megavoltage beams* of at least 1 MeV, backscatter is still smaller. However, with megavoltage beams such as cobalt 60 gamma rays and linac x rays there is a *buildup of electrons below the surface* so that ionization and dosage reach a maximum at some depth below the surface. In fact, this buildup depth is 5 mm for cobalt 60 gamma rays, and 15 mm for 6-MV x rays, and 4 cm for 25-MV x rays.

It should be noted that the source-skin distance has no effect on the backscatter factor or electron buildup, although it does influence percent depth dose (see below).

Given Dose. In practice we express radiation dosage in terms of absorbed dose. Knowing the entrance exposure (including backscatter) we can convert this value to absorbed dose in cGy. The conversion factor f published in *NCRP Report No. 69 (1981)* for various photon energies in muscle and bone are given for three qualities of radiation in Table 12.5. To obtain the absorbed dose with radiation having an energy of less than 3 MeV, we simply multiply the exposure by the appropriate f value:

TABLE 12.5
SAMPLE VALUES OF AVERAGE CONVERSION FACTOR f cGy/R*
(Based on Measurement of Exposure in Air)

HVL	Muscle	Bone**
mm Cu		
1.0	0.929	1.95
1.7	0.943	1.46
Cobalt 60	0.958	0.927

*Adapted from NCRP Report No. 69 (1981).
**Cellular elements in compact bone.

$$exposure \times \quad f \quad = absorbed\,dose$$
$$R \qquad cGy/R \qquad cGy$$

Since we are dealing with beams that always enter the body through soft tissue, we use the f factor for muscle in converting entrance exposure to entrance dose. The latter is often referred to as the **given dose**.

given dose = in-air exposure $\times f \times$ backscatter factor

Depending on the energy of the photon beam, the location of the given dose may be at the skin surface or at some depth below. Given dose is often referred to as D_{max}. There being no electron buildup with orthovoltage x rays, maximum ionization (and dosage) occurs at the skin surface where the beam enters, and that is where the given dose is specified. But with megavoltage beams, depending on their energy, D_{max} is located at some depth below the skin surface because of electron buildup, and so the given dose is specified at the appropriate depth. However, it should be realized that buildup occurs rapidly before D_{max} is reached.

In summary, then, in the orthovoltage region, D_{max} is the sum of the entrance dose and the backscatter dose, there being no electron buildup. In the megavoltage region, D_{max} includes the entrance, electron buildup, and backscatter doses, the last being much less than that in the orthovoltage region.

Tumor Dose. As a beam of photons traverses matter the dose rate gradually **decreases**. There are three factors responsible for the weakening or **attenuation** of the beam.

1. Absorption
2. Scattering
3. Inverse square law

Absorption of radiation accompanies the various types of interactions between its photons and the atoms in the irradiated material (or tissues) as described earlier. Photoelectric interaction with true absorption predominates with radiation energy up to about 140 kVp. Compton interaction with partial absorption predominates in the energy region between 150 kVp and 10 MV. Above this limit, pair production with partial absorption begins to assume major importance.

Scattering occurs in matter (tissues included) mainly by the Compton process for radiation ranging in energy up to several million volts. In the pair production process radiation seems to be scattered, but this actually results from the annihilation reaction.

The *inverse square law*, a geometric factor, applies because the beam, on passing through the body, reaches points that are progressively farther from the radiation source. The exposure rate is inversely proportional to the square of the distance between the source of radiation and the point of interest, such as the tumor.

The *tumor dose* may be defined as the absorbed dose in cGy (rad) at the location of the tumor. Ordinarily, this is determined for a point along the central axis (central ray) of the beam at the center of the tumor. Table 12.6 contains depth dose data for cobalt 60 gamma rays, and Table 12.7 for 6-MV linear accelerator x rays, obtained by measurement in a water phantom (a substitute for the human body). Refer back to Figure 12.8 which explains the concept of percentage depth dose. This is also expressed by the following equation:

$$\% \text{ depth dose at } x \text{ cm} = \frac{\text{dose rate at } x \text{ cm depth} \times 100}{\text{given dose rate}} \quad (12)$$

For example, suppose we are to treat a tumor whose center lies 10 cm below the skin surface where the beam enters the body. The field size measures $10 \times 10 \text{ cm}^2$ and we are to use a cobalt 60 beam (av. energy 1.25 MeV). In Table 12.6 for ^{60}Co at an 80-cm source-skin distance (SSD) we locate the 100 cm^2 column and find that at a depth of 10 cm the depth dose is *56.8 percent.* A total tumor dose of 6000 cGy would require a given dose (GD) computed as follows:

TABLE 12.6
PERCENT DEPTH DOSES*
COBALT 60

Square Fields				HVL 1.1 cm Pb			Source-skin Distance 80 cm	
Area (cm²)	25	50	80	100	150	200	225	300	400
Dep. Factor	0.942	0.970	0.990	1.000	1.019	1.035	1.042	1.055	1.070
Depth (cm)									
0	23.5	26.8	30.1	31.9	35.4	39.0	40.8	44.5	50.1
0.5	100.0	100.0	100.0	100.0	100.0	100.0	100.0	100.0	100.0
1	98.2	98.1	98.3	98.3	98.4	98.5	98.5	98.5	98.5
2	92.3	92.7	93.1	93.2	93.4	93.6	93.6	93.7	93.9
3	86.7	87.6	88.2	88.3	88.6	88.9	89.0	89.2	89.5
4	81.3	82.7	83.4	83.6	84.1	84.4	84.6	84.9	85.2
5	76.2	78.0	78.9	79.2	79.7	80.2	80.4	80.7	81.2
6	70.9	72.7	73.9	74.3	75.0	75.6	75.8	76.2	76.8
7	65.7	67.7	69.1	69.6	70.5	71.2	71.4	71.9	72.6
8	60.8	62.9	64.5	65.1	66.2	66.9	67.3	67.8	68.6
9	56.1	58.4	60.1	60.9	62.0	62.9	63.3	63.9	64.7
10	51.6	54.1	56.0	56.8	58.0	59.0	59.4	60.1	61.0
11	47.9	50.3	52.2	52.9	54.2	55.3	55.7	56.5	57.5
12	44.4	46.6	48.5	49.2	50.6	51.7	52.2	53.0	54.0
13	41.1	43.3	45.1	45.8	47.2	48.3	48.8	49.6	50.7
14	38.0	40.0	41.8	42.6	43.9	45.1	45.5	46.4	47.6
15	35.0	37.1	38.9	39.7	41.1	42.2	42.7	43.5	44.7
16	32.2	34.3	36.2	36.9	38.3	39.4	39.9	40.8	41.9
17	29.8	31.8	33.6	34.3	35.7	36.8	37.3	38.2	39.3
18	27.5	29.4	31.1	31.8	33.2	34.3	34.8	35.7	36.8
19	25.4	27.3	29.0	29.6	31.0	32.1	32.6	33.5	34.6
20	23.4	25.2	26.9	27.5	28.9	29.9	30.4	31.3	32.4

Dep. factor is dependency factor, used to correct the given dose rate for field areas relative to 100 cm².

*Adapted from original data by Frederick Hager, Physicist, East Texas Cancer Center, Tyler, Texas (by permission).

$$\text{tumor dose} = \text{given dose} \times \% \text{ depth dose} \qquad (13)$$

and rearranging,

$$\text{given dose} = \frac{\text{tumor dose}}{\% \text{ depth dose}} \qquad (14)$$

$$GD = \frac{6000 \text{ rads}}{0.568} = 10,563 \text{ cGy}$$

For linear accelerator 6-MV x rays, we use Table 12.7 where we find the 10-cm depth dose for a 100 cm² field and an SSD of 100 cm to be **68.8 percent,** so the GD would be

$$\frac{6000 \text{ cGy}}{0.688} = 8721 \text{ cGy}$$

Note that for the same tumor dose, the given dose is 17 percent smaller with 6-MV x rays than with cobalt 60 gamma rays at the usual SSD for these modalities.

The given dose in each example just cited is too large to be delivered through a single field. For this reason, as well to provide more uniform dose distribution within the tumor volume, the given dose is divided among two or more crossfiring beams centered on the tumor. For example, if we select a plan using three equally spaced beams around the body, and the tumor happens to be at a 10-cm depth for each beam, the total given dose for 6-MV x rays (for a 6000 rad tumor dose) is divided by three, giving

TABLE 12.7
PERCENT DEPTH DOSES*
6 MV X RAYS

Square Fields	HVL 1.55 cm Pb					Source-skin Distance 100 cm				
Area (cm²)	16	25	50	80	100	150	200	225	300	400
Output Factor	0.927	0.939	0.965	0.989	1.000	1.019	1.035	1.041	1.051	1.064
Depth (cm)										
1.5	100.0	100.0	100.0	100.0	100.0	100.0	100.0	100.0	100.0	100.0
2	99.9	99.9	99.9	99.9	99.9	99.7	99.6	99.5	99.3	99.1
3	95.7	96.1	96.1	96.0	96.0	96.0	96.0	96.0	95.9	95.8
4	90.9	91.5	92.0	92.0	92.0	92.0	92.0	92.0	91.8	91.8
5	86.0	87.0	87.9	88.0	88.0	88.0	88.0	88.0	88.2	88.3
6	81.5	82.4	83.4	83.7	83.8	84.0	84.1	84.2	84.2	84.4
7	76.6	77.6	78.9	79.7	79.9	80.4	80.6	80.7	80.8	81.0
8	72.2	73.3	74.7	75.8	76.2	76.8	77.0	77.1	77.1	77.4
9	67.9	69.0	70.8	71.9	72.4	73.1	73.6	73.7	73.8	74.0
10	63.9	65.1	66.9	68.2	68.8	69.4	70.0	70.1	70.2	70.5
11	60.6	61.8	63.7	65.0	65.4	66.2	66.8	66.9	67.1	67.5
12	56.9	57.3	60.2	61.7	62.1	63.0	63.5	63.7	63.9	64.3
13	53.7	54.7	56.7	58.0	58.6	59.5	60.2	60.6	61.1	61.6
14	50.5	51.7	53.7	55.1	55.6	56.6	57.3	57.8	58.2	58.9
15	47.8	48.8	50.8	52.1	52.8	53.8	54.5	54.9	55.4	56.1
16	44.8	45.9	47.8	49.2	49.9	50.9	51.8	52.0	52.7	53.3
17	42.3	43.4	45.1	46.5	47.1	48.1	49.0	49.4	50.0	50.8
18	39.5	40.8	42.6	43.9	44.5	45.6	46.4	46.9	47.5	48.2
19	37.5	38.6	40.4	41.9	42.4	43.5	44.3	44.7	45.4	46.1
20	35.4	46.4	38.0	39.4	40.0	41.0	41.8	42.3	42.9	43.7

Output factor is used to correct given dose for field areas relative to 100 cm².

*Adapted from original data by Frederick Hager, Physicist, East Texas Cancer Center, Tyler, Texas (by permission).

$8721/3 = 2907$ rads *for each field*

The total given dose for each field is fractionated over a period of about six weeks, all three fields being treated daily. Thus, for a 5-day per week schedule, the daily given dose to each field would be $2907/30 = 97$ cGy. The daily tumor dose from all three fields would be $6000/30 = 200$ cGy.

Applying similar calculations for a three-field plan with a ^{60}Co beam, the daily given dose would be 117 cGy for each field. Although this daily given dose is larger than that with 6-MV x rays, it is well within skin tolerance. Actually, with megavoltage radiation the tolerance of normal organs in the vicinity of a tumor is a greater limiting factor than skin tolerance.

Now let us examine the absorbed dose in *bone* lying in the path of the beam. In the case of *megavoltage* radiation such as ^{60}Co gamma rays, Table 12.5 shows the f values to be essentially the same in bone and soft tissue, so the absorbed dose in bone is practically the same as in soft tissue for a particular dose at a certain depth. On the other hand, with *orthovoltage* radiation the absorbed dose in bone is higher than in soft tissue, depending on the energy of the beam. Thus, in Table 12.5, with HVL 1.0 mm Cu, f is 1.95 for bone and 0.929 for soft tissue, so the absorbed dose in bone is about twice that in soft tissue under identical conditions.

Besides the central axis depth dose, we must know the doses at other points in the tumor to insure uniform irradiation so as to avoid undesirable over- or under-dosage. We must also know the doses to nearby normal organs to minimize radiation injury. Dosage to points away from the axis can be derived from *isodose curves,* which are computer-plotted percentage depth doses at various points in the beam (see Figure 12.17).

Factors Affecting Percentage Depth Dose

We have already shown how tumor dose varies with percentage depth dose (%DD). What conditions affect %DD?

The *percentage depth dose* increases with increasing (1) *half-value layer or beam energy,* (2) *treatment field area,* and (3) *source-surface distance.* The %DD decreases with increasing *depth of tumor.* These factors will now be discussed.

1. **Beam Quality.** As already noted, this is specified by the HVL and the kV or MV. The larger the HVL, the greater will be the penetrating ability of the beam and the %DD. In Tables 12.6 and 12.7, compare the %DD for ^{60}Co gamma rays (1.25 MeV) with 6-MV linac x rays for equal field areas. Examples of the application of %DD data in computing tumor dose are given on pages 192–195.

2. **Treatment Field Area.** As the area of the field increases, there is a corresponding increase in the volume of the irradiated tissue with resulting increase in the percentage of scattered radiation. The latter is added to the primary radiation reaching a given point below the surface and therefore contributes to an increased %DD.

3. **Source-Surface Distance** *(SSD).* As the distance between the source (x-ray target; ^{60}Co) and surface (skin) is increased, the dose rate at the surface decreases according to the inverse square law. However, due to less divergence of the rays at the greater SSD (see Figure 12.18) a larger fraction of the surface exposure rate reaches a given depth below the surface; hence, the greater the SSD, the larger the %DD. This depends on the inverse square law and can be shown by an example based on Figure 12.18, aside from the other factors that influence %DD:

Point P is located 10 cm below the surface

At SSD 70 cm
Point P is 10 cm + 70 cm = 80 cm from source F_1.
Dose rate at P, neglecting attenuation in tissue, is

$$\frac{70^2}{80^2} = 0.77 \text{ or } 77\% \text{ of given dose rate}$$

At SSD 100 cm
Point P is 10 cm + 100 cm = 110 cm from source F.
Dose rate at P, neglecting attenuation in tissue, is now

$$\frac{100^2}{110^2} = 0.83 \text{ or } 83\% \text{ of given dose rate}$$

4. **Depth of Tumor.** As the depth of the tumor increases, the %DD decreases because of the attenuation of the beam through absorption, scattering, and the inverse square law.

5. **Tissue Inhomogeneities.** The presence of tissues other than

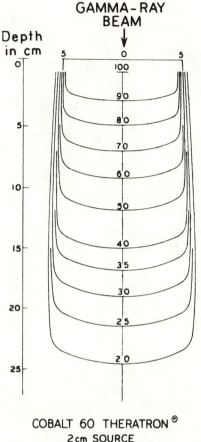

GAMMA - RAY
BEAM
Depth
in cm

COBALT 60 THERATRON®
2 cm SOURCE
1.25 MeV (av.) - HVL 1.2cm Pb - 80cm SSD
10 x 10 cm² FIELD

Figure 12.17. Isodose curves. A specific set of curves should be obtained for each combination of treatment factors. The curved isodose lines connect all points receiving identical dosage, expressed in this case as a percentage of the given dose.

those of water density will influence the %DD. For example, air filled lungs lying in the path of the beam will increase the %DD in the soft tissues beyond, due to the smaller attenuation per cm of path in the lung. On the other hand, bone will decrease the %DD because its greater density than soft tissue causes increased attenuation per cm of path; in other words, bone causes a "shadowing" effect on underlying soft tissue.

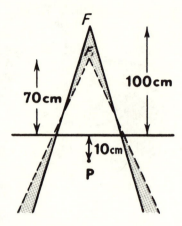

Figure 12.18. Percentage depth dose falls with decreasing source-surface distance, as shown by the greater divergence of the rays. This can be verified by the inverse square law.

Modification of Orthovoltage X-ray Beams by Filters

The use of filters with orthovoltage beams has been mentioned several times, and a more complete discussion will now be presented. A filter is simply a sheet of metal placed in the path of a polyenergetic x-ray beam with an HVL up to about 4 mm Cu, to increase its average penetrating ability; that is, *a filter increases the average hardness of a beam.* This is brought about through the interaction of the x-ray photons with electrons in the atoms of the filter material (see section on interaction of radiation with matter). Recall that these interactions consist of absorption and scattering of photons. As a result, a *greater fraction of low energy* (long wavelength) *than of high energy* (short wavelength) x rays is *removed by a filter,* although some higher energy photons are also removed because of the "edge effect"—photons with energy close to the binding energy of the K or L shells are more likely to be absorbed by photoelectric interaction with K or L electrons than are photons of lower energy. In general, though, an orthovoltage x-ray beam undergoes an increase in half-value layer on passing through a filter.

Filters are not used to harden the cobalt 60 gamma ray beam because this is already virtually monoenergetic, having only two discrete energies—1.17 and 1.33 MeV. However, the steel capsule enclosing the source filters out the beta particles. Beam-flattening filters are used with cobalt 60 gamma rays and megavoltage x rays to provide a more uniform dose rate over the cross-section of the beam.

Notice that filtration does not change the maximum energy of a polyenergetic beam (see Figure 12.19); this depends solely on the *peak kilovoltage* applied to the tube. However, the smallest energy a photon can have in an x-ray beam leaving tube depends on the *inherent filtration* of the tube; that is, the filtration afforded by the glass envelope and the cooling oil layer through which the beam must pass after leaving the target.

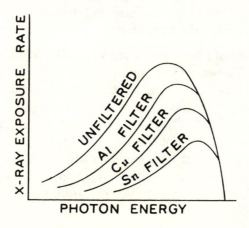

Figure 12.19. Family of curves for an orthovoltage x-ray beam filtered by various added filters, as compared with an unfiltered beam. Note that the maximum photon energy is not altered since it depends on the peak kV only. The peak of the curve is shifted toward a higher photon energy as the atomic number of the filter increases; at the same time, the total exposure rate (area under the curve) decreases.

The inherent filtration of a tube is usually expressed in the equivalent thickness of aluminum or copper. Any additional filtration one chooses to place in the beam is called *added filtration.* The *total filtration* of an x-ray beam obviously represents the sum of the inherent filtration plus the added filtration.

total filtration = inherent filtration + added filtration   (15)

Filtration affects both the *exposure rate* and the *quality* (half-value layer) of an orthovoltage x-ray beam:

1. **Exposure rate** is always decreased by filtration because some photons are removed by absorption and scattering. In fact, the thicker the filter and the higher its atomic number, the smaller will be the exposure rate.

2. **Quality** is changed by filtration. An increase in either the atomic number or the thickness of a filter increases the average penetrating ability of a beam; more specifically, the half-value layer increases. The type of filter should match the energy of the radiation. With radiation produced at 50 to 100 kV, aluminum is most suitable as a filter. Copper is used as the **primary filter** in the 120 to 400 kV range, but an aluminum filter must be placed between the copper and the patient as a **secondary filter** to absorb the characteristic radiation (8 keV) emitted by the copper. The resulting aluminum characteristic radiation (only 1.5 keV) is absorbed in a few centimeters of air.

A special type of filter often used with 250- to 400-kV x rays is known as the **Thoraeus filter** (named after its inventor), a variety of **compound filter** consisting of three metals—tin, copper, and aluminum—placed in that order from the tube side to the patient side (see Figure 12.20). The chief advantage of the Thoraeus filter is that it provides a beam of the same half-value layer as a copper and aluminum filter, but with significantly less attenuation of the beam. Thus, for the same half-value layer, the Thoraeus filter provides about a 25 percent greater exposure rate than an equivalent copper and aluminum filter.

Comparison of Megavoltage and 250-kV Therapy

A number of years ago irradiation therapy of deep-seated tumors with 200 to 250 kV x rays was replaced by higher energy megavoltage beams such as cobalt 60 gamma rays (1.25 MeV) and 6 to 25 MV x rays. Several outstanding advantages relate to the use of radiation in this high energy range.

1. **Skin-sparing Effect.** As described before, the maximum dose with the higher energy radiation is located beneath the entrance skin surface. For example, cobalt 60 gamma rays deliver the maximum dose 5 mm, and 6-MV x rays 15 mm, below the surface. This shifts the skin reaction, ordinarily visible with 200-kV x rays,

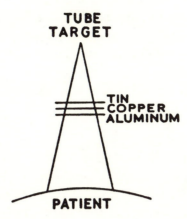

Figure 12.20. Schematic diagram of a Thoraeus compound filter, showing the relation of the component filters.

to a level below the skin, although the dose buildup occurs rapidly between the skin surface and the depth of maximum dose (D_{max}). While this permits a considerable increase in the total skin dose, still we must avoid excessive injury to the underlying tissues where it is not so readily apparent. In addition, we must avoid placing material in the path of the megavoltage beam any closer than about 15 to 20 cm from the skin, since Compton electrons liberated in such material and reaching the skin can significantly raise the dosage level there.

2. **Bone-sparing Effect.** An important advantage of megavoltage radiation is the nearly *equal absorption* occurring in bone and soft tissue, gram for gram. This permits larger doses to be delivered safely when there is bone in the path of the beam. However, there is "shading" of soft tissue beyond bone amounting to about 3.5 percent per cm of bone, up to 3 cm, for cobalt 60 gamma rays. This results from the greater attenuation of high energy photons in bone than in soft tissue for equal *thicknesses*. With 200-kV x rays and HVL of 1 mm Cu, absorption in bone is about two times that in soft tissue of the same mass, increasing significantly the likelihood of radiation injury to bone. At the same time, the lower energy radiation is severely shaded by the bone, so soft tissue beyond may be seriously underdosed.

3. **Greater Percentage Depth Dose.** The higher energy radiation provides a substantially higher percentage depth dose at a

given treatment distance. In fact, this may be its greatest advantage over lower energy x rays. For example, according to Tables 12.6 and 12.7, at a 10-cm depth and with a 100 cm^2 field:

Depth dose with cobalt 60 ....58.8% (SSD 80 cm)
Depth dose with 6-MV x rays........68.8% (SSD 100 cm)

This represents an increase in percentage depth dose of

$$100 \left(\frac{68.8 - 58.8}{58.8} \right) = \frac{100 \times 10}{58.8} = 17\%$$

with the higher energy radiation at the usual treatment distance for each modality. Even more striking is the 40 percent increase in percentage depth dose with megavoltage as compared with 200-kV x rays under similar conditions. At the same time, as we have just seen, the skin-sparing effect of megavoltage radiation further enhances the possibility of delivering an adequate dose to the tumor. It should be mentioned, however, that most authorities agree that the biologic effectiveness of cobalt 60 gamma rays relative to 250-kV x rays is about 85 percent; therefore, such gamma ray therapy requires a compensating increase in dosage of about 15 percent.

4. **More-forward Scatter.** Scattered radiation with high energy radiation is predominantly in a forward direction; that is, in the same direction as the primary beam. This results in less side scatter and a better defined beam edge, especially in the 6 to 18 MV range. Furthermore, the isodose curves can be flattened by suitably shaped filters, so there is more uniform irradiation of the selected volume.

QUESTIONS

1. Which properties of x rays led to their discovery?
2. What are roentgen rays? What are the two apparently contradictory theories as to their nature?
3. How are the frequency and wavelength of radiation related? What is the speed of x rays? Gamma rays?
4. How do soft rays differ from hard rays?
5. Describe the four essential components of a modern x-ray tube. Discuss the principles of x-ray production.
6. Discuss in detail the production of x rays when the electron stream hits the target.

7. What is characteristic radiation and how does it arise?
8. Discuss the differences between characteristic and general radiation. What is their relative importance in radiography?
9. Name ten properties of x rays.
10. What is meant by the quality of roentgen radiation? What factors influence it and in what manner? How can it be measured and specified (ICRU recommendations)?
11. Explain fully why a beam of x rays emerging from the target is heterogeneous. What are monoenergetic x rays?
12. What determines the maximum photon energy in an x-ray beam? State the equation that derives photon wavelength from energy.
13. Define and explain half-value layer and filters. What purpose does HVL serve?
14. Why is tungsten so widely used as the target material?
15. How is the quantity of radiation measured?
16. What factors influence exposure rate?
17. Discuss the four types of interactions of ionizing photon radiation and matter, showing how the corpuscular theory radiation applies.
18. What is the relative importance of the various interactions of photons with matter in radiography and radiotherapy?
19. Of what do scattered and secondary radiations consist?
20. How are tumor exposure and absorbed dose calculated?
21. Define absorbed dose. Name and define its units.
22. What is the relationship between exposure rate and exposure? Exemplify this relationship by means of appropriate units.
23. Prove the statement in Figure 12.18 by application of the inverse square law.
24. Define percentage depth dose and explain the factors that influence it.
25. Define backscatter factor. What conditions affect it and how?
26. Describe a Thoraeus filter and discuss briefly its advantages in orthovoltage therapy.
27. What are isodose curves? Why are they used?
28. How does a filter influence the maximum photon energy in an orthovoltage or radiographic beam? The exposure rate?
29. Discuss the advantages of megavoltage therapy over 250-kV therapy.

Chapter 13

X-RAY TUBES AND RECTIFIERS

THERMIONIC DIODE TUBES

IN CHAPTER 12 we learned that x rays are produced whenever a stream of fast-moving electrons undergoes rapid deceleration (loss of speed). These conditions prevail during the operation of a special type of *thermionic vacuum tube* — the hot filament or Coolidge x-ray tube. Before describing the detailed construction of such a tube, let us review its underlying principles:

1. *A hot metal filament gives off electrons* by a process called *thermionic emission.* The filament is heated by a separate *filament current.* The rate of electron emission increases with an increase in the temperature of the filament which, in turn, is governed by the filament current (measured in amperes).

2. *If no kilovoltage is applied,* the emitted electrons remain near the filament as an electron cloud or *space charge.*

3. *If kilovoltage is applied* between the filament and target so as to place a negative charge on the filament (cathode) and a positive charge on the target (anode), *space charge electrons are driven over to the anode at high speed by the large potential difference.* The maximum and average speeds of the electrons increase as the peak kilovoltage is increased. The electron stream crossing the gap between the cathode and anode constitutes the *tube current,* measured in milliamperes (mA).

4. *If the applied kilovoltage and resulting electron speed are high enough,* x rays are produced when the electrons enter the target, their *kinetic energy* being converted to heat (ordinarily more than 99 percent) and x rays (less than 1 percent).

A typical x-ray tube is a thermionic diode consisting of a tungsten filament cathode, a tungsten target anode, an evacuated glass tube enclosure, and two circuits to heat the filament and to drive the space charge electrons to the anode. However, various important modifications of these tubes adapt them to special purposes.

The structural and other details of radiographic and therapy tubes will now be described.

RADIOGRAPHIC TUBES

In this section we shall examine in detail the essential features of radiographic tubes used in medical x-ray diagnosis.

Glass Envelope

The working parts are enclosed in a *glass tube* or *envelope* containing as perfect a *vacuum* as possible. In fact, the tube is baked during manufacture to expel air and other gases that may have been trapped in the glass or metal parts, a process called *degassing.* As already noted, the vacuum offers an unobstructed path for the electron stream (tube current) and also prevents burning-out of the tube. Radiographic tubes are usually cylindrical in shape. Immersion in *insulating oil* in a suitable metal housing prevents high voltage sparkover between the terminals, thereby making possible the design of smaller tubes. The window through which the x rays leave is thinner than the rest of the tube.

Cathode

The negative terminal or *cathode assembly* consists of (1) the *filament,* (2) its *supporting wires,* and (3) the *focusing cup.* Loosely speaking, the terms "cathode" and "filament" are used interchangeably. The filament itself is a small coil of tungsten wire, the same metal that is used in electric light bulb filaments. In most radiographic tubes the filament measures about 2 mm in diameter and 10 mm or less in length, and is mounted on two stout wires which support it and also carry electric current. These wires lead through one end of the glass tube to be connected to the proper electrical source (see Figure 13.1). A low voltage *filament current* is sent through the wires to heat the filament, and one of these wires is also connected to the *high voltage source* which provides the high negative potential needed to drive the electrons toward the anode at great speed. A negatively charged concave metal cup behind the filament confines the electrons to a narrow beam and focuses them on a small spot on the tungsten target, known as the *tube focus* or *focal spot.*

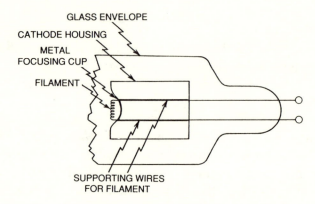

Figure 13.1. Details of the cathode of an x-ray tube (side view).

The filament current, serving to heat the filament and provide a source of electrons within the tube, usually operates at about 10 volts and 3 to 5 amperes. An increase in the filament current raises the temperature of the filament and the rate of electron emission causing a sharp rise in tube current (mA) as shown in Figure 13.2.

Some tubes, such as those used in older mobile equipment, have a single filament. However, most modern radiographic tubes, in both stationary and mobile equipment, are provided with two filaments mounted side by side. The filaments usually differ in size, producing focal spots of two different sizes on the target. Such tubes are called *double focus* tubes, although they are really double filament tubes. Note that only one filament is activated for a given exposure, being selected at the control panel. Supporting the two filaments are three stout wires, one of which is connected in common with both filaments. In Figure 13.3, if terminals 1 and 3 are connected to the low voltage source, the large filament lights up. If, instead, terminals 2 and 3 are connected to the low voltage source, the small filament lights up.

The high voltage, applied through wire 3 which is common to both filaments, gives either one a high negative potential.

Filament Evaporation. As the tube ages, the filament gradually evaporates as the result of heating during use. Consequently, a number of changes take place in the tube.

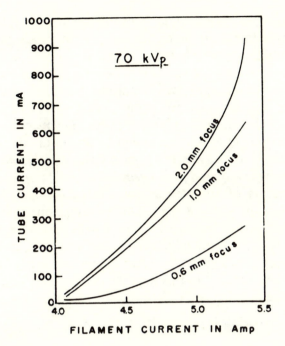

Figure 13.2. Relation of tube current (mA) to filament current. Note that a small increase in filament current produces a large increase in mA. Furthermore, mA is larger with a larger tube focus for a given filament current.

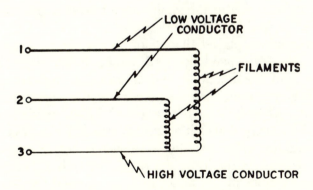

Figure 13.3. Simplified diagram of the connection of two filaments of an x-ray tube (double focal spot tube) to three wires, one of which is common to both filaments. The filaments are mounted side by side, facing the target.

1. *Thinning of the Filament.* Loss of metal (tungsten) by evapo-

ration causes the filament to become progressively thinner. Its electrical resistance therefore increases (*R* inversely proportional to cross sectional area). However, filament circuitry is designed to maintain constant current (*I*). Thus, *I²R* increases, thereby raising the filament temperature with resulting increase in tube current (mA). This may be corrected by a service person or by internal circuitry, which adjusts the filament current *downward* just enough to obtain the proper electron emission for the required tube current. It will be found that *as the tube ages, a progressively lower filament current setting is required for a desired tube milliamperage.*

2. *Deposition of Tungsten on the Glass Envelope.* As the filament evaporates, some of the vaporized tungsten deposits on the internal wall of the glass tube. When this occurs on the window of the tube, it acts as an additional filter, reducing x-ray output. Furthermore, the tungsten deposit impairs insulation so that electrons may be attracted to this metallic coating and possibly puncture the tube.

Anode

There are two main types of anodes: (1) stationary and (2) rotating.

1. **Stationary Anode.** A gap of a few centimeters separates the cathode from the anode which consists of a block of copper in which is imbedded the target, a small tungsten button. These metals are selected for definite reasons:

a. *Tungsten* — so placed that it forms the actual target of the x-ray tube, the high speed electrons striking it directly. The target is the area on the anode in which useful x rays are actually produced. Since more than 99 percent of the electrons' kinetic energy changes to heat at the target, the **high melting point** of tungsten — 3370 C — makes it especially suitable as a target metal. Furthermore, **tube efficiency**, that is, the percentage of the electrons' energy appearing in the form of x rays, is directly proportional to the atomic number of the target and the applied kV (see pages 345–346 for special mammographic tube with molybdenum target).

b. *Copper* — a much better conductor of heat than tungsten, carrying the heat away from the target more rapidly and thereby protecting it from overheating, within certain limits. Even greater heat-loading capacity can be achieved by circulating air, water, or oil around the anode, permitting larger exposures than would otherwise be possible.

Note again the electron stream bombards a limited area on the target called the *focus* or *focal spot*. X rays are emitted from the entire surface of the actual focus; in fact, there are innumerable points in the focal area from which x rays are projected in all directions.

As we shall see later, using the smallest practicable focus enhances the sharpness of the radiographic image. Therefore, radiographic tubes are provided with a small filament and a metal focusing cup. However, as the size of the focus decreases it experiences a greater concentration of heat with the attendant danger of overloading and melting the target. Tubes are so designed that the focus is small enough to give satisfactory radiographic detail, but not so small as to restrict their practical usefulness. Most stationary anodes have focal spot widths of 2 mm and 4 mm.

Anodes of radiographic tubes are constructed on the ***line-focus principle*** to provide a focus which, when projected toward the film, is smaller than is its actual area on the target. The line-focus principle can best be explained by a diagram, as shown in Figure 13.4. Here, the actual focus is a rectangle as in *B*. Assume this rectangle, *T*, to measure about 2 mm × 6 mm. In *A*, the side view of the tube, note that the anode is inclined, making an angle of 17° with the vertical. Since the film will be placed at some distance directly below the target, the ***effective*** or ***apparent focal area*** as projected toward the film is only *D*, which is about 2 mm × 2 mm. Perhaps it would become clearer if you were to imagine yourself lying supine in the position of the film and looking upward at the target—the focus would resemble a small square rather than a rectangle. In other words the focus is foreshortened, just as a pencil appears shorter when it is held obliquely in front of the eyes than it is when held parallel to the eyes. Furthermore, the effective size of the focal spot varies with the ***direction*** in which it is projected. As shown in Figure 13.5, it becomes smaller toward the anode end of the tube.

Another important relation is shown in Figure 13.6; the effective size of the focal spot diminishes as the anode angle becomes smaller (anode steeper). However, this also results in a more pronounced heel effect (see pages 394–397).

2. **Rotating Anode.** In 1936 there became available a radically

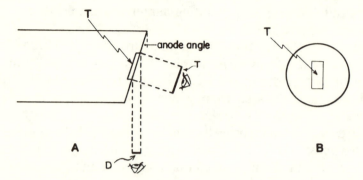

Figure 13.4. Line-focus principle. In *A* the target is shown in side view, and *D* is the *projected, apparent,* or *effective focus.* In *B* the target is shown in face view with the actual rectangular focus, *T,* at its center. The same principle is used in rotating anodes.

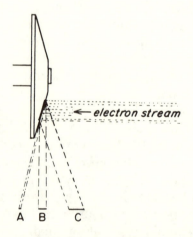

Figure 13.5. Effect of angle of emission of x rays on apparent or effective focal spot size. An anatomic detail at *A* (toward the anode end of the tube) will have a sharper image because in this direction the effective focal spot size is smaller than at *C* where the effective focal spot size is larger. *A, B,* and *C* represent the effective sizes of the same focal spot projected at different angles, that is, in different directions with respect to the anode surface. Rotating anode shown.

different kind of anode of ingenious design—the *rotating anode.* As its name indicates, this anode, attached to the shaft of a small induction motor, *rotates during x-ray production.* Rotating anodes were previously constructed of solid tungsten in the form of a disc. Later, an improved type of rotating anode was introduced, consisting of a *molybdenum disc* coated with a *tungsten-rhenium alloy*. The addition of rhenium diminishes roughening of the focal area, thereby maintaining a high x-ray output (emissivity). The usual diameter of rotating anodes is about 7.6 cm (3-in.) although 12.7 cm (5-in.) models have been introduced for extra heavy duty. Speed of rotation is ordinarily 3300 to 3600 rpm (revolutions per min), but special high-speed models capable of 10,000 rpm are available for special procedures. A diagram of a rotating anode tube is shown in Figure 13.7.

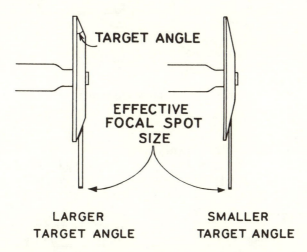

Figure 13.6. Effect of target (anode) angle on apparent or effective focal spot size. Note that a decrease in the target angle gives a smaller effective size when the actual focal spot size remains unchanged. Rotating anodes shown.

As already mentioned, rotating anodes utilize the *line-focus principle;* they therefore have a beveled edge as shown in Figure 13.7. The cathode is so placed that the electron beam strikes the beveled edge of the anode, as shown in Figure 13.8, the degree of beveling being called the *target* or *anode angle.* Anodes are available with targets of various angles such as 10°, 12°, and 17°. These influence the size of the exposure that can be used in a very short exposure time without damaging the tube (see Table 13.1).

Outside the tube is the stator of the induction motor, the rotor being inside the tube, connected to the anode. Self-lubricating bearings coated with *metallic barium* or *silver* reduce friction, thereby prolonging tube life.

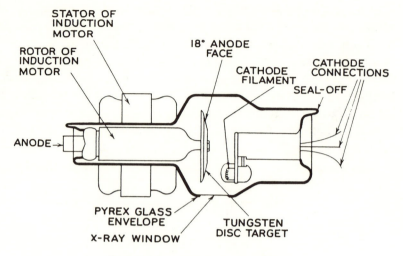

Figure 13.7. Rotating anode tube.

TABLE 13.1
EFFECT OF TARGET ANGLE OF A ROTATING ANODE
ON SHORT-EXPOSURE RATING RELATIVE
TO THAT OF A STANDARD TARGET ANGLE OF 17°

Target Angle	Approximate Increase in Short Exposure Rating
12°	30%
10°	60%

How does such an anode operate? During exposure the anode rotates rapidly while being bombarded by the electron stream. Although ordinarily rated at 3600 rpm, this is more likely to be 3000 rpm due to slippage of the rotor. During rotation the anode constantly "turns a new face" to the electron beam, so that the heating effect of the beam does not concentrate at one point, as in the stationary anode, but spreads over a large area called the *focal track* on the beveled face of the anode as shown in Figure 13.8. However, the effective focus remains constant in position relative to the film because of the extremely smooth motion of the anode. As a result, it is possible to have an effective focus of 1 mm or even smaller—the 0.3 mm fractional focus tube. Furthermore, it becomes feasible to subject a tube to high milliamperage. For instance, with a stationary anode, the upper limit for a 4 mm focus is 200 mA, whereas with a rotating anode, the upper limit for a 2 mm focus is 500 mA, the maximum exposure time being approximately the same in both cases. Special tubes have been designed to tolerate a current as large as 1000 mA for very short exposures, operating on three-phase current.

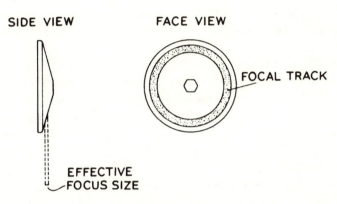

Figure 13.8. Line-focus principle with a rotating anode. The *focal track* is the actual area of impact of the electrons, or the actual focal area. Note how much smaller the apparent focal area is than the focal track. For example, a typical focal track is about 250 mm in length (circumference) and its width about 3 mm for a 1 mm² apparent focal spot. Thus, the area bombarded by electrons at any instant is $1 \times 3 = 3$ mm², while the heat is spread over an area of $3 \times 250 = 750$ mm².

Rotating anode tubes usually have two filaments to provide 1 and 2 mm focal spots; the smaller one is used to obtain fine radiographic definition, whereas the larger one can tolerate heavier loading in shorter exposure time to minimize motion in special radiographic procedures.

Special rotating anode tubes with 0.3 mm focal spots, and even smaller, have been designed for *magnification radiography*. Called *fractional focus tubes*, they permit placing the film at a distance from the part to permit direct magnification of the radiographic image. This is described on pages 352–359.

You can see from the above discussion that rotating anode tubes with a relatively high thermal capacity can successfully energize extremely small focal spots. This has made possible the superb recorded detail in radiography.

Space Charge Compensation

The low-voltage current supplied to the x-ray tube filament heats it to incandescence (white hot), causing the emission of electrons (thermionic emission). The rate of electron emission, or the number emitted per second, depends on the temperature of the filament which, in turn, is governed by the current supplied to it. In fact, *a small increase in filament current produces a large increase in the rate of electron emission and resulting space charge.*

If *no* high voltage is applied across the tube, the electrons remain in the vicinity of the hot filament as a *space charge.* This gives rise to the *space charge effect* which may be simply explained as follows: since the space charge consists of electrons, it is negative and therefore tends to hold back the further emission of electrons (like charges repel). In fact, some electrons are actually repelled from the space charge back into the filament. At any particular value of the filament current, an equilibrium is reached between the rate of electron emission by the filament and the rate of return to the filament from the space charge. Thus, *the space charge has a definite size for a given value of the filament current in a given tube,* and as just indicated, a small change in filament current produces a large change in the size of the space charge.

If a sufficiently high voltage (kV) is now applied across the tube, imparting a large negative charge to the filament and an equally large positive charge to the anode, the space charge electrons are driven toward the anode at a speed that depends on the kV (but *not* directly proportional). When tube current (mA) is plotted as a function of kV, a family of *characteristic curves* is obtained as shown in Figure 13.9. As expected, there is a different curve for each value of filament current (amp). Furthermore, the larger the filament current, the greater will be the resulting mA.

Let us now examine the characteristic curves. Note that the lowest curve (representing a filament current of 4.2 amp) has an initial curved portion and then flattens out at about 100 kV, after which a further increase in kV causes no additional increase in mA. We can explain this as follows: a progressive increase in kV drives more and more of the space charge electrons per second to the anode; that is, there is an increase in mA. This is called the *space-charge limited region* of tube operation. But above 100 kV all of the space charge electrons are driven to the anode immediately after emission from the filament, so that further increase in kV cannot increase mA; hence, this portion of the curve is flat, representing *saturation current* or the *temperature-limited region.* Here the mA can be increased only by increasing the filament current and temperature.

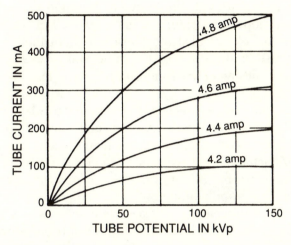

Figure 13.9. Family of characteristic curves of a typical rotating anode tube (*Machlett Dynamax "60"®, Dunlee*). Note the progressive rise in mA as kV is increased. A space charge compensator is used to flatten the curves so that kV can be changed without a change in mA as explained in the text.

The remaining curves in Figure 13.9 do not have a flat portion below 150 kV, so at the corresponding filament currents the tube operates only in the space-charge limited region. Operation of an x-ray tube in the space-charge limited region is undesirable in radiography because kV cannot be changed independently of mA—when the technologist changes the kV setting there is an unavoidable change in mA. To correct this situation a *space charge compensator* introduced into the filament circuit automatically lowers the filament current just the right amount, as the kV is raised, to keep the mA constant over the useful kV range. Now the technologist has full control of tube operation, being able to change kV without producing any appreciable effect on mA.

In modern automatic equipment with monitor-type mA selection, the space charge compensator has an additional function; it also selects the correct filament current to provide the desired space charge and resulting mA.

Factors Governing Tube Life

Overheating the filament or anode disc inevitably shortens tube life. While there are specific differences between the filament and anode in this regard, they have in common the tendency to vaporize and deposit tungsten on the glass wall of the tube. This leads to internal electrical stresses resulting in instability of operation and possible puncture of the tube wall. Furthermore, the tungsten deposit increases inherent filtration and so decreases x-ray output.

We shall evaluate the factors that influence tube life as they pertain to the filament, anode, and housing. The x-ray tube itself, as distinct from the housing, is often called the *insert.*

Filament Factors. The large tube currents (mA) often demanded in radiography require a large filament current and temperature. Failure of a radiographic tube may result from "burning out" of the filament due to application of excessive current which raises filament temperature beyond the rated limit. But tube failure can also occur from prolonged heating of the filament at normal current loads, just as an electric light filament burns out after a period of normal use. To extend tube life, a special *booster circuit* limits filament current to a low or standby level until the x-ray exposure switch is closed. At this instant, the booster circuit automatically raises the filament current sufficiently to provide the desired tube current. At the same time, the technologist must avoid using the "boost-and-hold" method of making exposure (as in radiography of infants) unless it is absolutely necessary; excessive boost time unquestionably shortens tube life.

Undue heating of the filament, whether by current or time, or both, causes evaporation of the filament metal with resulting decrease in its diameter. A diameter reduction of only 10 percent will cause the filament to break and so end tube life. Furthermore, as mentioned earlier, the vaporized tungsten coats the glass wall of the insert and causes instability in the operation of the tube; eventually, the insert may be *punctured* by electrical spark-over, at present the most frequent cause of tube failure.

Anode Factors. The ability of the anode disc to accumulate, store, and discharge heat, limits the maximum allowable techniques. These three anode factors will now be taken up individually.

1. *Radiographic Tube Rating Chart.* If the rate of temperature rise in the anode disc exceeds a limiting value, localized melting or cracking occurs as a result of thermal stresses set up in the metal by the difference in temperature between the focal spot or track at the surface, and the depths of the disc. With a slower rate of temperature rise, enough time may elapse to permit heat conduction to equalize the temperature throughout the disc. Melting of the metal, short of actually destroying the disc, causes it to vaporize and coat the glass wall of the insert, eventually resulting in its puncture. Furthermore, roughening of the focal track area and cracking of the anode disc reduces x-ray output because, as they leave the anode, the photons have to penetrate the uneven surface of the anode. Figure 13.10 shows the appearance of heat-damaged focal tracks and anode discs. Finally, warping of the anode disc causes noisy operation and decreased x-ray output.

The safe limits of exposure factors for a "cold" anode can easily be found by referring to the radiographic tube rating chart provided by the manufacturer for each type of tube and for a given kind of operation. Such a chart indicates the maximum "safe" exposure time for any selected combination of kV and mA for a single exposure and a relatively cool tube. The chart in Figure 13.11 is fictitious and is shown only by way of an example. It includes a series of curves representing various mA values of tube current. The vertical axis of the graph represents kV, and the horizontal axis maximum allowable exposure time in sec. Such a chart is quite simple to use. Assume that it applies to a particular tube of *given focus size,* and with a *given type of rectification.* Suppose we are to make an exposure of 300 mA and 65 kV, and we wish to determine the maximum allowable exposure time. Starting at 65 on the vertical axis, follow the horizontal line to its intersection with the 300 mA curve. Then from this point of intersection, follow vertically downward to the horizontal line, crossing it at three seconds, the maximum rating or tolerable exposure time with these factors.

It must be emphasized that the correct tube rating chart must be selected to avoid overheating the focal track, depending on the following conditions:

a. *Type of Rectification.* The maximum tube rating is less with half-wave (single-pulse) than with full-wave (two-pulse) rectification.

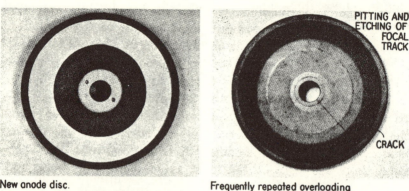

New anode disc.   Frequently repeated overloading

Repeated overload with surface melts   Overloading resulting in stopped anode
                                         and localized melt.

Figure 13.10. Surface appearance of anode disc under various conditions. *Adapted from Cathode Press, courtesy of the Machlett Laboratories, Inc.*

b. *Size of Tube Focus.* The maximum tube rating is less with a small focal spot or track than with a large one.

c. *Tube Design.* This affects the maximum tube rating; therefore the appropriate tube rating chart supplied by the manufacturer for the particular tube. For example, a tube with high-speed disc rotation has a higher rating than a conventional tube. Furthermore, smaller anode angles permit higher tube ratings for very short exposures.

d. *Cold versus Hot Tube.* The chart applies only to a relatively cold tube, one that has not been subjected to heavy loading just before use.

e. *Type of Power Supply.* Tube rating will be different for single phase and three phase, and for short or long exposures.

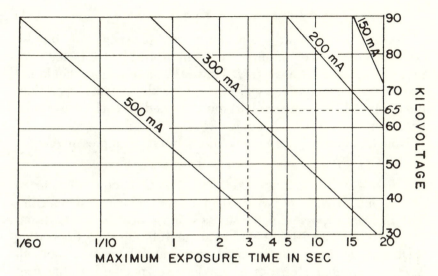

Figure 13.11. Tube-rating chart, arbitrary values. The chart for any particular tube must be obtained from the manufacturer.

f. *Application.* A special chart must be used for rapid sequence radiography, as in angiography and tomography.

The tube rating chart for the particular tube in use should be mounted in the control booth, especially if the equipment lacks automatic exposure limiters. These devices make it impossible to activate the x-ray tube if it were to be overloaded by the selected technical factors.

2. *Anode Thermal Capacity.* With multiple exposures the anode assembly gradually accumulates heat. (This is not the same as the maximum exposure factors for a "cold" anode, as determined from a tube rating chart.) We must therefore deal with the capacity of an anode to store heat without cracking, melting, or warping, any of which may reduce x-ray output (see above). Overheating and vaporization of disc metal result in coating of tungsten on the glass wall of the insert with the attendant harmful effects described earlier. Furthermore, excessive accumulation of heat may reach the anode stem and rotor bearings. The latter may become sufficiently deformed to slow down or stop the rotor, a condition that is obviously fatal to the insert. The thermal capacity of an anode is measured in *heat units,* defined by the following equation for single-phase operation:

$$\text{Heat Units (H.U.)} = \text{kVp} \times \text{ma} \times \text{sec} \qquad (1)$$

Because kV is nearly constant in a 3-phase circuit, 6-pulse operation requires equation (1) to be multiplied by the factor *1.35;* and for 3-phase 12-pulse *1.41.* However, these factors should *not* be used to convert single-phase charts to three-phase operation; the applicable tube rating chart must be used for each type of tube and equipment as noted on page 217.

Note that heat units are not fictitious, but really represent energy:

1 kVp × 1 ma × 1 sec = 1 watt-sec (*nominal*)

which is a unit of energy. (Actually, 1 H.U. is equivalent to an energy of 0.71 watt-sec, where 0.71 is the RMS conversion factor.) The *thermal capacity* of the anodes of various diagnostic tubes ranges from about 70,000 H.U. to about 400,000 H.U., depending on the *size of the anode* (ie, anode mass) and the cooling method. The cooling characteristics of anodes are specified by the *heat dissipation rate,* which must be determined from cooling curves supplied by the manufacturer of that particular tube (see Figure 13.12). If the anode has been heated to its full thermal capacity by a large number of exposures, additional exposures are limited by the cooling rate of the anode.

The equation defining heat unit storage also demonstrates that an x-ray tube operates more efficiently at higher kilovoltages. For example, a given radiographic technic calling for an exposure of 60 kVp, 100 mA, and 1 sec would develop the following number of heat units in the anode:

$$\text{H.U.} = 60 \times 100 \times 1$$
$$\text{H.U.} = 6000$$

If the kV_p is increased by 10 (ie, about 15 percent) and the time reduced one-half, the exposure in terms of visible radiographic effect would remain essentially the same, but now let us see how many heat units would be developed in the anode:

$$\text{H.U.} = 70 \times 100 \times \frac{1}{2}$$
$$\text{H.U.} = 3500$$

Thus, with an increase in kVp the heat units in the anode have been decreased almost one-half, even though the exposure has remained virtually unchanged.

The *short-exposure rating,* or the ability of a focal track to withstand a large heat input in a small fraction of a second, does not necessarily follow the heat storage capacity of the anode. For example, as described on page 256, the short-exposure rating is higher with three-phase than with single-phase current for a tube with the same thermal capacity. Furthermore, as shown in Table 13.1, the short-exposure rating of a rotating anode tube increases significantly as the target angle decreases. Still a third factor in this problem is the speed of rotation of the anode; thus, with high speed anodes—10,000 rpm—the short-exposure rating is increased about 70 percent over that with 3000 rpm.

In general, *an increase in the actual area of the focal spot or track increases the instantaneous rating of the anode, whereas an increase in the anode mass increases its thermal capacity.*

After prolonged use, or with minor overloading, the anode disc becomes roughened by *pitting* or *etching*. When severe, this reduces the x-ray output of the tube. The use of *rhenium-tungsten-molybdenum* instead of pure tungsten as the anode material aids in minimizing this problem.

In the ordinary oil-immersed tube, adequate *cooling* results from the transfer of heat to the oil, tube, housing, and surrounding air by simple convection and conduction. For heavier loads, as in special procedures and high voltage radiography, cooling is improved by mounting a small fan outside the tube housing. Some specially designed tubes are cooled by a forced oil circulating system, the oil flowing into the hollow anode and then between the tube and housing; this provides a very high rate of heat dissipation. Thermal capacity may be enhanced by the use of rotating anodes with surface areas 100 percent greater than usual. The oil incidentally provides efficient insulation, while the metal housing helps make the tube rayproof by shielding out unwanted radiation.

3. *Anode Cooling Curve.* The rate at which the anode cools, starting at any particular value of heat storage (heat units), is given by the anode cooling curve appropriate to the tube under consideration. Figure 13.12 shows a typical anode cooling curve. At no time should another exposure be made if the additional heat produced in the anode would increase the stored heat beyond the instantaneous rating. Thus, starting with a known amount of heat already present in the anode, we must allow enough time to elapse for it to cool down sufficiently to permit the additional heat load. An example is given in the legend to Figure 13.12.

Housing Factors. Often overlooked, the ability of the metal

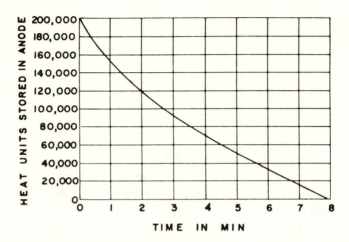

Figure 13.12. Anode cooling curve. With this tube, if the anode has the maximum 200,000 heat units already stored, then at the end of two minutes the stored heat will have decreased to 120,000 H.U. It would then be safe to load the anode with additional 80,000 H.U. for a total maximum of 200,000 H.U. For complete evaluation, a housing cooling curve should also be consulted. (*Based on data furnished by courtesy of Machlett Laboratories, Inc.*)

housing to store and dissipate heat contributes to tube life. Thus, the temperature of the housing must be kept below 90°C (200°F), both to prevent damage to the housing itself and to permit adequate cooling of the insert. In fact, excessive heat may bring the oil to the boiling point and even cause the housing to **explode.** This is a real hazard with rapid serial exposures in angiography and cineradiography. An air blower may help cool the housing, but is effective only if the housing cools more slowly than the anode disc. Regular servicing of the blower is mandatory to assure proper function, especially for an under-table tube. Special cooling curves, similar to those for anode cooling rate, are available for the housing.

Summary of Practical Steps in Extending Useful Tube Life. A number of precautions in the use of radiographic tubes contribute significantly to prolongation of their useful life. These have been described in detail but will now be summarized.

1. *Filament life* is shortened by application of large filament currents (and resulting high temperatures). Therefore the "boost" or preparation time during which the filament temperature is held at full value should be kept as short as possible.

2. *Instantaneous ratings,* as derived from tube rating charts,

should never be exceeded. Be certain to use the correct chart for the particular tube, focal spot size, type of rectification, and type of current (single or three-phase).

3. The *anode should be warmed up* before being subjected to a large load; use warmup times suggested by the manufacturer. Heat produced by a large power load on a cold anode causes uneven expansion that may crack the anode disc.

4. The *life of the bearings* in a rotating anode tube is shortened as the total rotation time is increased. Therefore the anode should not be run unnecessarily. This becomes a special problem in radiography of infants when the rotor is kept activated while the technologist awaits the opportune instant to make the exposure. In addition, the bearings may be damaged by heat overload of the anode.

5. *Excessive temperature of oil in the tube housing may shorten tube life.* Therefore adequate cooling of the tube housing must be provided; for example, by use of an auxiliary fan which may be provided at the time of installation.

6. *Rapid sequence exposures* as in angiography and cineradiography may easily exceed the tube rating. Therefore the proper rapid sequence rating charts and anode cooling curves should be consulted. In fact, it is good practice to work up a chart showing the maximum number of exposures in the time intervals for various procedures which may pose a hazard for the x-ray tube; this chart should be posted in the control booth for ready reference.

7. *Avoid overheating the filament and anode.* Even minor overheating, if repeated a sufficient number of times, causes evaporation of tungsten which is deposited on the glass wall of the insert. Eventually the insert may be punctured by spark-over, especially with application of high kV. In fact, tube puncture ranks high as a cause of tube failure in ordinary routine radiography, whereas anode warping or cracking occur more often in rapid sequence radiography.

X-RAY THERAPY TUBES

These are x-ray tubes designed exclusively for use in clinical radiation therapy. Such tubes, operating at relatively low mA values of tube current, are manufactured in various models for the desired kV range.

In x-ray therapy, four main kV energy ranges are available for treatment at different depths in the body.

1. **Low Voltage.** This includes a range of about 50 to 120 kV and is used to treat lesions in the skin. It is also known as *superficial therapy.*

2. **Intermediate Voltage.** This is about 130 to 150 kV and is used mainly in treating lesions lying just below the skin.

3. **Orthovoltage.** The usual range of 160 to 300 kV provides x rays of relatively higher penetrating power. Many years ago this was used in so-called "deep" therapy; now replaced by megavoltage therapy.

4. **Megavoltage.** Operating at potentials of mV and more, the specially designed tubes for this type of equipment, the *linear accelerator* or *linac,* will not be described here.

The first three types of tubes differ from a radiographic tube in having a larger filament, a stationary anode, and a larger focal spot; the last permits greater heat loading for continuous operation. The anode consists of a tungsten mass imbedded in a large block of copper, its face usually being inclined at an angle of about 30° to generate a beam of reasonably uniform intensity. As is true of radiographic tubes, therapy tubes have a separate filament current to heat the filament and produce a space charge; and a high voltage to drive the electrons to the target at high speed. Such tubes operate in the space-charge limited region of the characteristic curve.

Since there may be a tremendous amount of heat liberated in the anode, particularly in the orthovoltage region, special means of cooling the tube must be provided. Superficial therapy tubes are usually cooled in the same manner as radiographic tubes, depending on the anode heat load. On the other hand, intermediate and orthovoltage therapy tubes most often have hollow anodes cooled by circulating oil between the tube and housing. A pump then transfers the hot oil to a special tank where it is cooled by water circulating in coils.

The oil serves another purpose. Because it is highly refined and moisture free, it has superb insulating ability. Such insulating oils permit a reduction in the size of the tube housing, thereby making it more easily maneuverable.

RECTIFIERS

At one time, specially designed vacuum diode tubes served as rectifiers. These have been replaced by *solid state diode rectifiers*, which employ a special class of materials known as *semiconductors.*

To understand its basic principles we must look into the conduction of an electric current in solids from a slightly more advanced point of view than in Chapter 6.

One of the ways in which we can classify matter is on the basis of its *conductivity,* defined as the ease with which electrons move about within it. Accordingly, there are three classes of solids: (1) *conductors* (excellent conductivity — electrons can be made to drift readily from point to point), (2) *nonconductors* or *insulators* (very poor conductivity — high resistance to movement of electrons), and (3) *semiconductors* (intermediate conductivity).

What accounts for the difference in conductivity of these three classes of matter? Energy-state diagrams come to our aid. An *energy state* is the energy of a particular electron, corresponding to its energy level in the atom. In Figure 13.13 we see the energy states that an electron may have in a single atom. Since a piece of matter is an aggregation of a fantastic number of atoms usually arranged in a crystal lattice, that is, a regular pattern, their energy levels merge into a series of *energy bands.*

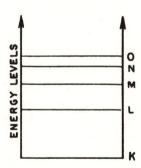

Figure 13.13. Possible energy states of an electron in a single atom. The higher energy states are in the direction of the arrows.

There are three types of energy bands — *valence, forbidden,* and *conduction.* No electron can remain in the forbidden band; this band represents the energy needed to raise an electron from the

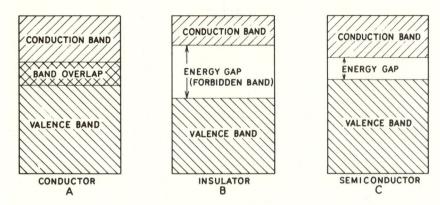

Figure 13.14. Energy level diagrams of three types of matter. In a conductor (A) electrons move freely from the valence band to the conduction band. In an insulator (B) there is a large energy gap between the valence and conduction bands preventing free movement of electrons. In a semiconductor (C) there is a small energy gap between the valence and conduction bands making it possible to raise electrons to the conducting band, thereby controlling resistance.

valence band to the conduction band where the electron can move freely. The various bands are shown in Figure 13.14.

In a *metallic conductor*, shown in Figure 13.15A, the outermost or *valence* electrons are very loosely bound; in fact, the valence band actually overlaps the conduction band. Thus, an extremely weak electric field (applied emf) causes the electrons in the valence band to drift in a particular direction as an electric current. Because electrons can be made to flow in a metal with the application of so little energy, electric current in a conductor is not easily controllable (as it is in a semiconductor).

In an *insulator*, shown in Figure 13.14B, all electron-containing bands are completely filled. Since a very large energy gap exists between the valence band and the conduction band in insulators, an extremely large electric field would be needed to raise an electron to the conduction band, so that, for all practical purposes, an insulator does not conduct an electric current (we may say that it has virtually infinite resistance).

Now we come to the third class of solids—*semiconductors.* These are intermediate between conductors and insulators in their ability to conduct an electric current. In Figure 13.14C you can see that a small energy gap exists between the valence band and the conduction band. Thus, only a small amount of energy is needed in semiconductors to raise the valence electrons to the conduction band. Note that this energy gap is much smaller in semiconductors than in insulators (compare Figure 13.14B with 13.14C). Because the valence and conduction bands do not overlap in semiconductors, current flow can be regulated; for example, the addition of carefully measured amounts of certain impurities to a semiconductor modifies its conductivity. In fact, its conductivity is controllable by the kind and amount of impurity.

Of the available semiconductors, we shall limit our discussion to *silicon* (an element in ordinary sand—silicon dioxide) because it is now being used in the manufacture of solid state rectifiers for x-ray equipment.

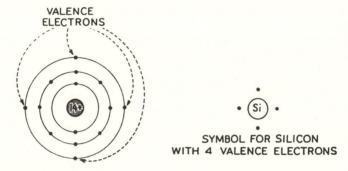

Figure 13.15. The silicon atom has four valence (outer shell) electrons as indicated by the broken lines.

Basic to an understanding of semiconductors is the rule that, in combining chemically, most atoms share their valence electrons in such a way that the outer shell of each contains eight electrons. This is known as the *octet rule.*

The silicon atom has four valence electrons (in its outermost shell) as shown in Figure 13.15. Silicon atoms aggregate into crystals by sharing these valence electrons with each other—socalled *covalent bonding* —as shown in Figure 13.16. Such bonds, in effect, saturate the outer or valence shells, thereby satisfying the octet rule and enhancing the stability of the crystal structure. Pure silicon is an example of an *intrinsic* semiconductor.

If a small amount of *arsenic* is added to silicon, it contributes to the crystal lattice an *extra* electron, not needed for bonding silicon and arsenic atoms (see Figure 13.17). This is a loosely bound

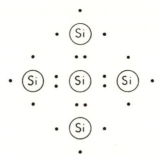

Figure 13.16. Diagram of the crystal lattice of silicon atoms which share valence electrons in pairs by means of *covalent bonds.* Each silicon atom is surrounded by four others (actually, in three dimensions). Note that the central Si atom, by electron sharing, now has an *octet*—eight valence electrons. In a real crystal there are tremendous numbers of Si atoms forming covalent bonds.

electron which can easily be raised to the conduction band by a small applied electric field. Arsenic is known as a **donor** atom because it furnishes electrons. Silicon "doped" with arsenic is called **n-type silicon** (*n* stands for negative). Its conductivity can be controlled by the amount of arsenic doping—the more arsenic, the more electrons available and the greater the conductivity.

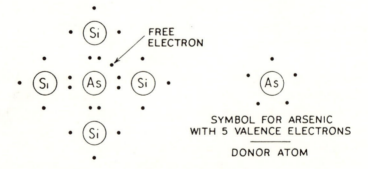

Figure 13.17. *N*-type silicon. Addition of a minute amount of arsenic with its five valence electrons introduces a free electron into the crystal. Under the influence of an applied emf, this free electron can be raised to the conduction band and made to drift through the crystal as part of an electric current.

On the other hand, if **gallium** with its three valence electrons is added to the silicon as shown in Figure 13.18, there is a shortage of one electron in the crystal structure called a **"hole."** Such a hole

behaves like a positive charge, gallium-doped silicon being known as ***p-type silicon*** (*p* stands for positive). Since the hole can accept an electron, gallium is an ***acceptor*** atom.

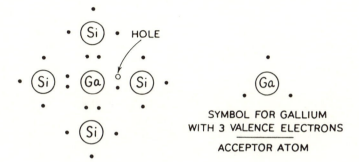

Figure 13.18. *P*-type silicon. Addition of a minute amount of gallium with its three valence electrons results in a deficit of one shared electron in the crystal lattice. An absent electron is equivalent to a positive charge, here called a "hole"; this can accept an electron.

How do these two types of semiconductors behave? In Figure 13.19A, representing an n-type semiconductor, electrons drift in an applied electric field just as they do in a conductor. But in a p-type semiconductor, shown in Figure 13.19B, an applied electric field causes electrons to move toward holes, at the same time leaving new holes which are then, in turn, filled by other electrons. Thus, the holes, in effect, drift in the opposite direction from that of the electrons as though the holes were positive charges.

Suppose, now, we were to combine n-type and p-type silicon at a ***junction*** as in Figure 13.20. Some electrons from the n-type silicon combine with holes of the p-type near the junction, leaving positive and negative ions, respectively, as shown in the figure. A potential difference exists between these oppositely charged ions, called the ***barrier voltage*** or ***potential hill.***

Let us see how an n-p junction rectifies—that is, ***conducts current in only one direction.*** Suppose we apply an emf across this junction in the direction shown in Figure 13.21A, with the negative terminal of the source at the n-end, and the positive terminal at the p-end. Electrons in the n-type silicon move toward the junction and, at the same time, holes in the p-type silicon also move toward the junction where the electrons and holes combine.

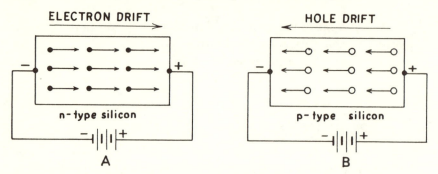

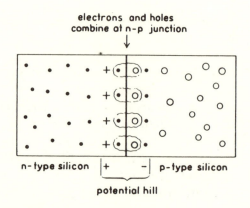

Figure 13.19. A, behavior on *n*-type silicon when an electric field (emf) is applied. Free electrons are raised to the conduction band and drift through the crystal just as in a conductor. B, in *p*-type silicon, electrons enter from the negative terminal of the battery and combine with holes which may be regarded as drifting toward the negative terminal.

Figure 13.20. Behavior of an *n-p* junction. When *n*-type and *p*-type silicon are brought into contact, holes and electrons combine near the junction leaving, respectively, negatively and positively charged ions. A potential "hill" (potential difference) is built up at the junction.

As a result, the n-type silicon accepts electrons from the negative terminal of the battery, while the p-type silicon passes electrons to the positive terminal. Thus, the diode (ie, the n-p junction) continues to conduct current as long as the battery terminals are connected in the manner shown. This is called a *forward biased diode.*

Now if the battery terminals are reversed, as in Figure 13.21B, electrons leave the n-type silicon, and holes leave the p-type silicon.

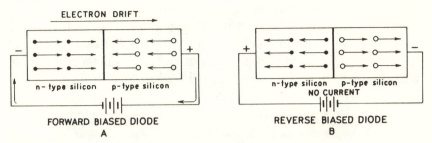

FORWARD BIASED DIODE
A

REVERSE BIASED DIODE
B

Figure 13.21. Effect of an applied electric field (emf) on an *n-p* junction. In *A*, with the negative terminal connected to the *n*-type silicon, electrons drift toward the junction where they combine with holes. The net effect is that the electrons enter the *n*-type silicon and leave the *p*-type silicon. In *B* the battery terminals have been reversed; now electrons are withdrawn at the positive end, and holes at the negative end. Both are immediately depleted and no current flows. When *AC* is applied, current flows only during the forward-biased half of the cycle; in other words, the *n-p* junction behaves as a *rectifier.*

This causes an increase in the number of positive ions on the n-side of the junction, and an increase in the number of negative ions on the p-side of the junction, thereby increasing the barrier voltage. As a result, the barrier voltage instantaneously becomes equal and opposite to the applied voltage and no current flows across the diode. This is now a **reverse biased diode.** Thus, if an alternating current is applied to an n-p junction it undergoes rectification, being conducted across the junction only when the negative half of the AC cycle is directed from *n* to *p*. The symbol for a solid state diode rectifier is shown in Figure 13.22.

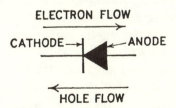

Figure 13.22. Symbol of a solid state diode rectifier.

A practical silicon rectifier for x-ray equipment consists of a stack of individual diodes, each containing an n-p junction. These diodes, called **modules,** are connected in series. The complete assembly is enclosed in a sealed ceramic tube containing an insulating liquid. Each module is capable of withstanding a maximum of 1000 volts, so that for a 150-kV generator, 150 modules are required. The rectifier stack is of the **controlled avalanche type** to minimize breakdown of individual modules. A silicon rectifier for x-ray equipment is shown in Figure 13.23.

Figure 13.23. Silicon rectifier for x-ray equipment. Various sizes are available, from about 8 to 10 inches in length.

What are the advantages of the silicon rectifier over the valve tube? They may be listed as follows:

1. Compact size.
2. No filaments, hence no filament transformers needed for solid state rectifiers. This is especially important in three-phase equipment where twelve rectifiers are ordinarily used.
3. Low forward voltage drop—less than 300 volts at maximum current rating of 1000 mA.
4. Low reverse current—0.002 mA at maximum voltage.
5. Long life due to rugged construction.

Because of these advantages, solid state rectifiers have replaced valve tubes in modern radiographic units.

RECTIFIER FAILURE

The technologist should form the habit of checking for satisfactory operation of the rectifier diodes during each exposure. This can be done by simply watching the milliammeter, which indicates **average mA**. With four diode (full-wave) rectification, the **peak mA** is actually about one and one-half times the reading of the milliammeter. Whenever one rectifier diode in such a system fails, the system operates as a two-diode (half-wave) rectifier, producing a drop in the milliammeter reading to about one-half the normal value, and causing radiographic underexposure. With pushbutton selection of mA there is no compensation for half-wave rectified operation, and so the average mA drops to one-half the normal value.

The simplest device for testing the competence of any type of

Figure 13.24. Diagram of a lead spinning top used to determine the integrity of the rectifiers in a full-wave rectified circuit.

rectifier diode is the **spinning top.** This is a flat metal top which has a small hole punched near one edge (see Figure 13.24). It is placed on a film exposure holder containing an x-ray film, and made to spin during an x-ray exposure of one-tenth second. If all four diodes are operating, the circuit is fully rectified and there should be 120 pulses per sec with 120 corresponding peaks of x-ray output on 60-cycle current. Therefore, in one-tenth second there will be twelve peaks and the image of the spinning top on the radiograph will show twelve spots, as in Figure 13.25A. If only six dark spots appear in the radiograph of the top, as in Figure 13.25B, there were only six pulses in one-tenth second or sixty

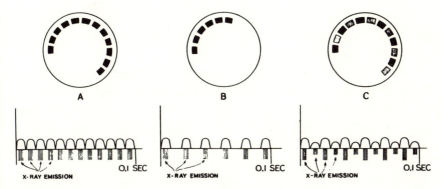

Figure 13.25. Spinning top test. In *A*, the radiograph of the spinning top shows twelve spots in 1/10 sec, representing 120 kV peaks per sec; this is full-wave single-phase rectification. In *B*, there are six spots in 1/10 sec, representing 60 kV peaks per sec; this is half-wave single-phase rectification, which may occur when one diode rectifier is completely inoperative. In *C*, one rectifier diode is malfunctioning (ie, partly operative), so there is less voltage left for the x-ray tube; therefore, every other spot appears lighter (data kindly furnished by Willis R. Preston).

pulses per second, indicating that the circuit is half-wave rectified, provided the exposure timer is known to be operating correctly.

The spinning top may be used to test the accuracy of the x-ray exposure timer if all four rectifier diodes are known to be functioning properly, in single-phase equipment. However, with 12-pulse three-phase equipment, the image produced by a spinning top is a dark band with intermittent indistinct darker pulses that do not permit accurate evaluation of the timer. An oscilloscope should be used with such equipment, as it provides an image of the actual pulses; in a properly operating 12-pulse circuit, there will be 12 pulses in $1/60$ sec.

QUESTIONS

1. Describe the main features of a thermionic diode tube.
2. Compare the structural details of radiographic, therapy, and valve tubes.
3. Describe, with aid of a diagram, a double focus, stationary anode tube.
4. Explain the principle of the rotating anode tube and show its important structural features by means of a simple diagram. Enumerate its advantages over a stationary anode tube.
5. As a radiographic tube ages, what adjustment must be made in the filament current, and why?
6. What effect does evaporation of the filament have on tube operation? tube life? x-ray output?
7. Describe the line focus principle and discuss its importance in radiography. To what types of tubes does it apply?
8. Discuss saturation current.
9. How does the space charge effect influence the operation of a radiographic tube? How does a change in filament current affect space charge?
10. To what part of the tube does heat storage refer? In what units is it expressed?
11. Using the chart in Figure 13.11, determine the tube rating ("safe" exposure time) at 70 kV and 500 mA.
12. Discuss the importance of anode heat storage; anode cooling rate; housing cooling rate; anode pitting; anode evaporation.
13. Why does a radiographic tube anode have a greater heat

loading capacity with full-wave rectification than with half-wave rectification?

14. Summarize the steps that should be taken to prolong tube life.
15. Compare the values of peak milliamperage and milliameter readings in a full-wave and half-wave rectified circuit.
16. How are radiographic and deep therapy tubes cooled?
17. Describe the spinning top method and state its purpose. What is used instead of a spinning to check timer accuracy in twelve-pulse three-phase equipment, and why?
18. What are the main causes of x-ray tube failure? How can they be minimized?
19. What are semiconductors? How is their conductivity controlled?
20. Describe the principle of the n-p junction.
21. What intrinsic semiconductor is most often used in solid state rectifiers in medical radiography?
22. Discuss solid state rectifiers on the basis of principle and construction. List their advantages and disadvantages relative to valve tubes.
23. Prove that kV × mA = watts.
24. Show how heat units are related to energy according to equation (1).

Chapter 14

X-RAY CIRCUITS

THE MAJOR ITEMS of an x-ray unit have been covered in some detail, but they have been discussed individually without relation to each other. There now remains the task of combining them to create a functioning x-ray machine. A number of auxiliary devices are necessary for the proper operation of x-ray equipment, and these will also be described.

Source of Electric Power

The electric power company supplies a high voltage alternating current to the *pole transformer* outside the building. The transformer, which is mounted atop a pole, is a step-down transformer. It reduces the high voltage to 120 to 240 volts, depending on the type of equipment to be served. The voltage supplied to the radiology department is called the *line voltage.*

A *three-wire system* conducts the current into the building, the two outer wires being "hot" and the middle one "grounded" (neutral), as shown in Figure 14.1. At the instant that one hot wire is 120 volts above ground, the other is 120 volts below ground, thereby producing a potential difference of 240 volts. Thus, connection to the two hot wires provides a 240-volt source, whereas connection to either hot wire and the neutral wire provides 120 volts.

Most x-ray units now operate on 240 volts, although some of the less advanced equipment, and especially the older mobile units, require only 120 volts.

Note that the current delivered to the Radiology Department is *alternating current* (AC) because it is necessary for the operation of the x-ray transformer. Under unusual circumstances when only direct current (DC) is available, it must first be converted to AC so that it can be stepped up by the transformer.

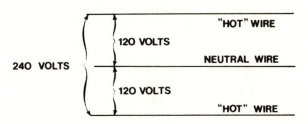

Figure 14.1. Three-wire system which brings electric power into the building. By connecting the electrical appliance or equipment to the appropriate wires as indicated, one may obtain 120 volts or 240 volts.

The Main X-ray Circuits

To simplify the discussion of the electrical connections of an x-ray unit, we may regard the circuit as being divided at the x-ray transformer. The part of the circuit connected to the primary coil of the transformer is the *primary* or *low voltage* circuit, whereas the part connected to the secondary coil is the *secondary* or *high voltage* circuit. This scheme is perfectly natural, since the two coils of the transformer are electrically insulated from each other, and since the profound difference in voltage in the two circuits requires separate consideration. However, some parts of the equipment are connected to both the low and high voltage sides.

1. **Primary Circuit.** Included in this portion of the circuit is all the equipment that is connected between the electrical source and the primary coil of the x-ray transformer. Each essential component will be described in proper sequence.

a. *Main Switch.* This is usually a double-blade, single-throw switch, illustrated in Figure 14.2.

b. *Fuses.* Each wire leading from the main switch is provided with a *fuse* — an insulated cylinder with a metal cap at each end, in the center of which a thin strip of metal of low melting point connects the caps. Overloading the circuit with excessive amperage causes the metal strip within the fuse to melt because of the heating effect of the current, thereby breaking the circuit. Figure 14.3 shows the construction of such a fuse and its electrical connection. Fuses protect the equipment and reduce fire hazard. A blown fuse is easily replaced, but the underlying cause should first be ascertained and corrected.

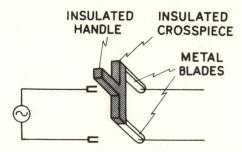

Figure 14.2. Double-blade single-throw switch. When the switch is closed toward the left, the metal blades contact the corresponding wire and the circuit is completed to the x-ray machine.

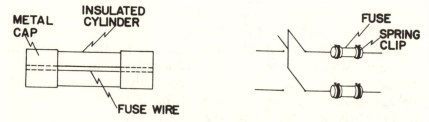

Figure 14.3. Diagram of a fuse (left), and its connection in a circuit. When the circuit is overloaded (excessive amperage) the wire in the center of the fuse melts, breaking the circuit.

c. *Autotransformer.* Discussed in detail in Chapter 10, the auto-transformer changes the voltage supplied to the primary of the transformer, thereby providing various kilovoltages for the x-ray tube.

d. *Prereading Kilovoltmeter.* Since kV selection must be made (by means of an autotransformer) before a radiographic exposure, we must have a metering device that can indicate the desired kV. This function is served by the prereading kilovoltmeter which is an AC voltmeter connected in parallel with the autotransformer, that is, across the primary circuit between the autotransformer and the transformer (see Figure 14.12). How, then, is it possible for a voltmeter in the primary circuit to indicate kV across the second-ary circuit? The answer lies in the calibration (standardization) of the prereading kilovoltmeter in the primary circuit by the manu-facturer, against actual peak kV values in the secondary circuit by means of spark gap measurements. These depend on the fact that a given kV will cause a spark to jump between two metal spheres separated by a particular distance or *air gap* (at standard temperature, atmospheric pressure, and relative humidity). The kilovoltages obtained by such spark gap measurements are recorded directly on the scale of the prereading kilovoltmeter for each corresponding autotransformer setting. In other words, the prereading kilovoltmeter, since it is in the primary circuit, does not measure kV directly, but rather indicates what the kV will be in the secondary circuit because it has been initially calibrated to do so at the factory. The main advantage of the prereading kilovoltmeter is that *slow* fluctuations in line voltage can easily be corrected. If, for instance, the line voltage should drop, the autotransformer controls are adjusted manually until the correct kV shows on the prereading kilovoltmeter.

Some types of x-ray equipment do not employ a prereading kilovoltmeter. Instead, the desired kV is obtained by means of kV-labeled *pushbuttons* which select the appropriate transformer settings. The calibration of these settings is similar to that of a prereading kilovoltmeter. But these fixed pushbutton settings indicate kV accurately only if the line voltage is the same as it was during the actual calibration. Since the line voltage may fluctuate slowly, depending on the other electrical equipment in use, a means of detecting such fluctuation is provided by a *compensator voltmeter* connected in parallel across a portion of the primary side of the autotransformer. A mark on the compensator voltmeter scale indicates the correct line voltage. If the pointer is above or below this mark on the meter, a *line voltage compensator* permits adjustment of the line voltage to the correct value. The compensator varies the number of turns on the primary side of the autotransformer until the compensator meter needle indicates that the correct line voltage has been restored; only then will the correct kV be obtained.

e. *Timer and X-ray Exposure Switches.* These switches control the current to the primary coil of the transformer, serving to complete the x-ray exposure. The switches themselves are modifications of an ordinary pushbutton, shown in Figure 14.4, and can be used for either hand or foot operation. But these are not designed to withstand the high amperage in the primary circuit (as high as 90 A with a 200 mA unit), or to prevent electrical shock.

Therefore, these switches operate a *remote control switch,* which in turn closes the primary circuit.

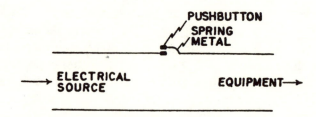

Figure 14.4. Pushbutton switch. Depressing the button brings the metal contactors together, thereby closing the circuit.

f. **Remote Control Switch.** Operated by either hand or foot switch, its basic design can best be appreciated by studying Figure 14.5.

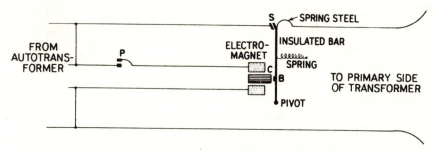

Figure 14.5. Remote control switch. The pushbutton, *P,* when depressed, completes the circuit through the coil of the remote control switch. This magnetizes the core, *C,* which attracts the metal button, *B,* on the insulated bar. This closes the primary circuit by bringing together the contacts at *S.* When the pushbutton is released, the reverse occurs, breaking the circuit.

g. **Timer.** Included in the exposure switch circuit is a timer which can be set to start and stop an exposure of preselected duration. There are four types of exposure timers. They include (1) mechanical, (2) electronic, (3) milliampere-second meter, and (4) phototimer.

(1) **Mechanical Timer.** As the name implies, such a hand-held timer has a clock mechanism with a pointer that can be rotated, by means of a knob, around a scale calibrated in seconds. Turning the knob to the selected time winds up a spring; and activating the

exposure button releases the spring and terminates the exposure at the preselected time interval. Mechanical timers are only accurate to about 0.1 sec and are limited to dental x-ray machines where a high degree of accuracy is not required.

(2) *Electronic Timer* — provides extremely accurate exposures down to 1 millisecond ($\frac{1}{1000}$ sec). Its circuitry, shown in Figure 14.6, consists essentially of a rheostat (variable resistor), a capacitor, and a *thyratron* which is a special gas-filled triode that becomes conductive at a particular, critical voltage. When the x-ray exposure switch is closed, switch *S* is simultaneously opened, allowing current to build up a charge on the capacitor. When the capacitor charge reaches a critical value, discharge suddenly occurs through the thyratron, activating the exposure-terminating switch and thereby ending the exposure. The time required to build up the necessary charge on the capacitor governs the time of the x-ray exposure; it depends on the product RC (resistance × capacitance) and can be varied by adjusting the resistance of the circuit by means of the rheostat. The capacitance C remains constant in this particular circuit. In six-pulse and twelve-pulse three-phase equipment the thyratron, and the electromagnetic relay with its mechanical contactor, are replaced by a *thyristor*, which is a solid state device that functions like a thyratron but can provide much shorter and more accurate exposures. These may be as short as 1 millisecond with a 1-millisecond delay (1 millisec = $\frac{1}{1000}$ sec) between activation of the exposure switch and the actual start of the exposure.

(3) *Milliampere-second (mAs) Meter.* This is, strictly speaking, not really a timer since it reads out mAs — the product of mA × exposure time. It is described on pages 247–248.

(4) *Automatic Exposure Control* — includes a phototimer circuit. With this system, originated by Morgan and Hodges, a film is exposed to the x-ray beam and as soon as it has received the correct amount of radiation for the desired degree of film darkening (density), the exposure terminates automatically. This requires a special type of diode known as a *phototube;* its cathode is coated with an alkali metal such as potassium or cesium, which has the peculiar property of *giving off electrons when struck by light.* If the anode of the phototube has, in the meanwhile, been given a positive charge from an outside source, the electrons emitted by the cathode are attracted to the anode, constituting a current in

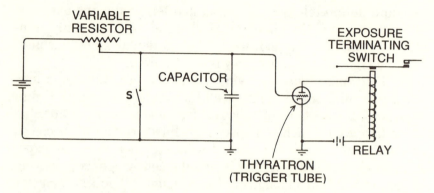

Figure 14.6. Simplified circuitry of an electronic timer. When switch S is closed, charge cannot build up on the capacitor. Activation of the exposure switch (in another part of the circuit) starts the exposure and opens switch S, allowing a charge to build up on the capacitor. When this reaches a preset value, the capacitor suddenly discharges through the thyratron tube, activating the exposure termination switch. This may be either an electromagnetic switch, as shown, or a *thyristor*, a solid state device functioning like a thyratron but allowing exposures as short as 1 millisec, with a 1-millisec delay. The time for buildup of capacitor charge depends on RC, which is the product of resistance by capacitance. Since C is constant, we select the exposure time by adjusting R by means of a rheostat.

the phototube. By placing a fluorescent screen between the phototube and the x-ray source, we can make the phototube indirectly sensitive to x-rays, the brightness of the screen depending on the intensity of the x radiation. As shown in Figure 14.7, the phototube-fluorescent screen combination is placed behind the cassette whose back is radiotransparent (x-ray transmitting). When a predetermined quantity of radiation has reached the fluorescent screen, depending on the part being radiographed, the resulting current in the phototube operates a capacitor, thyratron, and relay circuit which activates a contactor to terminate the exposure.

Automatic exposure control in modern x-ray units has abandoned the phototube, replacing it with a thin, radiolucent, *parallel-plate ionization chamber* positioned between the patient and film cassette. This also eliminates the need of a fluorescent screen. Because the charge liberated in the ion chamber varies directly with the desired radiographic density, the chamber is calibrated initially (upon installation) to terminate the exposure via the remainder of the circuit as in Figure 14.7, when the proper amount of radiation has been delivered.

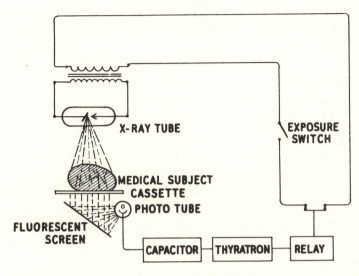

Figure 14.7. Automatic exposure control. When the exposure switch is closed, x rays passing through the patient and reaching the fluorescent screen of the phototimer assembly cause the screen to emit light which activates the phototube. Current then flows in the phototube and charges the capacitor while the exposure is in progress. When the predetermined voltage is built up on the capacitor, the thyratron becomes conductive, activating the relay which opens the exposure circuit and terminates the exposure even while the manual exposure switch is still closed. In most automatic exposure systems available now, the fluorescent screen has been replaced by a parallel plate ionization chamber located between the patient and cassette.

Automatic exposure control entails extremely accurate reproduction of radiographic density, provided the anatomic part is carefully centered. The most useful application of automatic exposure control is in spotfilm and chest radiography, and in photofluorography, although it is also available now for general radiography.

h. *Backup Timer and Minimum Exposure Time.* Modern equipment designed for automatic exposure control has a special manual backup timer that is set to terminate the exposure in the event of failure in the automatic circuit, to avoid the hazard of overloading the tube or overexposing the patient. According to the Code of Federal Regulations (21CFR), the maximum allowable exposure time shall be equal to or less than $1/60$ sec, or the time interval to deliver 5 mAs, whichever is greater. The maximum allowable mA × time shall be limited to 600 mAs per exposure, except when the tube potential is less than 50 kVp, in which case the limit shall be 2000 mAs per exposure. In equipment with pushbutton selection of the anatomic part, the backup times is set automatically for each selection by internal circuitry, but HEW regulations still apply as above.

A manual *backup timer* is set for a specific time, usually about 0.5 sec. As an example, if the exposure called for 140 mAs at a 200 mA setting, the time would be 140 mAs/200 mA = 0.7 sec, and the exposure would be terminated at 0.5 sec by the backup timer. Thus, the film would be underexposed.

Radiographic equipment timer circuits have a *minimum exposure time*, which varies with the type of equipment, ranging from about 0.05 sec (50 millisec) in older equipment, to 0.005 sec (5 millisec) in later models. (With the introduction of thyristors, minimum exposure times will be in fractions of milliseconds.) For example, with a equipment having a 0.05 sec minimum exposure time, if the exposure required 2 mAs at 200 mA, the time would be 2 mAs/200 mA = 0.01 sec. Since this is less than the minimum exposure time (0.05 sec), the exposure would terminate at 0.05 sec and the film would be overexposed. But if the exposure required 12 mAs, the time would be 12 mAs/200 mA = 0.06 sec; as this is longer than the minimum exposure time (0.05 sec), the exposure time would terminate correctly after a total of 0.06 sec.

i. *Circuit Breaker.* Additional protection against overloading the circuit is provided by a circuit breaker which can easily be reset, while a blown fuse has to be replaced. It is usually connected in series with the exposure switch, timer, and remote control switch. Figure 14.8 illustrates schematically the operation of a magnetic circuit breaker, shown in the exposure circuit, 2. When the exposure switch is closed, circuit 2 is completed through the timer, the electromagnet of the remote control switch, and the circuit breaker contacts which are in the closed position. This activates the electromagnet of the remote control switch closing the primary circuit, 1, that leads to the primary side of the x-ray transformer. If there should be a momentary surge in the primary current, this will, through circuit 3, increase the strength of the electromagnet in the circuit breaker to the point where it will open circuit 2, thereby interrupting the current to the electromagnet of the remote control switch, opening the remote control switch, and breaking the primary circuit, 1. The circuit breaker must then be reset manually before another exposure can be made.

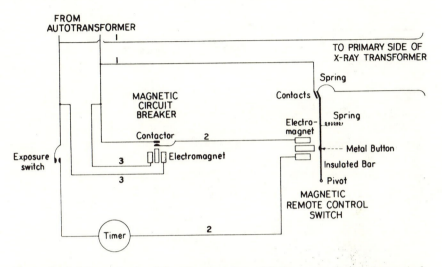

Figure 14.8. Magnetic circuit breaker together with the magnetic remote control switch, as they are connected in the primary circuit.

j. *Filament Circuit of X-ray Tube.* The primary circuit supplies the *heating* current for the filament of the x-ray tube, but this current must first be reduced to 3 to 5 amp and 6 to 12 volts by a *rheostat* (variable resistor); in accordance with Ohm's law, the greater the resistance, the smaller the current (amp). Most radiographic units are equipped with *pushbutton type* mA selectors which also choose the desired focal spot size. The pushbuttons operate through individual preset resistors which provide the correct filament current for each tube current, such as 100, 200, and 500 mA. Other types of filament current regulators include the *saturable reactor* and the *high-frequency* control, described on pages 132–134. In series with this is an *oil immersed step-down transformer,* which further reduces the voltage to the required value, as shown in Figure 14.9. Because this *filament transformer* has its secondary coil in the high voltage circuit and its primary coil in the low voltage circuit of the x-ray unit, it must be oil immersed to provide sufficient insulation between these two circuits.

There is often included in the filament circuit a *filament stabilizer* for the x-ray tube because a relatively small change in the

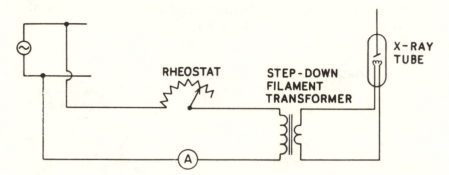

Figure 14.9. Filament circuit. The rheostat varies the filament current (and voltage), thereby controlling filament temperature, electron emission, and tube current (mA). Thus, it is an *mA selector.* A saturable reactor is being used instead of the rheostat in advanced design equipment. The step-down transformer further helps reduce the voltage and also insulates the primary circuit from the high voltage in the secondary circuit. Instead of a rheostat, modern equipment uses a saturable reactor or a high-frequency circuit.

filament voltage or current causes a large change in electron emission and consequent tube current (mA). The filament stabilizer corrects for instantaneous fluctuations in line voltage that may be caused by momentary demand elsewhere on the line, such as in the starting of an elevator or air conditioner. It may be so effective that a variation in line voltage of 10 percent will cause no greater change than 1/2 percent in the filament voltage. The stabilizer consists of a capacitor and a small, modified split transformer which are so arranged that they compensate for a rise or fall in line voltage and provide a more uniform filament voltage, maintaining the filament current more nearly constant.

k. *Filament Ammeter.* In order to measure the filament current, and hence the amount of heat developed in the filament, an ammeter is connected in series in the *filament circuit* of old style equipment. By previous calibration, we can establish the readings of this meter that correspond to desired mA in the x-ray tube at a given kV. In modern radiographic units a *space charge compensator* automatically adjusts the filament current to maintain constant mA over a wide range of kV; furthermore, pushbutton selection of mA automatically provides the correct filament current for the desired tube current.

l. *Primary Coil of X-ray Transformer.* Although the high voltage transformer operates as a unit, we may regard its primary winding (coil) as a component of the primary circuit. Both the primary and the secondary coils are immersed in special transformer oil for added insulation.

In general, the wires in the primary circuit must be relatively large, because of the high amperage. Circuit breakers and fuses should be conveniently located for easy resetting or replacement.

2. **Secondary Circuit.** This includes the secondary coil of the transformer and all devices to which it is connected electrically. The wire conductors in the secondary circuit have a smaller diameter (than those in the primary circuit) because they carry a smaller current (mA).

a. *Secondary Coil of X-ray Transformer.* As we have indicated earlier, this consists of many turns of electrically insulated wire that is thinner than the wire in the primary coil because of the very small current in the secondary circuit. The transformer *steps up* the primary voltage to provide the high voltage required to operate the x-ray tube, the step-up ratio being about 500:1.

b. *Milliammeter.* In order to measure the mA in the x-ray tube, a *milliammeter* is connected in series in the high voltage circuit. Since the milliammeter is grounded together with the midpoint of the secondary coil of the x-ray transformer, it is at zero potential and can therefore be safely mounted in the control panel without hazard to personnel.

Attention must be called to the fact that the *tube current is measured in milliamperes by the milliammeter placed in the high voltage circuit, while the filament current is measured by an ammeter placed in the low voltage filament circuit.* The milliammeter measures average values and gives no indication of the peak values of tube current.

c. *Milliampere-second or mAs Meter.* It is desirable to have in series with a regular ammeter, a mAs meter. Due to the large mass of the rotating mechanism, it turns slowly so that the attached indicator needle registers the product of mA and time, that is, it indicates *milliampere-seconds* (mAs). This device is needed because the ordinary milliammeter does not have time to register the true mA at exposures of less than 1/10 sec. The ballistic meter therefore serves to measure mAs at very short exposure times, and because of its great sensitivity, it must be used in conjunction with an impulse timer or an electronic timer. At exposure times longer than 1/10 sec, the ordinary milliammeter will register the correct average mA.

d. *Rectifier.* In all but self-rectified units, a system of rectification is included to change AC, supplied by the transformer, to DC. As we have already noted, rectification increases the heating capacity of the x-ray tube, safely permitting larger exposures.

e. *Cables.* These conduct the high voltage current from the rectifier to the x-ray tube. The ***midpoint of the secondary coil of the transformer is grounded.*** However, the total kV across the ends of the transformer secondary remains unchanged because one-half is below, and the other half above, ground potential (see Figure 14.10). For example, if there is 90 kV across the transformer and the center is not grounded, the kV will fluctuate between 90 kV above ground and 90 kV below ground according to the AC sine wave. If the center of the secondary coil is grounded, the peak kV will still be 90, but in this case one-half or 45 kV will be above ground and one-half will be 45 kV below. The difference between 45 above zero and 45 below zero is still 90. Under these conditions each cable has to be insulated for only 45 kV (rather than 90 kV), representing a significant saving in weight and cost.

The cables are designed to eliminate the danger of shock, provided the insulation remains intact. Ordinarily, very heavy insulation would be required to prevent the high voltage from sparking over to the patient or other objects near the cable. However, the grounded woven wire sheath surrounding the cable obviates the danger of sparkover. Figure 14.11 shows a shockproof cable in cross section with its standard three conductors. The same type of cable is used for both terminals of the x-ray tube. When the ***cathode*** cable is connected to the cathode end of a double-focus x-ray tube, all three conductors make contact with the two filaments (see Figure 13.3). The other end of this cable makes contact with both the filament circuit and one end of the transformer secondary. When the ***anode*** cable is connected to the anode, connection is made with only one of the conductors within the cable. At the other end of the cable, the three conductors join a single conductor which connects with the opposite end of the transformer secondary. Thus, the three conductors in the anode cable serve as a single conductor, which is all that is needed for the anode.

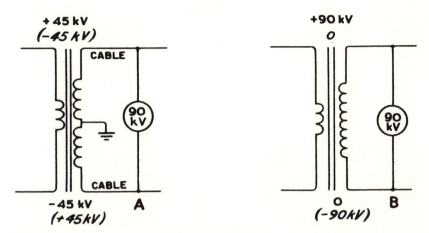

Figure 14.10. In *A* the transformer secondary has been grounded at its midpoint. With the upper end of the secondary coil positive, and the lower end negative, *final kV = 45 kV − (−45 kV) = 90 kV.* Each cable needs insulation for only 45 kV above or below ground. A similar situation prevails when the polarity of the secondary reverses in the next half cycle (values in parenthesis).

In *B* the secondary has not been grounded at its midpoint. Total secondary voltage is again 90 kV, but it now represents the difference between 90 kV at one end of the secondary, and 0 at the other end. Each cable must therefore be insulated for the full 90 kV above or below ground.

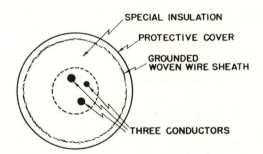

Figure 14.11. Cross section of a shockproof cable.

f. *X-ray Tube.* The ultimate goal of the x-ray equipment is the operation of an x-ray tube, the last piece of equipment connected in the high voltage circuit.

Completed Wiring Diagram

In order to visualize clearly the relationship of the various parts of the x-ray unit, we must connect them correctly in the circuit. Figure 14.12 presents schematically a full-wave rectified x-ray unit employing four rectifier diodes (these may be valve tubes or solid state diodes). Most of the auxiliary items, though desirable or even necessary for proper operation, have been purposely omitted to avoid complicating the diagram. The hook-up of these accessory devices has been indicated in the preceding sections. We can readily modify this basic diagram; thus, for self-rectification we simply omit the rectifier and connect the transformer to the tube terminals.

Simple X-ray Control Panel

An x-ray machine requires not only suitable control devices for easy selection of kV, mA, and time, but also readily accessible meters to check the operation of the equipment. The controls and meters are mounted compactly in a *control panel* — a separate unit which is connected electrically to the x-ray equipment and is comparable in function to the dashboard of an automobile.

X-ray control panels vary greatly in complexity, depending on the design of the individual x-ray machine. If we realize that all control panels, regardless of their intricacy, are similar in their basic design, the problem of operating unfamiliar equipment becomes a relatively simple matter. The more complicated units are variations of the basic pattern.

In Figure 14.13 is shown a simple, nonautomatic control panel comprising all the essential items; familiarity with this diagram will make it much easier to understand the operation of the more complex types of equipment. The three meters shown in the figure include the milliammeter, filament ammeter, and kilovoltmeter. The milliammeter registers, in mA, the current in the secondary circuit, including the x-ray tube.

A filament ammeter indicates filament (heating) current, which ultimately determines mA. Turning the control knob in the lower left hand corner of the diagram, labeled "filament control," varies the filament current. This control actually operates a rheostat in the filament circuit to vary resistance.

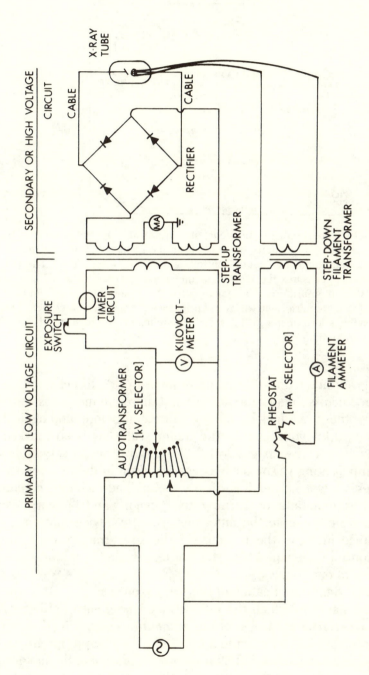

Figure 14.12. Simplified wiring of a single phase x-ray unit, with full-wave rectification.

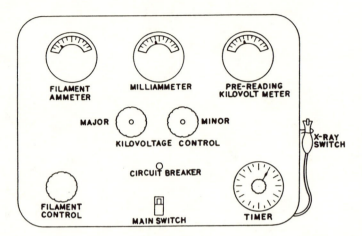

Figure 14.13. The essential components of a primitive x-ray control panel. The filament control knob regulates the rheostat in the filament circuit, varying the filament current and consequently controlling the tube current (mA). The filament ammeter measures the current in the filament circuit. The kV control operates the autotransformer in steps of 10 kV (major) and 1 kV (minor). Actually, the autotransformer varies the primary voltage which is registered on the prereading kilovoltmeter in "kilovolts." The milliammeter indicates the average tube current in mA.

A kilovoltmeter does not actually measure kV directly, but is connected across the primary circuit and therefore measures volts. However, these readings have been calibrated beforehand against known tube kilovoltages and the meter scale has been marked accordingly in kV (see pages 238–239). A kilovoltmeter serves also as a compensating voltmeter because any drop in the line voltage produces a lower kV reading on the meter. The technologist can adjust this manually by turning the "major" and "minor" kV controls which operate the autotransformer, a device that varies the voltage input to the primary of the x-ray transformer. The major control varies the kV in steps of ten, while the minor varies it in steps of one.

An important modification of this basic plan is found in the type of control panel in which the kilovoltmeter is replaced either by *kilovoltage selector* control knobs or *pushbuttons* operating through the autotransformer to provide a range of kV. The appropriate kV settings have been arrived at by previous calibration at the factory. In other words, when the settings read "56," it means that if the kV were actually measured it would be 56 kVp, provided the line voltage corresponds to that at the time of original calibration. With this system a line voltage meter and a line voltage compensator must be included to correct for variations in line voltage.

Modern radiographic equipment includes a *milliamperage selector.* Instead of a filament control knob, there is an array of pushbuttons which activate the proper resistor to obtain the correct filament current for the desired mA. The pushbuttons also select the focal spot. For example, the pushbutton setting for large focal spot and 200 mA should automatically activate the large focal spot and deliver a tube current of 200 mA.

The timer, indicated in the lower righthand corner of the diagram, controls the duration of the x-ray exposure which is initiated by the pushbutton x-ray switch. In addition, there is a main switch; when this is in the "on" position, the primary voltage is applied to the autotransformer, and also to the kilovoltmeter which is connected in parallel across the primary circuit. At the same time, the current also flows through the *filament only* of the x-ray tube and registers on the filament ammeter. Thus, *when the main switch is closed, only the filament ammeter and the kilovoltmeter should normally be activated.* Then, when the x-ray switch is closed, the primary voltage applied to the primary side of the x-ray transformer is stepped up by the transformer, and current finally passes through the x-ray tube as indicated on the milliammeter. Thus, *the milliammeter does not register until the x-ray switch is closed.*

The construction and operation of the circuit breaker have already been discussed.

Most diagnostic equipment is supplied with a special radiographic-fluoroscopic *changeover switch,* enabling the radiologist to do spotfilm work automatically. Shifting the spotfilm device into the radiographic position activates the *changeover switch* and automatically selects the exposure factors for radiography. When the spotfilm is withdrawn from the radiographic position, the exposure factors automatically return to fluoroscopy. Phototiming is essential for timing spotfilm exposures.

You will find it advantageous to correlate the sections on the various items of x-ray equipment with the operation of the basic control panel, thereby developing a much clearer concept of the function of the various parts.

THREE-PHASE GENERATION OF X-RAYS

Available for a number of years in Europe, three-phase generators are being installed in the United States in ever increasing numbers. This trend has paralleled the growth of angiography with its need of large tube currents and very short exposure times.

The circuitry of three-phase equipment is necessarily more complex than that of single-phase. At present, there are three main types of three-phase circuits: six-pulse-six-rectifier, six-pulse-twelve-rectifier, and twelve-pulse-twelve-rectifier. All use *solid state diode rectifiers*.

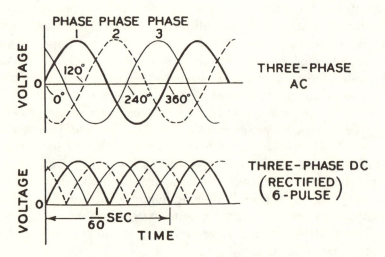

Figure 14.14. Voltage curves for three-phase power supply. In the upper diagram, phase *1* lags phase *2* by 120°, and phase *2* lags phase *3* by 120°. In the lower diagram the three-phase current has been rectified to give six pulses in 1/60 sec, with a 13 percent variation in voltage, or *ripple*. Special circuitry (not shown) provides twelve pulses in 1/60 sec with a ripple of only 3 percent, giving virtually constant potential.

A three-phase generator operates on three-phase current, which consists of three single-phase currents out of step with each other by one-third cycle or 120° (see Figure 14.14). Therefore, the transformer primary and secondary coils must each have three windings. These are arranged in either a *delta* or a *star* configuration (see Figure 14.15). Three-phase circuits all have delta-wound primary coils, but differ in the form of the secondary windings. In

Figure 14.16 is shown a wiring diagram of a twelve-pulse-twelve-rectifier generator having a balanced circuit.

With three-phase equipment, three autotransformers are needed for kV selection, one for each phase. Furthermore, because of its nearly constant value, the high voltage circuit cannot be closed and opened at or near zero potential as in the conventional single-phase unit. This introduces the danger of power surge which can damage the equipment. Therefore, special thyratron timer circuits have been designed to make the contactors open and close sequentially instead of simultaneously. Power surge is further suppressed by the solid state rectifiers themselves. In more modern equipment thyratrons and mechanical contactors have been replaced by solid state *thyristors* to improve further the accuracy of exposure timers (see pages 241–243).

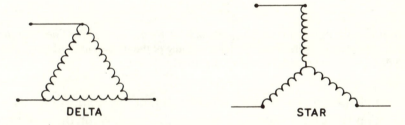

DELTA **STAR**

Figure 14.15. Types of windings used in transformers for three-phase equipment.

At present, three-phase generators are available with ratings up to 1600 mA at 150 kV (ie, 240 kW) and with exposure times as short as 1 millisec ($^{1}/_{1000}$ sec). Furthermore, as shown in Figure 14.17, the voltage is the resultant of three out-of-phase single-phase voltages and never reaches zero. In fact, the voltage shows a small fluctuation or *ripple* and may be regarded as nearly *constant potential*. Although the effective kV at the transformer is about 95 percent with a 12-pulse, and 87 percent with a 6-pulse generator, the effective kV during operation is actually 95 percent with either generator because of the voltage-smoothing effect of the capacitance in the high tension cables. The resulting ripple is about 3 to 5 percent in both instances, contrasted with about 70 percent in single-phase full-wave rectification.

Comparison of radiographic technics with the two systems shows that with the same mAs, 85 kVp three-phase is equivalent to 100

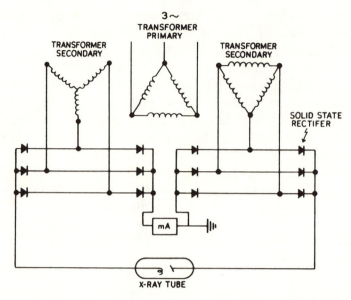

Figure 14.16. Simplified wiring diagram of a twelve-pulse-twelve-rectifier generator in a balanced circuit. Note the delta winding of the transformer primary, and
the delta and star windings of the secondaries.

kV single-phase generated x rays. On the other hand, if one uses
100 kVp with both types of equipment, the mAs for three-phase
will be about 0.4 that of single-phase for equal radiographic density.
However, in this case the contrast will be less with the three-phase
generator because of the greater half-value layer of the x-ray beam.
When technics are adjusted to obtain the same density and contrast,
the *radiation exposure of the patient is the same with both types of
equipment.* Hence, from the standpoints of radiographic quality
and patient exposure, there is no appreciable improvement with
the three-phase generator.

Probably the single most important advantage of the three-
phase system is the higher tube rating at *very short exposures*
allowing the use of very high mA with exposure times as short as
one millisecond. Figure 14.18 shows tube rating curves comparing
the maximum mA that can be used with three-phase and single-
phase equipment at various exposure times (Dynamax® "61" tube,
1 mm focus, 100 kV). Note that at $1/120$ sec the maximum ratings
are 1000 mA with three-phase, and 860 mA with single-phase. On
the other hand, at five seconds the situation is reversed; now the

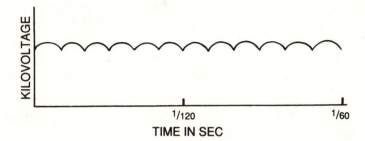

Figure 14.17. Rectified voltage curve with three-phase twelve-pulse x-ray generator. The voltage remains nearly constant throughout the alternating current cycle, the slight variation being called *ripple;* in this case, ripple is 3.5 percent. Note the twelve peaks or pulses per cycle of 1/60 sec.

maximum ratings are 190 mA with three-phase, and 230 mA with single-phase. The reason for this is that with very short exposures, heating occurs mainly at the anode surface and is readily dissipated. But with long exposures with three-phase equipment the large power input heats the anode to a greater depth and thereby decreases tube rating.

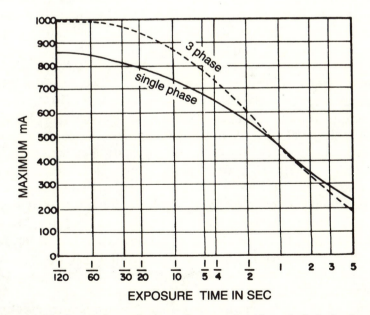

Figure 14.18. Comparison of tube rating curves with typical single-phase and three-phase x-ray generators. Note the higher rating with three-phase for short exposures, and the slightly higher rating with single phase for long exposures.

Because of the complex circuitry of three-phase equipment, changeover from fluoroscopy to spotfilm radiography is slow – 1.5 to 2 sec. This can be overcome by introducing a device known as a Q stator into the circuit; it reduces changeover time to 0.5 sec.

In summary, then, the advantages of the three-phase generation of x rays are as follows:

1. High mA at very short exposures, especially useful in angiography, spotfilm radiography, and tomography.
2. Nearly constant potential characteristics.
3. Higher effective kV.

HIGH-FREQUENCY X-RAY GENERATORS

For many years attempts were made to design x-ray equipment that would generate x rays at constant potential to eliminate the kV fluctuations inherent in 60-cycle (hertz or Hz) single-phase machines. You should recall that with half-wave rectification there are 60 voltage peaks/sec, and with full-wave rectification 120 peaks/sec; between these peaks, x-ray energies range downward to zero and are increasingly absorbed in the tissues without contributing to the x-ray image. It would therefore be advantageous to reduce the tissue-to-film dose ratio.

In the 1970s an x-ray *therapy* unit called the *Vanguard* (Picker X-Ray Corp.) was introduced, generating 280-kV x rays at more nearly constant potential than other x-ray machines. The *Vanguard* converted a three-phase current to a high-frequency (HF) 1.2 kHz (1200 cycles/sec) current; but this system lost out to the already growing popularity of the cobalt-60 therapy units. At about the same time, 6 to 70 kHz generators were being devised for general radiography, but fell victim to the imported three-phase units which, although expensive, were widely accepted for angiography and later for general radiography.

With the advent of high-frequency (H–F) generators, it became possible to obtain nearly constant voltage output with ripple of less than two percent. Such H–F units operate on 115–120 VAC (AC voltage) or 220–240 VAC° single phase, depending on their design. For example, the Bennett dedicated *mammographic* unit

---

°volts alternating current

requires 115 VAC, whereas General Electric, Acoma, and Lorad units operate on about 220 VAC. Lorad also has a 115-VAC unit.

The advantages of H–F mammographic equipment include their compactness, increased mobility, and reliability. In fact, they are being used not only for fixed installations but also in special vans with an on-board technologist for breast screening in outlying communities where no such equipment exists. Film processing may be done on-board, or the exposed films taken to home base for processing. Because of developer and fixer cross-contamination during transport, the preferred method at present is to return the exposed films to home base where a dedicated mammography processor should be available. This should be done each evening after the day's run because saving the films for several days leads to deterioration of the image. Mobile mammography has unquestionably extended the reach of breast screening with its attendant early detection of more breast cancers.

From the standpoint of exposure, H–F generators result in reduction of breast tissue dose by about 25 percent relative to 60-Hz generators. Moreover, they greatly reduce exposure time to about 30 to 40 msec (0.03 to 0.04 sec), thereby minimizing the effect of voluntary and involuntary motion and improving recorded detail. Finally, tube life is increased because the very small ripple distributes the heat in the anode more evenly than 60-Hz operation.

High-frequency generators for dedicated mammographic units are available at 20 or 100 kHz (kilohertz) operation. Only one manufacturer — Bennett — is manufacturing 100 kHz units at the present time. Claims and counterclaims are being made as to the superiority of one H–F over the other, and only time will answer this problem. Here, discussion will center, in simplified terms, on the general aspects of H–F circuitry (see Figure 14.19). A power supply of 115 or 220 VAC (single phase) is fed to a *battery charger.* The battery serves as a reserve source of electric power because of the large drain by the x-ray tube. An *H–F converter* (oscillator) changes the battery DC to H–F AC that supplies the full-wave rectified high-voltage *transformer* to step up the potential to 22 to 28 kV (in one case, to 49 kV). For some 220 VAC and in three-phase units the *battery is omitted,* so the incoming AC is fed directly to the H–F converter; the remainder of the circuit remains the same as in Figure 14.21. The secondary transformer voltage undergoes "smoothing" by a *filter capacitor* whose output supplies

260   *The Fundamentals of X-ray and Radium Physics*

the x-ray tube (see pages 345–346 for tube details). To minimize random voltage fluctuations, during x-ray exposure an auxiliary circuit samples the kV output at the transformer secondary 20 or 100 times per msec (20 kHz or 100 kHz generators, respectively) and, via a *feedback loop* to the H–F converter, adjusts from instant to instant the voltage to the transformer primary.

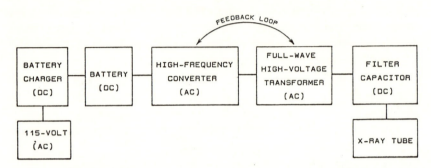

Figure 14.19. Block diagram of a 100-Hz generator for a mammographic unit supplied by 115-volt AC (*Bennett X-Ray Corp.*).

High-frequency generators are being introduced into general radiography because of their compactness and reliability. They include features of automation that are not available with conventional radiographic equipment. Another advantage of the H–F over the 60-Hz generator is the greater mR/mAs output, which results in shorter exposure time, lower patient dose amounting to about 25 percent, and increased generator accuracy and consistency in image quality.

It must be emphasized that the information about these mammographic units was kindly furnished by the respective manufacturers, is not necessarily complete, and does not in any way imply the author's preference of one unit over another. In addition to complete technical information, the purchaser must, *as always,* consider equipment reliability, exposure consistency, auxiliary devices, user friendliness, availability of service, survey of present users, and comparative cost before making a final decision.

FALLING–LOAD GENERATOR

An ingenious method of reducing exposure time and simplifying the selection of exposure techniques is provided by the *falling-load* *generator.*[*] Unique circuitry, controlled by a microprocessor, automatically reduces mA in tiny steps closely following the selected kV curve of the pertinent tube rating chart (see Figure 14.20). In this process, the tube operates at near-maximum rating along the required initial segment of the kV curve, to produce the optimal mAs at each point.

Power supply is three-phase, converted to high-frequency current as described for H–F mammographic units. This permits a feedback loop from the secondary side of the full-wave rectified high-voltage transformer to the H–F converter to adjust for random voltage fluctuations in the primary circuit. A filter capacitor then "smooths" the kV before it is applied to the x-ray tube.

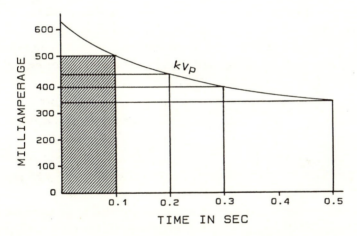

Figure 14.20. Tube rating chart to show principle of falling load generator. The crosshatched area represents the exposure factors of kVp and mA, allowable for 0.1 sec. The generator follows the course of the upper right-hand corner of the rectangle at tiny intervals along the curve until the selected exposure has been delivered.

The falling-load generator permits either of two kinds of technique selection, depending on the circuitry of the machine:

1. *One-knob selection* allows the technologist to set *kV* by means of an "up-down" touch pad, like channel selection on a modern TV set. This requires a phototiming circuit, which automatically sets the optimum mAs for the desired examination.

---

[*]*Medical Systems Division, North American Philips Corp.*

2. *Two-knob selection* is available in the absence of phototiming. As before, kV is selected by means of an up-down touch pad, but now the mAs must be set by a separate up-down touch pad.

Thus, technique selection may be greatly simplified while exposure reliability compares favorably with that of other types of x-ray generators.

SPECIAL MOBILE X-RAY EQUIPMENT

Battery-Powered Mobile X-Ray Units

Self-contained mobile ("portable") apparatus energized by powerful storage batteries eliminates the need of an external power supply, except for recharging the batteries. An example of the updated version° of the original model utilizes a sealed lead-acid minimum-liquid battery that does not leak even if punctured. It actually comprises nine 12-volt packs connected in series, operating both the x-ray tube and the drive motor for the mobile unit. This maintenance-free battery carries a 5-year warranty.

A message center contains an *automatic voltage sensor,* which flashes a warning message indicating the need of a recharge when 90 percent of the stored mAs has been exhausted. A second message gives the warning that no charge remains; whereupon the unit shuts down automatically, but still allows a short interval in which the motor still operates to allow time to move the unit to a wall outlet. Recharging involves plugging the electric cord into a 110- or 220-V line for all models. The charging process stops automatically when the battery attains full charge. This procedure must be conducted in an open area, never in a closet or operating room.

The circuitry is not too complex. An *inverter* changes the battery DC voltage to 1 kHz (1000 Hz) AC. The high-voltage *transformer* with full-wave rectification steps this up to 125 kV at nearly constant potential, and yields a usable storage of 20,000 mAs.

Microprocessor control provides highly accurate output as to kV and mAs, the kV being available as follows: 50–76 kV in 2 kV, and

<hr>

°*GE Medical Systems*

80-125 kV in 5-kV steps. A touch panel allows two-point selection of kV and mAs; no separate timer is needed.*
The options for technique selection are as follows:

kV	mAs
50-90	0.4-320
95-105	0.4-250
110-125	0.4-200

This unit operates a rotating anode tube with a 0.75 mm nominal focal spot and a thermal capacity of 275,000 heat units.

Another battery-powered mobile x-ray unit** has essentially the same general circuitry, but there are several differences in details. The battery has twenty 12-volt gel type battery packs, giving a total of 240 VDC, rechargeable automatically by either a 100-120 or a 200-240 VAC source by plugging the cord into the pertinent wall outlet. An inverter changes the battery DC to high-frequency (HF) tube current at 20 kHz, and filament current at 0.5 kHz. The 50 to 125 kV output is nearly constant potential. The rotating anode tube has dual focal spots — 0.6 and 1.2 mm — and a thermal capacity of 300,000 HU. With a fully charged battery, 20,000 mAs are stored at 100 kV. Techniques can be set in 1-kV steps from 50 to 125 kV, and in 30 steps between 0.4 and 320 mAs by means of touch pads. No timer is needed because mA is automatically programmed according to the mAs setting.

Capacitor (Condensor)-Discharge Mobile X-ray Units

One of the early methods of generating x rays utilized the principle of storing a quantity of electric charge in a capacitor (condenser) and then discharging it through an x-ray tube. At this point review the capacitor on pages 82-84. For our present purpose, we may say that the *capacitor stores mAs.* (Note that mA is a measure of charge per sec, so mA times sec should represent charge.)

The voltage and charge on a capacitor are related by the following equation:

Important Note: Please see pages 82-84 about recommendations before purchasing equipment.

**Acoma X-Ray Industry Co., Ltd.*

$$Q = CV \qquad (1)$$

where Q = charge in coulombs
   V = potential in volts
   C = capacitance in farads (constant for a particular capacitor)

By selection of appropriate units, equation (1) is readily adapted to x-ray equipment; thus, with a capacitance of 1 microfarad (μF),

$$mAs = \mu F \times kV$$

In other words, if the capacitor is designed to have a capacitance of 1 μF, then 1 kV of potential is acquired for every mAs quantity of electric charge stored in the capacitor. For example, at a potential of 75 kV, it will have a charge of 75 mAs, so a total of 75 mAs will be available to the x-ray tube. To charge the capacitor in ten seconds requires only 75 mA/10 sec = 7.5 mA, a small current, indeed.

The time it takes a *capacitor* to discharge to 37 percent of its initial charge is called the *time constant*, τ (Greek letter, *tau*), which is determined by the equation

$$\tau = RC \qquad (2)$$

where R and C are the resistance and capacitance of the circuit, respectively. By selecting the proper capacitance for a particular circuit we have a means of controlling the discharge time.

In practice, we should avoid too great a drop in kV during exposure, a situation that would exist if the capacitor were allowed to discharge completely. Instead, some type of interval timer must be used to control the duration of the exposure. The term *wave-tail cutoff* refers to the process of stopping the discharge of the capacitor at some preselected point on the discharge curve (see Figure 14.21). The first portion of the diagram shows the charging curve as the capacitor is charged to a desired peak kV, V_τ, in time, T_p. At time T_0 the exposure is initiated and the capacitor discharges along the discharing curve which, in this case, is cut off at time T_r. The tube current decreases along a similar curve. Wave-tail cutoff is represented by the dotted portion of the discharging curve.

If we were to start with 100 kV and a tube current of 500 mA, and used 30 mAs for an exposure, the exposure time, t, would be obtained from

$$mAt = mAs$$
$$500t = 30$$
$$t = 30/500 = 0.06 \text{ sec}$$

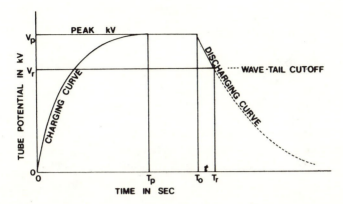

Figure 14.21. Charging and discharging characteristics of a capacitor discharge radiographic unit. V_p is peak kV; V_r is residual kV after capacitors have discharged during exposure time $(T_r - T_0)$; and T_p is time required to charge capacitors to peak kV.

During this exposure time the tube potential will have dropped from 100 kV to 70 kV (a loss of 30 kV accompanying a loss of charge of 30 mAs), when the total capacitance is 1 μF.

How do we start and stop the discharge of the capacitors, thereby activating and deactivating the x-ray tube? This requires a *grid-controlled x-ray tube* (see Figure 14.22). An ordinary x-ray tube is a diode, having two electrodes. On the other hand, a grid-controlled tube has a third electrode or "grid," making this type of tube a triode. The grid, which in ordinary vacuum triodes consisted of a metal wire mesh, is actually the *focusing cup* itself, completely insulated from the filament in a triode x-ray tube. When a *negative potential* or *grid bias* of 1 to 2 kV is applied to the cup relative to the tube kVp, it "breaks" the tube current (mA).

In the capacitor discharge unit, grid bias functions as a switch: activation of the exposure switch instantaneously reduces grid bias virtually to zero, permitting the applied kV to drive the space charge electrons to the anode as in a conventional x-ray tube. Once the selected mAs value has been attained (determined either by an mAs control device or an automatic exposure system), a grid bias of −1 to 2 kV is automatically applied and wave-tail cutoff achieved.

Figure 14.23 shows a simple block diagram of a capacitor-

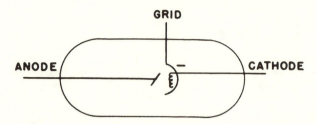

Figure 14.22. Simplified diagram of a grid-controlled x-ray tube. Exposure is started by removing the negative charge on the grid (nearly to zero); it is stopped by restoring a negative charge to the grid.

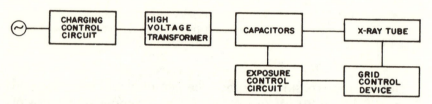

Figure 14.23. Block diagram of a capacitor discharge radiographic unit.

discharge unit. For those who may be interested in greater detail, Figure 14.24 contains a wiring diagram of such equipment. As shown, one capacitor is charged during each half cycle of the alternating current to one-half the desired kV. The two capacitors in series discharge almost continuously through the tube during exposure, with a relatively small ripple in kV. Note the summation of the kV of the two capacitors in series.

A grid-controlled tube requires a cathode cable with a fourth connector or pin at its tube terminal, through which the negative bias voltage can be applied to the focusing cup (what are the other three connectors?)

In actual radiography with capacitor discharge apparatus, after the main switch is turned on, the line voltage is adjusted and the kV selected. Next, the capacitors are charged by depressing the charger button. When the preselected kV has been acquired by the capacitors, charging automatically stops. The mAs is selected by another control, although this step is omitted with phototiming (optional). Finally, the exposure button is activated, causing the capacitors to discharge through the x-ray tube.

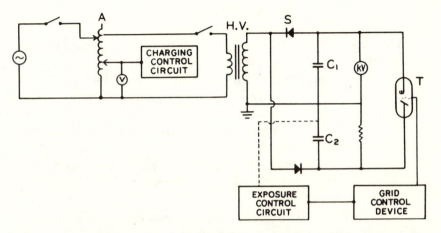

Figure 14.24. Simplified wiring diagram of a capacitor discharge radiographic unit. *A* is autotransformer. *H.V.* is high voltage transformer. *S* is solid state diode (rectifier). C_1 and C_2 are capacitors.

Note that after completion of the exposure, kV may not return immediately to zero because of a residual charge on the capacitors. Because of electric shock hazard the high voltage cables cannot be disconnected while any charge remains. Another warning should be sounded here—if a charge remains on the capacitors, leakage may cause generation of x rays without activation of the exposure switch.

Some Advantages of Newer Mobile Type Apparatus

Unlike ordinary mobile x-ray units, battery-powered equipment is not subject to troublesome line voltage fluctuations. Another advantage is their constant potential output. Radiographs are equivalent in quality to those produced at high kV; for example, 85 kV at constant potential is equivalent to about 100 kV with conventional full wave rectification.

The decision as to which type of mobile equipment to purchase depends on several factors, such as use, cost, and availability of service, just as in the case of other equipment.

QUESTIONS

1. Describe with the aid of a diagram the overload circuit breaker.
2. What is the function of a fuse and how does it work?
3. Discuss the principle of the prereading kilovoltmeter.
4. Why is a remote control switch necessary? How is it constructed?
5. Describe the three main types of exposure timers. Describe automatic exposure control.
6. What is the advantage of a circuit breaker over a fuse?
7. Show by diagram the filament circuit and its important components.
8. Why can a milliammeter be safely mounted on the control panel? What is a ballistic mAs meter and under what conditions is it an essential part of the control apparatus?
9. Explain why the center of the transformer secondary coil is grounded.
10. Explain how the kilovoltage control knob varies the kilovoltage. Which device does it operate in the primary x-ray circuit?
11. Which device in the filament circuit is operated when the filament control knob is manipulated? What function does it serve?
12. Describe a shockproof cable with the aid of a diagram.
13. Prepare a simple wiring diagram of an x-ray machine, including four-diode rectification.
14. Of what does a three-phase current consist?
15. How does the design of a three-phase transformer differ from that of a single-phase? What type of rectifiers is used with three-phase equipment?
16. What is meant by ripple? Discuss its importance in radiography.
17. What are the advantages and disadvantages of the three-phase over the single-phase generator?
18. Describe, with the aid of a block diagram, a high-frequency mammographic unit. A H–F filament voltage controller.
19. List the advantages of high-frequency current in mammographic equipment. In filament voltage control.
20. The minimum exposure time with an x-ray unit is set at 0.04 sec. A particular technique calls for 80 kVp and 3 mAs at 200 mA. Would the radiograph be under- or overexposed? Why?
21. A manual backup timer is set for 0.5 sec. If a technique calls

for an exposure of 100 mAs at 500 mA, would the radiograph be over-, under-, or correctly exposed? Why?

22. Explain the principle of a capacitor discharge unit. How is exposure time controlled? What is wave-tail cutoff?

23. Describe the construction and operation of a grid-controlled x-ray tube.

24. Describe a storage-battery operated mobile (portable) x-ray unit and list its advantages.

Chapter 15

X-RAY FILM, FILM HOLDERS, AND INTENSIFYING SCREENS

ON PASSING THROUGH THE BODY, an x-ray beam interacts with the constituent atoms by photoelectric and Compton processes. As a result, the beam emerging from the body—the *exit beam* —consists of a pattern in which different areas have different numbers of photons corresponding to the pattern of tissue thicknesses, atomic numbers, and densities through which the beam has passed. The term *aerial image* refers to such an exit beam, since it contains information about the tissues through which the beam has passed. This information pattern can be revealed by various image receptors such as film, film-screen combinations, magnetic tapes and discs, and xerographic plates.

In the early days of roentgenography, glass photographic plates coated with an emulsion sensitive to light were used to record the radiologic image. The disadvantages of plates included the danger of breakage, the hazard of cutting one's hands, the difficulty in processing, and the inconvenience in filing these plates for future reference. With the introduction of flexible films, these disadvantages were eliminated. This chapter deals with the use of films and intensifying screens—one type of image receptor system whose purpose is to display the aerial image.

Composition of X-Ray Film

There are two essential components: the *base* and the *emulsion.* Figure 15.1 shows the configuration of a typical double-emulsion x-ray film.

1. **The Base.** Modern safety film has a base measuring about 0.2 mm thick, consisting of a transparent sheet of *polyester plastic* that is usually tinted blue. It is called "safety" because it is no more flammable than an equal thickness of paper.

2. **The Emulsion.** This consists of microscopic crystals of *silver bromide* suspended in *gelatin.* A small amount of sodium iodide may also be present. The gelatin, a transparent material similar to that used in cooking, is obtained mainly from cattle skins and bones treated with mustard oil to improve the sensitivity of the emulsion by providing *traces of sulfur.* To the gelatin dissolved in hot water are added, in total darkness, silver nitrate and potassium bromide to produce *silver bromide,* and a much smaller amount of silver nitrate and potassium iodide to produce *silver iodide.* Silver bromide and silver iodide are examples of *silver halides.* Heating the mixture to a temperature of 50 to 80°C—a process called digestion—further improves the sensitivity of the emulsion. After cooling, it is shredded, washed, reheated, and mixed with additional gelatin. Not only does the gelatin keep the silver halides uniformly dispersed in the emulsion, but it is also readily penetrated by the processing chemicals and by the water used in the final wash.

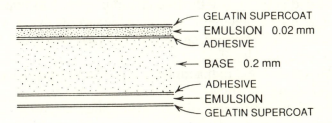

Figure 15.1. Cross section of a typical x-ray film.

Finally, the emulsion is spread in a layer about 0.007 to 0.02 mm thick on both sides of the polyester base which has previously been coated with a special adhesive agent. Cooling is then carried out under strictly controlled conditions. Such double-coated film is the type most widely used in routine radiography. Special single-coated film is available for special purposes such as mammography. It must be emphasized that from the moment the ingredients are brought together in the emulsion until the finished sheet film is packed in boxes, the entire process must be carried out in *total darkness.*

Film Emulsion Characteristics. Emulsions have been designed to respond to certain regions of the color spectrum such as ultraviolet, blue, or green light emitted by intensifying screens, as well as to x rays. Furthermore, an emulsion should display the radiologic image of a variety of tissues with optimum recorded detail, density, and contrast, qualities that will be discussed later.

An ordinary film contains the silver halide crystals clustered in the form of irregular microscopic pebble-shaped grains. The speed of such emulsions can be increased only by increasing the size of the grains and the thickness of the emulsion. This inevitably impairs recorded detail or image sharpness because of the coarser overall granular pattern. Such pebble-shaped grains vary in size and shape, measuring about 1 to 1.5 microns in width.

A type of emulsion called *T-grain* (Eastman Kodak Company) contains silver halide grains of various sizes and shapes, which are *flat*, with the flat side facing the x-ray tube. Their size is similar to that of conventional silver halide grains. The flatness of the grains, as well as the addition of special sensitizers, allows them to capture more light without the need of thickening the emulsion. Consequently, the films display increased speed without sacrificing recorded detail. Since the T-grain emulsion is sensitive mainly to the light from green-emitting screens, a magenta dye is added to the emulsion to minimize crossover (see page 287) and further improve recorded detail.

To be suitable for radiography, an emulsion should have two important characteristics:

1. *Speed or Sensitivity* — the relative ability of an emulsion to respond to radiation such as light and x rays. An emulsion is said to be fast or have high speed if a small exposure produces a radiograph of adequate density (darkening).

2. *Latitude* — the ability of an emulsion to display a radiographic image with a reasonably long tonal range, from white, through various shades of gray, to black. This is *long scale* or *low contrast*, to be discussed later in more detail. Another aspect of latitude is *exposure latitude* — the range of exposure factors that will produce an acceptable radiograph. Obviously, an emulsion should have sufficient latitude to allow a reasonable degree of error in exposure without serious impairment of radiographic quality. Both aspects of latitude—long tonal range and permissible margin of exposure error—are closely related. However, excessive film latitude may impair image visibility, that is, the ability to see fine recorded detail.

Another type of latitude is *subject latitude*, which depends on subject contrast (see pages 337–341). This has to do with the range of densities resulting from contrast in the subject being radiographed; with higher subject contrast, latitude decreases and so does the margin of exposure error.

Types of Films

There are three main types of x-ray films in use at the present time in medical radiography.

1. **Screen Film** is the type most often used in radiography. Films are designed to have maximum sensitivity (response) to the principal light-color emitted by the screen with which they will be used. Such film-screen combinations have made possible the high speed and excellent quality of radiography. Manufacturers produce screens and films having a variety of speeds and contrasts to suit the fancy of any radiologist, both for general radiography and for special examinations.

2. **Nonscreen or Direct-exposure Film** had a thicker emulsion than screen-type film and was used without intensifying screens to obtain fine recorded detail. This type of film is no longer available, but ordinary film can be used in cardboard holders to obtain superb recorded detail of small anatomic parts such as fingers or toes.

3. **Mammography Film** is a single-coated, fine-grain film, designed to be used with a single intensifying screen. The combination must be fast enough to deliver a minimal dose to the grandular tissue of the breast.

4. **Duplicating Film** is used to copy radiographs. The original radiograph is inserted into a cassette whose opaque front has been replaced by a pane of clear glass. A sheet of special duplicating film is then placed, with emulsion side down, onto the radiograph and the cassette lid closed. An exposure is made to an ordinary illuminator at a distance of 12 in. (30 cm) for 6 sec and the film is processed as usual. When necessary, the density of the copy may be improved by adjusting the exposure time; a longer exposure will decrease image density, and conversely. Duplicating films, when properly exposed, produce excellent copies of the original.

Practical Suggestions in Handling Unexposed Film

1. Films deteriorate with age; therefore, the expiration date stamped on the box should be observed, the older films being used first.
2. Since moisture and heat hasten deterioration, films should be stored in a cool, dry place.
3. Films are sensitive to light and must be protected from it until processing has been completed.
4. Films are sensitive to x rays and other ionizing radiations, and should be protected by distance and by shielding with protective materials such as lead.
5. Films are marred by finger prints, scratches, and dirty intensifying screens, and by crink marks due to sharp bending.
6. Rough handling causes black marks due to static electricity. These appear as jagged lines, black spots, or tree-like images after development.

Film Exposure Holders

Each x-ray film must be carried to the radiographic room in a suitable container which not only protects it from outside light but also allows it to be exposed in radiography. Film holders are available in standard film sizes.

There are two types of film holders used in general radiography: *cardboard holders* and *cassettes.*

1. **Cardboard Film Holder** — an enclosed light-proof envelope into which the film is loaded *in the darkroom.* The film is placed in the folder and the long flap of the envelope is folded over it, followed by the two shorter side and end flaps. After being closed and secured by the hinged clip, the holder is ready to be taken to the radiographic room for exposure. During exposure, the front of the film holder must face the tube, the back cardboard being lined with lead foil to prevent fogging of the film by x rays scattered back from the table. Cardboard holders are seldom used, except for extremely fine recorded detail of small parts.

2. **Cassette** — a case measuring about one-half inch in thickness and having an aluminum, stainless steel, or plastic frame, a hinged lid with one or more flat springs, and a bakelite or light metal front. One of a pair of intensifying screens is mounted on the

inside of the cassette front, and the second screen is mounted on the inside of the lid. As will be shown below, these screens convert the energy of x-ray photons to light photons, thereby amplifying the photographic effect on the film. This entails not only a marked reduction in the exposure needed for satisfactory radiographs but also a similar decrease in patient exposure. Moreover, it makes possible the use of grids for improved contrast.

The front of the cassette faces the x-ray tube during exposure. We load the cassette by raising the hinged lid *in the darkroom,* slipping a film gently into the cassette of the same size, and closing the lid by means of the springs. The film is thus sandwiched between two screens and in close contact with them. Figure 15.2 shows a cassette in cross section. Lead foil cemented to the lid behind the back screen helps absorb radiation scattered back toward the screen, but this must be omitted in phototimed exposures.

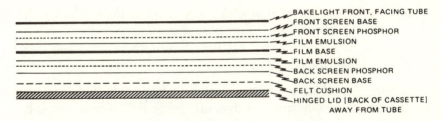

Figure 15.2. Cross section of a cassette with intensifying screens and film.

Intensifying Screens

We have already mentioned the use of intensifying screens and shall now describe their structure and function.

1. **Composition.** An intensifying screen (see Figure 15.3) consists of microscopic crystals of a *phosphor* (luminescent material), incorporated in a binding material, and coated on one side of the white reflecting surface of a sheet of high grade cardboard or Mylar® plastic, 0.25 mm thick, called the *base.* The phosphor coating is the *active layer;* it usually measures about 0.01 mm thick in medium speed screens. It is, in turn, coated with a thin, smooth, abrasion-resistant material, its edges being sealed against

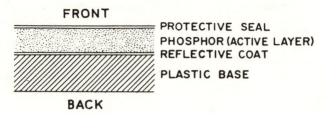

Figure 15.3. Magnified cross-sectional diagram of an intensifying screen.

moisture. For years *calcium tungstate* was the most widely used phosphor because it:

a. Emits blue and violet light to which ordinary x-ray film is particularly sensitive.
b. Responds well in the kV range ordinarily used in radiography.
c. Does not deteriorate appreciably with use or with age.
d. Does not have significant afterglow or lag (see pages 277, 278).
e. Can be used to manufacture screens of uniform quality. As will be seen below, other phosphors (eg, rare earth types) have largely replaced calcium tungstate because of their greater efficiency.

2. **Principle.** Phosphors are crystalline materials that have the unique ability to *luminesce* or give off visible light when struck by x rays. While the process is incompletely understood, we do know that it involves the transition of electrons between *energy bands* in the crystal lattice, shown diagrammatically in Figure 15.4. (Note the similarity to the concept of semiconductors, pages 225–227). The *valence band* contains atomic valence electrons and is usually filled, the *forbidden band* normally remains unoccupied by electrons except at certain points, and the *conduction band* allows free movement of electrons. An imperfect crystal results from faults or metallic impurities which serve as *traps* for electrons in the forbidden band; in fact, the minute amount of impurities added to the phosphor crystals in intensifying screens produce traps that usually contain electrons.

There are two kinds of luminescence—*fluorescence* and *phosphorescence* —classified, respectively, according to whether immediate or delayed emission of light occurs following exposure to x rays. A brief, simplified discussion of these processes will now be given.

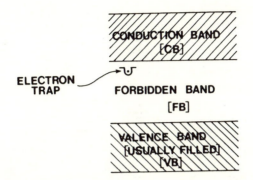

Figure 15.4. Schematic concept of a phosphor crystal. Note the similarity to a semiconductor crystal.

a. *Fluorescence* — light emitted *promptly*, within 10^{-8} sec after absorption of an x-ray photon by photoelectric or Compton interaction. The resulting primary electron raises one or more valence electrons to the conduction band, leaving a "hole" in the valence band (see Figure 15.5). Immediately, in less than 10^{-8} sec, a trapped electron drops from the forbidden band into the hole in the valence band, giving off visible light. The trap is reset, so to speak, when it is refilled by transition of an electron from the conduction band. A much less common event in radiographic phosphors is the direct transition of an electron from the conduction band to a hole in the valence band without preliminary trapping. As a result of fluorescence, such as that occurring in a pair of medium speed calcium tungstate screens, *one incident x-ray photon ultimately causes the emission of about 300 light photons*, about one-half of which leave the screen and reach the film (see Ter-Pogossian); so the *screen efficiency* of this particular screen is 50 percent. The ability of a given phosphor to change x-ray photon energy to light energy is called its *conversion efficiency.*

b. *Phosphorescence* — light emitted after a delay of 10^{-8} sec or more after the absorption of an x-ray photon. Such *delayed emission* of light is also known as *afterglow.* Here, absorption of x-ray photons causes elevation of electrons to the conduction band, leaving many holes in the valence band (see Figure 15.6). However, in crystals subject to intense phosphorescence, such as zinc sulfide, the traps in the forbidden band are normally empty of electrons.

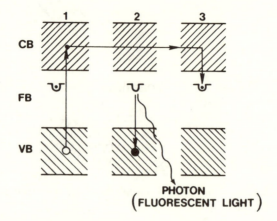

Figure 15.5. Process of fluorescence. An incoming photon (not shown) initiates a photoelectric or Compton interaction in the valence band (*VB*), releasing a primary electron which, in turn, imparts energy to a number of secondary electrons and raises them to the conduction band (*CB*). In *1*, a secondary electron has been raised to the *CB*, leaving a hole in the *VB*. In *2*, any electron previously trapped in a flaw in the forbidden band (*FB*) immediately falls into a hole in the *VB* and a photon is emitted as fluorescent light. In *3*, an electron from the *CB* finally drops into the empty trap, resetting it. Note that fluorescent light emission occurs *promptly,* in less than 10^{-8} sec.

When the newly arrived electrons drop from the conduction band into traps in the forbidden band, they remain there for a variable length of time. Heat from the kinetic energy of the atoms then lifts the trapped electrons back to the conduction band, whereupon they may drop into holes in the valence band, accompanied by the emission of visible light.

Thus, there are two types of luminescence occurring in certain crystals following the absorption of x rays. One is the *prompt emission of light* by the process of *fluorescence.* The other is the *delayed emission of light* through *phosphorescence* (lag). In both types, the tiny phosphor crystals absorb energy in the form of x rays and convert this energy to visible light which is emitted in all directions. Phosphorescence is virtually nil in intensifying screens and in cesium iodide image intensifier screens. The total amount of light emitted in a particular region of the active layer depends on the quantity of x rays striking that region, and is a summation of the light given off by the innumerable, closely packed crystals.

In the object being radiographed, those parts which are readily

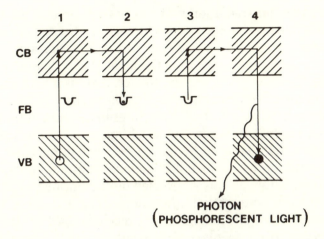

PHOTON
(PHOSPHORESCENT LIGHT)

Figure 15.6. Process of phosphorescence. Here, traps are normally empty. In *1* we have the same sequence of events as in fluorescence (Figure 15.5). But in *2*, electrons drop from the conduction band (*CB*) into empty traps, one of which is shown here. Electrons remain in the traps for variable lengths of time before a small amount of energy, such as the heat present in the crystal, raises them back to the *CB* as in *3*. Finally, they drop into holes in the *VB*. Note that *delayed* emission of light occurs in phosphorescence.

penetrated by x rays will appear on the *screen* as brighter areas than those parts which are poorly penetrated. Thus, the screen registers, although temporarily, the *aerial image* and includes zones of various degrees of brightness. These, in turn, ultimately produce corresponding differences in darkening of the radiograph. Since the film emulsion is specially sensitive to the particular color of light emitted by the screens, this *photographic effect* is of major importance. In fact, *as much as 98 per cent of the recorded density (blackening) in a film exposed with intensifying screens may be photographic in origin,* that is, due to light emitted by the screens; the remaining 2 percent resulting from direct x-ray exposure. Thus, screens greatly intensify the effect of x rays on the film emulsion, thereby reducing the exposure needed to obtain a particular degree of film blackening and, at the same time, substantially decreasing patient exposure. Intensifying screens made possible the use of grid radiography of thick anatomic parts such as the abdomen, skull, and spine, and led to the development of high speed radiography.

To summarize this discussion, we may say that the *exit* or

remnant radiation (radiation that has passed through the patient and carries the aerial image) passes through the front of the cassette and impinges on the front intensifying screen. This emits light which varies in brightness depending on the amount of radiation reaching any particular area of the screen. In fact, there is a light and dark pattern on the screen corresponding to the relative transmission of x rays through the various body structures in the path of the beam; the screen image represents the aerial image in the form of light. The image in light emitted by the screen photographically affects the film emulsion with which it is contact, recording the light and dark pattern, although this is reversed on the finished radiograph. X rays also pass on directly to the film causing a variable, though small, photographic effect. Some x rays reach the back screen which fluoresces and affects the film emulsion nearest it in the same manner as the front screen. Thus, a film is coated on both sides with a sensitive emulsion so that one side receives light from the front screen, and the other from the back screen, as shown in Figure 15.7.

3. **Screen Speed.** A screen is said to be fast or to have high speed when a relatively small x-ray exposure produces a given output of light and causes a certain degree of blackening of a film. Conversely, a screen is said to be slow when a relatively large exposure is required for a given amount of film blackening. *The speed or intensifying factor of a pair of intensifying screens may be defined as the ratio of the exposure required without screens to the exposure required with screens to get the same degree of blackening of x-ray films.* Another name for intensifying factor is *speed factor.*

$$\text{intensifying factor} = \frac{\text{exposure without screens}}{\text{exposure with screens}}$$

Since the denominator is always less than the numerator, the intensifying factor of a pair of screens always exceeds unity, which means simply that the exposure with screens is less than that without screens for the same amount of film blackening.

At present, screen speed is rated according to **arbitrarily** assigned numbers as follows:

Screen Speed Number	Generic Name
less than 100	detail
100	medium speed
more than 100	high speed

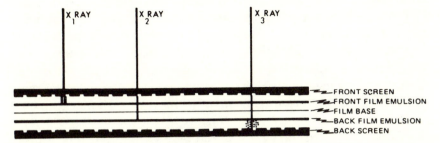

Figure 15.7. Schematic representation of the principle of the intensifying screen. X-ray photon *1* strikes the front screen causing it to fluoresce at that point. Another x-ray photon such as *2* may pass directly to the film and affect it at that point. Or a photon such as x ray *3* may pass through the front screen and the film, striking a crystal in the back screen and exciting fluorescence in it. Actually the screens and film are in close contact. Not shown here is *crossover* from one screen to the emulsion on the opposite side of the film (see Figure 15.11).

On this basis, screens with a speed of 400 require about one-fourth the exposure of screens with a speed of 100, for equal radiographic density. It is expected that the generic names for screens will eventually be discontinued, and screen speed will be identified by number alone. Note again that these numbers are arbitrary and relative; they are not the same as the intensification factors.

But we must take into account the speed of the *film-screen system,* since different speeds pertain to a particular type of screen depending on the speed of the film being used. Thus, high-speed screens with slow films may be no faster than slower screens with high-speed films. So, choice of radiographic image receptors for a particular application should be based on a *film-screen system* that provides the requisite speed and image quality, rather than on screen speed alone.

The *speed* of intensifying screens depends on two main classes of factors—intrinsic and extrinsic.

a. **Intrinsic Factors** are those inherent in the composition of the screens and include the phosphor, the thickness of the active layer, the size of the phosphor crystals, and the reflectance of the backing.

(1) *Phosphor.* The speed conferred on an intensifying screen by a particular phosphor depends on its *conversion efficiency* which is determined by two factors: (1) its ability to absorb x-ray photons,

and (2) its ability to convert the absorbed x-ray energy to light energy, but this is a very inefficient process. Actually, only about 5 percent of x-ray energy in ordinary calcium tungstate, and 20 percent in rare earth phosphors, is converted to light energy (see Curry and others). Conversion efficiency can be enhanced by the addition of special activators; for example, activated calcium tungstate has twice the conversion efficiency of ordinary calcium tungstate.

Note that not all the light emitted by the phosphor reaches the film—about 50 percent is absorbed in the screen itself. *Screen efficiency* refers to the ability of light given off by a phosphor actually to leave a screen and expose a film. Thus, the intensifying factor of a screen depends on both phosphor conversion efficiency and screen efficiency.

Calcium tungstate screens emit blue and violet light to which ordinary x-ray film is especially sensitive. As a rule, the maximum color sensitivity of a particular film should be matched to the color of the light emitted by the screens with which the film is used.

Some of the newer phosphors with intrinsically higher conversion efficiency include *barium strontium sulfate* which fluoresces at shorter wavelengths than calcium tungstate and gives screen speed about 1.5 times that of par speed screens. Another phosphor, *barium fluorochloride* provides screen speed about four times that of medium speed screens, but with moderate increase in mottle.

The rare earth phosphors, comprising salts of gadolinium, lanthanum, and yttrium tantalate, display much greater conversion efficiency than calcium tungstate while retaining excellent image quality. As shown in Table 15.1, these phosphors emit strongly at their characteristic light frequencies; for optimum efficiency, they should be used with films whose maximum sensitivity matches the color emitted by the particular phosphor. Screen-film combinations having a speed of 400 should ordinarily be used in general radiography because they provide optimum speed and image quality while substantially reducing patient exposure. While 800-speed systems reduce patient exposure even more, radiographic image quality suffers due to excessive quantum mottle (see pages 328–329).

(2) *Thickness of Active Layer.* For a given phosphor, the thicker the active layer the greater will be the speed. In fact, this contributes more to screen speed than does crystal size.

TABLE 15.1
PHOSPHOR COMPOSITION AND LIGHT EMISSION OF VARIOUS
INTENSIFYING SCREENS AND THEIR SPEED FACTORS
RELATIVE TO 100 FOR MEDIUM (PAR) SPEED SCREENS[a]

	Phosphor	Principal Light Emission	Relative Speed
Non-Rare Earth	Calcium tungstate[b]	Blue	60–400
	Barium lead sulfate[c]	Blue	20–50
	Barium strontrium sulfate[c]	Blue	100–300
Rare Earth Phosphors	Gadolinium oxysulfide[c]	Green	120–1200
	Lanthanum oxybromide and Gadolinium oxysulfide[d]	Blue/green	400–800
	Yttrium tantalate (high density)[d]	UV/Blue[e]	100–400
	Yttrium tantalate and Lanthanum oxybromide[d]	UV/blue	400–800

[a]depends on film speed
[b]Kodak; DuPont
[c]Kodak
[d]DuPont
[e]UV = ultraviolet

(3) *Size of Phosphor Crystals.* Larger crystals impart greater speed to intensifying screens than do smaller ones for two reasons. First, more light is emitted by larger crystals because they receive more x-ray photons. Second, larger crystals scatter less light within the screen's active layer, increasing screen efficiency.

(4) *Reflectance of Backing.* The more light reflected back to the active layer by the cardboard or plastic backing, the greater will be the speed. However, this may be undesirable with thick active layers because of the detrimental effect on image sharpness caused by the diffusion (spread) of light in the screen.

Thus, a faster screen would probably incorporate a more luminescent phosphor, such as specially activated calcium tungstate in a thicker layer, or some other phosphor that is inherently faster, such as one of the rare earths. But note that *as the thickness of the active layer is increased the radiographic image becomes more blurred* (ie, *less sharp*) because an increase in average distance the emitted light has to travel to the film results in greater diffusion (spread) of the light rays (see Figure 15.8). Consequently, there is a practicable limit to the thickness of the active layer.

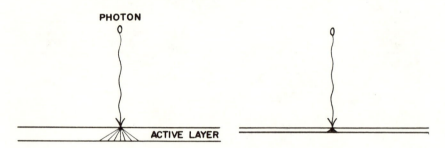

Figure 15.8. Effect of active layer thickness of an intensifying screen on image definition. Note that the light originating at points deep within the thinner active layer (right figure) spreads less before reaching the film, thereby producing a sharper image.

Theoretically, we should expect the size of the crystals to influence image sharpness because a larger crystal would produce a larger spot of light than a smaller crystal, but this turns out to be unimportant in actual practice. The crystals are extremely minute, averaging about 5 microns (0.005 mm) with a range of 4 microns (slow or "detail" screens) to 8 microns (faster calcium tungstate screens). This range of crystal size has a smaller effect on recorded detail than does active layer thickness.

b. **Extrinsic Factors** include the conditions under which the screens are used, namely temperature and kV.

(1) *Temperature.* As room temperature rises, screen speed decreases and film speed increases. These two tendencies nearly cancel each other, so that no temperature correction is necessary for screen speed under ordinary conditions. However, at the high temperatures prevailing in the tropics, basic mAs values have to be increased about 40 percent at 50 C (120 F). On the other hand, in extreme cold such as may be encountered in industrial radiography, a decrease of about 25 percent in the basic mAs is required at −10 C (15 F).

(2) *Kilovoltage.* The speed of rare earth screens (except for yttrium tantalate°) does not remain constant over the useful kV range, being maximal at about 100 kVp and gradually dropping to

---

°*Ultra-Vision*™ *(DuPont)*

about 75 percent at 70 kVp. These values depend on the phosphor. On the other hand, calcium tungstate and yttrium tantalate screens show negligible change in speed with change in kVp.

4. **Screen Contact.** The film must be sandwiched uniformly between the two screens, with perfect contact throughout. Even the smallest conceivable space between the film and screen at any point will permit the light rays emerging from the screen at that point to spread over a wider area, producing a blurred image of that particular point (see Figure 15.9). You can easily ascertain the uniformity of contact of the screens with the film by placing a piece of hardware cloth with 1/8 inch mesh over the front of the cassette and making a film exposure — 40 kV, 10 mAs, and 100 cm (40 in.) focus-film distance, with a 3 mm Al filter. With satisfactory screen contact, the image of the wire mesh is sharp over the entire radiograph; but in areas of poor contact the image appears blurred and patchy (see Figure 15.10).

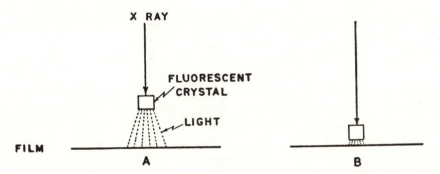

Figure 15.9. Poor screen contact causes a blurred image. In *A*, there is an appreciable space between the fluorescent crystal and the film (when screen contact is poor), so that a tiny but spreading bundle of light rays strikes the film, producing a blurred image instead of a fine point. In *B*, the crystal is in close contact with the film so that the image is about the same size as the crystal. (The crystal is magnified many times in the diagram.)

Screen contact is usually, though not universally, good in smaller cassettes. However, larger cassettes, especially 30 × 35 cm (11 × 14 in.) and 35 × 43 cm (14 × 17 in.) often exhibit poor contact even when new. Therefore, a layer of contact felt, fiberglass, or foam rubber should be cemented (preferably by an expert) as a cushion between the screen and the cassette. When the cassette is

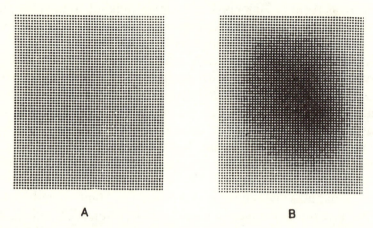

A   B

Figure 15.10. Tests for screen contact. In A, with excellent screen contact, the image of the wire screen is uniform. In B, with poor contact at the center, the image of the wire is blurred, producing a typical blotchy, dark area in the radiograph.

closed, the padding tends to equalize the pressure applied to the screens, squeezing them tightly against the film.

It is essential that all new cassettes and, periodically, old ones as well, be checked for contact as described above. Whenever a radiograph appears sharper in one area than in another (for example, edges sharper than the center), poor screen contact should be suspected and the cassette tested without delay.

The *causes of poor screen contact* include (1) warped cassette front, (2) cracked or twisted cassette frame, (3) loose hinges or spring latches, and (4) local elevation of a screen by a foreign body underneath. Rough handling and dropping of cassettes are mainly responsible for damage leading to poor screen contact. Another important cause is the placement of a cassette without a tunnel under a patient in bed often resulting in a warped cassette.

5. **Recorded Detail with Screens.** In previous sections we described the relationship of intensifying screens to radiographic image sharpness. These data may be summarized to good advantage under a single heading. In general, *recorded detail is never so sharp with screens as with direct radiographic exposure of x-ray film in a cardboard holder,* provided the patient can be adequately immobilized. In actual practice, image sharpness with screens becomes worse as the thickness of the active layer increases. As

noted before, a thicker active layer causes greater diffusion of light and blurring of the image. But remember that as the active layer thickness is decreased to improve recorded detail, as in detail screens, the screen becomes slower. Another fact to bear in mind is that regardless of the type of screen, recorded detail suffers when film-screen contact is poor. In nongrid radiography of small parts, detail screens give much better recorded detail and less mottle than medium speed or fast screens.

Another cause of impaired recorded detail with intensifying screens, related to diffusion of the image, is a condition known as *crossover* or *punch-through.* This is shown in Figure 15.11. Note that crossover occurs only with the double-emulsion films and two screens prevailing in general radiography. Light from one screen expands in the form of a cone as it passes successively through this screen, the nearer film emulsion, the film base, and the farther film emulsion where a slightly enlarged, less sharp image is formed. Recorded detail is thereby worsened. However, special dyes incorporated in the emulsion minimize crossover.

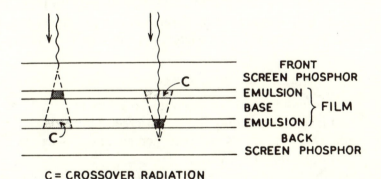

C = CROSSOVER RADIATION

Figure 15.11. Crossover or punch-through of light from one screen to the other in the form of an expanding small cone of light. This causes a blurring effect in the corresponding emulsion of the film, resulting in impaired recorded detail.

6. **Care of Screens.** These must be kept scrupulously clean, since dust and other foreign matter absorb light from the adjacent screen and cast a white shadow on the film. Screen cleaners should be of the type recommended by the manufacturer. Some may be cleaned with cotton *dampened* with bland soap and water and wiped with additional pieces of cotton moistened with water,

never dripping wet. They can then be air-dried by half-opening the cassette and standing it on its side. However, such washing must first be approved. Ultravision (yttrium tantalate) screens require special instructions.

Care must be taken not to nick, scratch, or chip the screens as, for example, by carelessly digging the film out of the cassette with the fingernail, or by scratching the screen with a corner of the film. In loading the cassette, one carefully slips the film into the cassette with the lid elevated about 5 cm (2 in.). Doing this gently helps avoid static marks. To remove the film after exposure, one carefully raises a corner of the film, being certain that the fingernails do not slip across the surface of the screen.

One must avoid using ultraviolet-blocking lotions before loading cassettes equipped with ultraviolet-emitting yttrium tantalate screens; any lotion adhering to the films or screens will block the ultraviolet radiation and create white patches on the processed film.

The cassette must always be kept closed, except while being loaded or unloaded, to prevent accidental damage to the screen surface and help keep out dust. Scratch marks and dust block the light from the screen, leaving a *white spot* on the finished radiograph.

QUESTIONS

1. Show by cross-section diagram the structure of x-ray film.
2. What is meant by safety film? Nonscreen film?
3. State two differences between screen and nonscreen film. What is mammography film?
4. Describe the structure of an intensifying screen. How does it differ from a fluoroscopic screen?
5. Define the intensifying factor of an intensifying screen. State the speed *numbers* of slow, medium, and fast screens; how are they related intensifying factor.
6. What are three differences in the composition of a slow screen and a fast screen? How do they differ in their effect on radiographic sharpness?
7. How does an intensifying screen intensify the x-ray image?
8. Why is an x-ray film coated on both sides with sensitive emulsion?

9. How do rare earth screens differ from ordinary intensifying screens?
10. Explain screen lag. Luminescence. Phosphorescence. Fluorescence.
11. How does active layer thickness affect image sharpness? Explain. What other factors cause impairment of image sharpness by screens? Explain.
12. Discuss the four intrinsic factors affecting screen speed.
13. How do temperature and kilovoltage influence screen speed?
14. In general, is recorded detail better with direct exposure radiography or with intensifying screen radiography? Why?
15. What is meant by the conversion efficiency of an intensifying screen? Screen efficiency?
16. Discuss rare earth phosphors on the basis of emitted color and conversion efficiency, as contrasted with calcium tungstate.
17. Explain crossover or punch-through related to film-screen systems. How can it be reduced?
18. Why should we avoid scratching or otherwise marring a screen surface?
19. What is the effect of poor screen contact on radiographic sharpness? Why?
20. Describe a test for film-screen contact.
21. What special precaution must be taken in handling Ultra-Vision™ film/screen systems?

Chapter 16

THE DARKROOM

THE IMPORTANCE of the darkroom in radiography cannot be exaggerated. Radiography unquestionably begins and ends in the darkroom, where the films are loaded into suitable light-proof holders in preparation for exposure, and where they are returned for processing into a finished radiograph. In general, a darkroom is a place where the necessary handling and processing of film can be carried out safely and efficiently, without the hazard of producing *film fog* by accidental exposure to light or x rays.

The term "darkroom" overstates the case, since complete blackout is unnecessary. In fact, as will be shown later, a great deal of safe illumination can be provided to facilitate darkroom procedures. Although the term "processing room" is more accurate, it is not widely used. We shall therefore continue to refer to it as the "darkroom," bearing in mind that it is dark only insofar as *it must exclude all outside light and provide "safe" artificial light.*

Although this chapter will deal only with the darkroom designed for the *manual processing of films,* the basic principles apply also to darkrooms for automatic processing, to be discussed on pages 310–317.

Location of the Darkroom

Because of its overriding importance, the location of the darkroom should be determined during the planning stage of the Radiology Department. Convenient placement of the darkroom with relation to the radiographic rooms should save time and eliminate unnecessary steps. Walls adjacent to the radiographic rooms should be shielded with the correct thickness of lead all the way to the ceiling; 1.6 mm (1/16 in.) lead usually suffices. Stored radioactive materials should be located as remotely as possible from the darkroom because even as little as 5 mR total exposure from x or gamma rays causes detectable fog. As a general rule, protection of films approximates that of personnel.

Windows should be avoided because they are extremely difficult to render lightproof. Furthermore, they serve no useful purpose, since much more satisfactory methods of ventilation and lighting are available.

Finally, the darkroom should be readily accessible to plumbing and electrical service.

Building Essentials

Walls between the darkroom and adjoining x-ray rooms should contain enough lead thickness, or its equivalent in other building materials, for adequate protection of the films. This is especially important because efficiency demands that the darkroom be close to the radiographic rooms (see above).

In a busy department, the efficiency of work flow can be improved by having *passboxes* built into the walls at appropriate locations. Typical passboxes have two light-tight and x-ray proof doors that are so interlocked that both cannot be opened at the same time. The cassettes, after radiographic exposure, are placed in the passbox through the outside door, and then removed through the inside door by the darkroom technician. The most suitable location for the passbox is obviously near the film-loading bench although this is not always possible.

Darkroom walls should be covered with chemical-resistant materials, particularly near the processing tanks. Such materials include special paint, varnish, or lacquer. Ceramic tile or plastic wall covering provides a more durable finish.

Floor covering should consist of chemical-resistant and stainproof material such as asphalt tile. Porcelain or clay tile may be used, but a nonskid abrasive should be incorporated to minimize the danger of slipping. Ordinary linoleum and concrete are unsuitable because they are readily attacked by the processing solutions.

Entrance

Access should be conveniently located with relation to the darkroom equipment. The simplest type of entrance is a single door which must be made absolutely light-tight by weatherstripping.

Such an entrance generally prevails in offices and hospitals, but it should have an inside lock to prevent opening while films are being processed.

Another type of protective entrance is a small hall with two electrically interlocked doors, so designed that one door cannot be opened until the other is completely closed, thereby preventing entrance of light (see Figure 16.1). A separate door should be provided for emergency use and for moving equipment into and out of the darkroom.

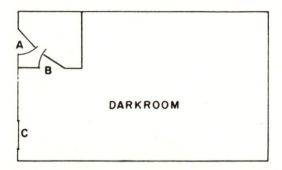

Figure 16.1. Darkroom floor plan with interlocked doors. If door *A* is opened, an electrical relay prevents anyone from opening door *B*, and vice versa. *C* is an emergency door, which is usually kept closed.

A more elaborate type of entrance is the ***maze***, shown in Figure 16.2. Note that one must execute a complete turn in going through the three doorways. Serving as a light trap, the maze requires no doors, especially if the walls are painted black. Because of the high cost of the required floor area, mazes are rarely used today.

For most installations, the preferred modern type of entrance is the revolving door, shown in Figure 16.3. Several sizes are available. For example, one measures about 3 feet in diameter and, built into the wall, it extends about 15 in. into each adjoining room so that it occupies relatively little floor area. At the same time, it offers convenient, lightproof access to the darkroom. Models with three-way and four-way doors are also available.

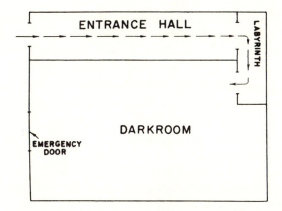

Figure 16.2. Darkroom with maze or labyrinth entrance.

DARKROOM

LIGHTROOM

Figure 16.3. Diagram (top view) of a two-way revolving darkroom door. The outer cylindrical chamber, built into the wall, has two openings, one into the darkroom and the other into the lightroom. The inner chamber has one door which, suspended at its top by a central bearing, can easily be rotated by means of a handrail until its opening coincides with either of the openings in the outer chamber. The entire unit is completely lightproof and takes up very little space.

Size

Darkroom size will vary, of course, with the size of the department, but it must be large enough to house conveniently all the necessary benches for loading and unloading films, a film storage bin, cupboards, processing tanks, refrigeration system, and dryer. On the other hand, the darkroom should not be too large, since the excessive distances between various units of equipment result in wasted time and steps. Placement of the loading bench across the room from the tanks minimizes accidental splashing of water and solutions on the films, benches, cassettes, or other equipment that may come in contact with the films. However, this is not a problem in the darkroom with an automatic processor.

Ventilation

Essential to efficient work and to the technologist's health are the adequate removal of stale humid air and the supply of fresh air. Air conditioning is definitely the preferred method of ventilating the darkroom. However, in a small office or department in areas where climate permits, an exhaust fan may provide adequate air circulation. The system should be absolutely light proof and should include a means of filtering out dust at the point of entry of the fresh air.

Lighting

A properly designed darkroom should have three types of illumination: safelight, general, and radiographic.

1. **Safelight.** We must have a source of light which will not fog films and still provide adequate illumination under processing conditions. *Safelight lamps* serve this purpose; they must have filters of the proper color, such as the Wratten Series 6B® filter for ordinary blue-violet sensitive film, but a special filter such as the Kodak GS-1 must be used with green-sensitive Ortho-G film. The *working distance* from the safelight should be no less than 1 meter (3 ft), and bulb wattage should correspond to that specified on the lamp housing. If a brighter bulb is used, light transmitted even through the correct filter may cause film fog. Safelights are designed for either indirect ceiling illumination or for direct lighting.

The safety of darkroom lighting may be tested as follows: subject a film in a cassette to a very small x-ray exposure, just enough to cause slight graying; screen film is more sensitive to fogging by light after initial exposure to the fluorescence of intensifying screens. Then remove the film from the cassette in the darkroom, cover one-half the film with black paper, and leave it exposed under conditions simulating as closely as possible those normally existing when a film is being loaded and unloaded. Process the film as usual. If the uncovered portion appears darker than the covered, you may conclude that the darkroom lighting is unsafe, and must then make every effort to eliminate the source of light responsible for the fogging. This may result from a cracked or an incorrect filter, a light leak in the safelight housing, or a leak around the entrance door to the darkroom.

Note that the darkroom walls do not have to be painted black. A light color enhances safelight illumination without increasing the risk of film fogging.

2. **General Illumination.** A source of overhead lighting is needed for general purposes such as cleaning, changing solutions, and carrying out other procedures that do not require safelight illumination.

3. **Radiographic Illumination.** A fluorescent illuminator for viewing wet radiographs should be mounted over the washing compartment to avoid contamination of processing solutions. In more elaborate darkrooms for manual processing the wash tank is installed partly through the wall into the light area to permit viewing the wet radiographs outside the darkroom, thereby mini-mizing traffic in the darkroom. With automated processing, a fluorescent illuminator helps only when the entire processor is located within the darkroom.

Apparatus and Equipment

Ready-mixed processing chemicals are available as liquid solutions and made up to the proper volume by the addition of water according to the instructions printed on the label. Two stirring paddles are needed, one for the developer and one for the fixer. A thermometer is used to determine the temperature of the developing solution and to check the temperature of water used to prepare the solutions. All apparatus coming in contact with the processing solutions must consist of materials such as plastic, hard rubber, enamelware, or stainless steel of the proper composition, to avoid corrosion and resulting contamination of the solutions.

Comprising the *major equipment* are the manual processing tanks. The simplest type consists of a three-compartment tank, one end-compartment being used for developing and the opposite one for fixing. The middle compartment serves both to rinse and wash the films and should be supplied with running water. Thermostatic temperature control produces optimal results.

Stainless steel of the proper alloy composition is the best material for processing tanks because not only does it resist corrosion but it also permits rapid equalization of the temperature of the processing solutions with that of the water in the middle compartment (stainless steel is an excellent conductor of heat in contrast to rubber or stone). Insulation of the outside walls of the tank prevents condensation of moisture and helps maintain the proper temperature within the compartments.

A more satisfactory arrangement is that shown in Figure 16.4, consisting of a large, insulated, stainless steel, double-compartment master tank. Two stainless steel insert tanks are placed in one of the compartments, the first insert being the developing tank and the other, the fixing tank. Water between the inserts in this compartment serves both to rinse the films and to control the temperature of the solutions. A fixing tank should have about twice the volume of a developing tank since the time required for fixing films is approximately twice that for development. The other main compartment serves as the washing tank and should be about twice the size of the fixing tank, since washing requires about twice as long as fixation.

Inserts should be measured when first installed to determine their correct volume. If the level of the solution is to be 2.5 cm (1 in.) from the top, then

$$\text{volume in liters} = \frac{(\text{height} - 2.5)\underset{cm}{} \times \text{length}\underset{cm}{} \times \text{breadth}\underset{cm}{}}{1000}$$

or

$$\text{volume in gallons} = \frac{(\text{height} - 1)\underset{in.}{} \times \text{length}\underset{in.}{} \times \text{breadth}\underset{in.}{}}{231}$$

The wash tank should be deep enough for the water to pass completely over the tops of the hangers for thorough washing.

A conveniently located water inlet facilitates washing the tanks and making up the solutions. Thermostatically controlled warm water and refrigerated water are essential and should be circulated around the processing tank inserts to maintain optimal temperature.

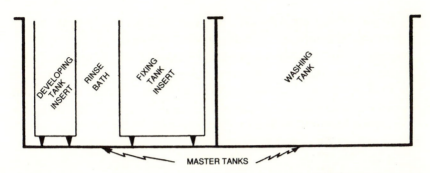

Figure 16.4. A convenient array of tanks for manual processing. The developing and fixing tanks are separate units placed in the master tank on the left. Note the relative sizes of the tanks.

Various types of *drying devices* help speed up the drying of films. Driers are available as enamel or stainless steel cabinets provided with a heating element and blower fan. Venting to the outside minimizes heat and humidity in the darkroom.

A *lightproof storage bin* for unexposed film should be placed under the loading bench. Vertical partitions subdivide the bin to accommodate film boxes of different sizes. Counterweighting the drawer of the bin makes it close automatically when released. A warning light must be posted on the front of the bin to prevent its being opened in white light. *Film hangers* of proper size and in sufficient number should be available to hold the films during processing, but these are unnecessary with automatic processing.

QUESTIONS AND PROBLEMS

1. Where is the best location for the darkroom? How should the walls be shielded?
2. Discuss the various kinds of darkroom entrances, including the advantages and disadvantages of each.
3. Describe interlocking doors. Passboxes.
4. What materials are preferred in the construction of processing tanks? Name the best material for tanks and dryers.
5. Describe a darkroom safelight lamp. What precautions are necessary to assure that the safelight is really safe?
6. How do you test a darkroom for safe illumination?
7. Which colors should be used in painting the walls of the darkroom? Of the maze?

8. Make a diagram of a convenient arrangement of the processing tanks, and indicate their comparative sizes.
9. A developing tank measures 8 in. wide × 16 in. long × 19 in. high. How many gallons of developer does it hold when filled to a level one inch below the top? Convert the data to the metric system and compute the volume in liters. Approximately how many liters equal 1 gallon?
10. How can you prevent x rays from entering the darkroom, and why?
11. Describe briefly a film drier.
12. Why is running water essential in the darkroom? Why do we need warm and refrigerated water?

Chapter 17

CHEMISTRY OF RADIOGRAPHY AND FILM PROCESSING

Introduction

THE BASIC PRINCIPLES of radiographic image formation in sensitive emulsions have been established for many years, evolving as they have from ordinary photography. However, as we shall see later, certain changes were required to adapt conventional film processing to automation, these changes being perhaps more *physical* than chemical. Therefore, a thorough understanding of radiographic image formation and photographic chemistry should precede the study of both manual and automatic processing.

Radiographic Photography

Fundamental to an understanding of the production of an image on an x-ray film, is the concept of the *latent image*. An ordinary radiographic film emulsion contains silver halides in the form of minute crystals that are invisible to the naked eye (1 to 1.5 μm micrometers). These halides consist mainly of silver bromide (AgBr) but also include a small amount of silver iodide (AgI) to enhance sensitivity.

Figure 17.1 presents a diagram of the cubic lattice structure of a silver halide crystal. In it, the positive silver ions alternate with the negative bromide ions (and an occasional negative iodide ion), all being held together by ionic bonds (see page 51), although some silver ions drift freely through the lattice. During the ripening phase of emulsion manufacture, *silver sulfide particles,* serving as *sensitivity specks* or *development centers,* appear on the surface of the crystals. Only those crystals having sensitivity specks can be affected by exposure to light or x-rays.

According to the Gurney-Mott hypothesis, a photon entering such a sensitized crystal may interact with a bromine ion, liberat-

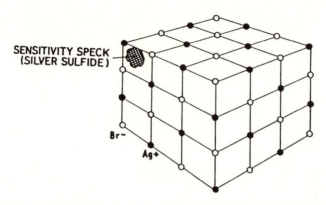

Figure 17.1. Lattice diagram of a silver bromide (AgBr) crystal. The straight lines joining the Ag+ and Br− ions represent the electrovalence forces holding the ions together in the crystal. An occasional iodide ion (I−) may also be present in place of a Br− ion. The sensitivity speck (silver sulfide) renders the crystal highly sensitive to light and x rays.

ing a loosely bound (valence) electron and leaving a neutral bromine atom. As this valence electron drifts through the lattice, it may be trapped by a sensitivity speck to which it imparts a negative charge. A migrant positive silver ion, now attracted to the negatively charged speck, picks up the electron, and becomes neutralized to a silver atom. Another electron soon becomes trapped by the speck, and another silver ion drifts to it and is neutralized. This process is repeated until several silver atoms are deposited, mainly on the surface of a sensitivity speck in a very short time interval. The silver atoms that have been deposited on the sensitivity specks during exposure remain invisible and constitute the *latent image.*

Radiographic Chemistry

When the latent image is acted upon by reducing agents known as *developers,* the process initiated by photon action is greatly speeded up; *the sensitivity speck serves as a development center for the entire crystal.* In fact, development amplifies the latent image about 10^8 times. The speck rapidly picks up electrons from the reducing agents and attracts more silver ions which become reduced to silver atoms, and grow into thread-like or flat (T-grain or tabular) particles of *metallic silver.* Thus, the *dark areas in a radiograph consist of metallic silver in a very fine state of subdivision,* the amount of silver deposited in a given area increasing with the amount of radiation received by that area. The basic concepts of photographic chemistry are shown in Figure 17.2.

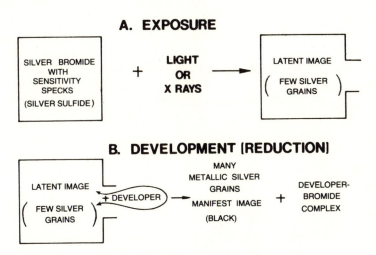

Figure 17.2. Scheme of the basic theory of photographic chemistry. In *A*, radiant energy in the form of light or x rays converts a silver bromide (or silver iodide) crystal containing a sensitivity speck (silver sulfide) into a *latent image center.* In *B*, the developer reacts with the altered crystal and reduces it to metallic silver, which constitutes the *manifest image.*

During development the bromine (and iodine) ions diffuse out of the developed crystals and into the solution. This, in addition to the gradual exhaustion of the reducing agents, eventually causes deterioration of the developer to the point where it must be discarded.

What happens to the portions of the film emulsion that are not affected by light or x-ray photons? Since the silver halides (AgBr and AgI) in these areas have not been altered, they are relatively unaffected by the developer. However, they must be removed in order to render the unexposed film areas **transparent,** and also to prevent **fogging** by subsequent exposure to light. The unexposed and undeveloped silver halides are eliminated from the emulsion by immersion in a **fixing agent,** ammonium thiosulfate, a process known as **fixation.** As a result, the areas from which the silver halides have been removed become clear, while the black areas remain black since metallic silver is not dissolved by the fixing agent in the ordinary course of processing. However, prolonged immersion in the fixing solution will cause bleaching of the image; this may be appreciable even in twenty-four hours.

The amount of blackening of a particular area of a radiograph depends on the amount of radiation it has received. This is the radiation that has passed through the various thicknesses, types, and densities of tissue interposed between the tube and the film and is called *remnant* or *exit radiation.* It comprises the *aerial image* (image in space) which, in cross section, consists of more or less closely spaced photons that will be displayed on the image receptor (film, intensifying screen, fluorescent screen) (see Figure 17.3).

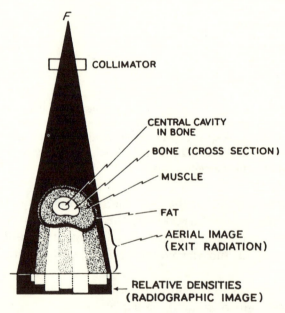

Figure 17.3. Formation of the aerial image. The x-ray beam, on passing through the body, undergoes absorption and scattering which depend on the kV and on the atomic number and density of any particular tissue. The beam emerging on the opposite side of the body contains information in terms of the number of photons per unit cross-sectional area of the beam; this comprises the *remnant* or *exit radiation,* which forms the *aerial image* or *image in space.* The resulting image on the film—the *radiographic image* —consists of the various densities corresponding to the aerial image.

Let us consider, for example, the simple case in which a radiograph of the hand is made by direct exposure to x-rays (ie, without intensifying screens). The light areas represent the bones; because of their high calcium content, the bones absorb a large fraction of the incident x rays so that very few x-ray photons pass through them to the silver halide crystals in the film emulsion underlying them. Therefore, the reducing action of the developer on these crystals produces almost no darkening. The soft tissues, on the contrary, absorb only a relatively small fraction of the incident x rays and therefore the areas of film emulsion beneath the soft tissues receive a relatively large amount of remnant or exit radiation. As a result, more silver halide crystals are affected in these regions and the developer causes considerably greater blackening. Notice that the finished radiograph is really a negative, corresponding to a negative in ordinary photography. It is actually the recorded aerial image of the tissues of different thicknesses and densities, which vary in their attenuation of x rays (mainly by photoelectric absorption). As was shown earlier (see pages 278–280), the radiographic effect of the aerial image is greatly enhanced by intensifying screens.

MANUAL PROCESSING

We shall now turn to the practical aspects of *film processing*, which consist of (1) conversion of the latent image to a visible or manifest image and (2) the preservation of that image. This includes primarily development, rinsing, fixing, and washing. Each of these important steps will be described in detail for manual processing as an introduction to automated processing. Optional steps such as fixer neutralization and detergent rinse will also be included. Cleanliness of hands, utensils, and darkroom are of utmost importance in both manual and automated processing.

Development

The function of development is to convert the latent image to a visible image by means of a *developing solution* which contains four essential ingredients.

Developing Solution. Composition:

1. *Organic Reducing Agents* — usually include a mixture of

hydroquinone and *metol* (or elon, a synonym for metol) for manual processing. The metol acts during the early stage of development, producing the basic gray image, whereas the hydroquinone acts more slowly, building up density and contrast to the desired level. Two such reducing agents in proper combination assure optimum development of the radiographic image. (A combination of hydroquinone and phenidone yields the high speed developer needed for automatic processing.)

2. *Preservative—sodium sulfite* —protects the organic reducing agents (hydroquinone and metol) from oxidation by the air, thereby prolonging the effective life of the developer. It also suppresses staining of the emulsion by decomposition products formed during development.

3. *Accelerator—sodium carbonate* or *sodium hydroxide* —swells the emulsion slightly, making it easier for the developing agents to enter the emulsion, and the products of the reaction to diffuse out. Furthermore, hydroquinone can act only in an alkaline medium. Thus, gel swell and alkalinization speed up development.

4. *Restrainer—potassium bromide* —preferentially holds back any action of the developer on the unexposed silver bromide grains without preventing the action of the developer on the exposed grains. Thus, it inhibits fogging of the lighter areas without interfering significantly with the development of the radiographic image.

Practical Factors in Development. The two most important factors in development are (1) the *temperature* of the developing solution and (2) the total *time* of development. A correctly exposed film undergoes complete development with a particular combination of temperature and time; in fact, it is almost impossible to compensate for underexposure by prolonging development time.

The optimum temperature for development is 20 to 22 C (68 to 72 F). With cold developer, that is, below 16 C (60 F), the *action of hydroquinone ceases* and the resulting radiograph lacks contrast and density. However, even at 16 C (60 F) quality is not impaired if developing time is sufficiently long. Developer that is too warm, that is, above 24 C (75 F), may soften the emulsion and also produce chemical fog. At a given appropriate temperature there is a definite, correct developing time, best determined from the data furnished by the manufacturer of the films and solutions being used. However, various manufacturers recommend nearly the same developing times so that for practical purposes one may use the graph shown in Figure 17.4. This graph averages the data supplied by the two leading film companies (Eastman and DuPont), for *medium-speed screen film* and *rapid developer.* From this graph, we can readily determine the correct developing times at various temperatures, based on a normal time of 3 min at 20 C (68 F). At 17 C (62 F) the developing time is 4½ min and at 24 C (76 F) it is 2 min. Notice that an increase in temperature of the solution increases the speed of chemical action, so that developing time must be decreased according to the graph. Processing of films on the above basis is called *time-temperature development.*

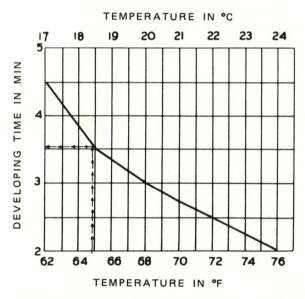

Figure 17.4. Time-temperature development chart for manual processing. This is a composite based on the data furnished by the two leading American manufacturers for medium speed film and rapid developer.

With manual development radiographic exposure is based on *five-minute development at 20 C (68 F).* This produces radiographs of superior contrast with about 5 kV less exposure than is required for three-minute development.

When films are immersed in the developer they should be *agitated gently* at first and about once every minute thereafter to insure uniform development and prevent streaking. During development, bromides are freed from the emulsion and if not removed by agitation, tend to inhibit development wherever they cling to the film. This is most frequently manifested by light streaks below the letters of the identifying labels on the film. Agitation of the film removes these waste products and allows fresh solution always to be in contact with the film emulsion, thereby assuring uniform development.

Replenishment. With continued use of the developing solution its activity diminishes as the hydroquinone and metol are gradually exhausted during the reduction of exposed silver halides to metallic silver. Furthermore, the level of developer in the tank drops progressively because some is carried out with each film. Because of this gradual weakening and loss of volume of the developer, a special solution—*replenisher*—is added periodically to maintain constant activity and volume and assure uniform results. Replenisher composition differs from ordinary developer in having (1) no bromide, because this ion is released from the film emulsion and accumulates in the developer, and (2) a higher concentration of hydroquinone, metol, and alkali, because these chemicals become exhausted with continued use of the developer. Films must be removed rapidly from the developer, without being allowed to drip back into the tank. Properly replenished developer should be usable for two to three months.

In small departments and offices, with very low film throughput, special replenisher is available containing more bromide and less alkali because oxidation from the air increases alkali concentration and produces no additional bromide.

Rinsing

After development, the film must be rinsed for about thirty seconds in running water to remove most of the adhering chemicals, thereby diminishing contamination of the next solution, the fixer. Rinsing can be made more efficient by immersing the films in a tank containing a dilute solution of acetic acid (about 1 percent), which neutralizes the alkali carried over from the developer. Otherwise, the alkali gradually neutralizes the acid in the fixing solution, decreasing its effectiveness as well as its useful life. *Without thorough rinsing, the fixer does not act evenly and the radiograph becomes streaked.* Rinsing is omitted in automatic processing.

Fixation

The purpose of this step is (1) to *remove the unexposed and undeveloped silver halides,* (2) to *preserve the film image,* and (3) to *harden the emulsion* so that it will not be easily damaged.

Fixing Solution. This consists of four essential ingredients:

1. *Fixing Agent—hypo* (*sodium thiosulfate* in powdered fixer; *ammonium thiosulfate* in liquid fixer)—*clears* the film by dissolving out the unexposed, undeveloped silver bromide and silver iodide, leaving the metallic silver in the exposed and developed areas of the film more readily discernible. Without fixation, the undeveloped silver halides would leave the radiograph nearly opaque. Moreover it would eventually turn black, completely hiding the radiographic image.

2. *Preservative—sodium sulfite* —protects the fixing agent from decomposition and helps clear the film.

3. *Hardener—chrome alum or potassium alum* —"tans" or hardens the gelatin in the emulsion, thereby protecting it against scratches.

4. *Acid—sulfuric acid or acetic acid* —serves two purposes: it neutralizes the alkali still remaining on the film, and provides an optimum medium for the fixer and hardener.

Fixing time depends on the age of the fixer and the number of films processed. A satisfactory fixer requires about one to four minutes to *clear* a film—that is, to remove all the unexposed silver halides—but it takes two or three times the clearing time to *harden* the emulsion. The solution should be discarded when it has become so exhausted that the fixing time is prolonged beyond ten minutes. When the films are immersed in the fixing bath, they should be *agitated* and *separated* to allow uniform action of the chemicals and prevent streaking. To avoid fog, white light should not be admitted until the films have been fixed for at least one minute. Prolonged fixation is to be condemned because the *hypo becomes so firmly bound to the emulsion that it cannot be removed in washing,* thereby causing eventual brown discoloration of the radiograph. Besides, prolonged fixation may cause *bleaching* of the image. The optimum temperature of the fixing bath is 18 to 24 C (65 to 75 F); at very low temperatures, chemical action is retarded, while at high temperatures the emulsion may be softened and easily damaged.

Replenishment requires the periodic addition of fresh fixer after an equal volume of older fixer has been discarded. Not only does this prolong the useful life of the fixing solution, but it also maintains its full strength. The level of solution in the tank remains constant (except for slight loss due to evaporation), because each film carries in about as much liquid as it carries out.

Washing

An important step in film processing is thorough washing to remove residual chemicals from the film emulsion. For example, hypo remaining in the emulsion will eventually change deposited black silver to brown silver sulfide, with gradual fading of the image. Ordinarily, if the volume of water flowing per hour is about eight times the capacity of the wash tank, washing is complete in 20 minutes at 20 C (68 F). This is the minimum washing time regardless of how fast the water changes. A slower flow rate makes for longer washing, but if excessive, this may soften the emulsion. Abnormally high temperature also softens the emulsion, whereas extremely cold water slows the washing process.

A *fixer neutralizer bath* significantly shortens washing time. After removal from the fixer, the films are rinsed briefly in the rinse tank and then immersed for two minutes in a tank containing a special solution which removes the fixer from the emulsion. Washing will then be complete (in running water) in five minutes.

Drying

Films can be dried by various methods. Although small numbers of films can be dried in wall-mounted racks, special film driers perform much more efficiently (see page 297).

A *wetting agent* such as household liquid detergent reduces washing time by about 50 percent. The washed films are immersed for two minutes in a tank containing one teaspoon of detergent per gallon of water, and then placed in the drier.

Darkroom Problems

It may be of some help to indicate the causes of the more common film defects.

1. **Fog.** There are many causes of film fogging, that is, a generalized darkening of the film.

a. *Exposure to Light.* This may occur when the darkroom is not light proof; the safelight contains too large a bulb; the safelight housing or filter is cracked; the safelight filter series is incorrect; or the exposure of the film to the safelight is prolonged, especially at short distances.

b. *Exposure to X Rays or Radionuclides.* Films should be shielded from these sources of radiation by distance and sufficient thickness of lead.

c. *Chemical Fog.* The many causes include overdevelopment or development at excessively high temperatures; oxidized, deteriorated developer, which may also stain the film (oxidized developer is brown); prolonged or repeated inspection of films during development; and contamination from corroded tanks.

d. *Age Fog.* Either mottled or uniform fogging due to outdated films, or films stored under conditions of high temperature and excessive humidity.

2. **Stain.** Various types of discolorations may appear on films at different intervals after processing. These can generally be avoided by the use of reasonably fresh solutions and correct processing.

a. *Brown.* Oxidized developer.

b. *Variegated Color Pattern.* Inadequate rinsing.

c. *Grayish-yellow or Brown.* Excessive fixation, or use of exhausted fixer.

d. *Grayish-white Scum.* Incomplete washing.

3. **Marks and Defects.** There are several different kinds of characteristic markings which appear when films are not handled gently.

a. *Crinkle Marks* are curved black or white lines about 1 cm in length which result from bending the film acutely over the end of the finger.

b. *Static Marks* are lightning or tree-like black marks on the film, caused by static electricity due to friction between the film and other objects such as intensifying screens and loading bench. To avoid this, films should always be handled gently. In addition, the loading bench should be grounded in order to prevent the buildup of static electricity.

c. *Water Marks* are caused by water droplets on the film surface, which leave round dark spots of various sizes because of migration of silver particles.

d. *Cassette Marks* are caused by foreign matter such as dust, hair, fragments of paper, etc, or by screen defects, which leave a corresponding white mark on the radiograph.

e. *Air-bell Marks* result from formation of air bubbles in the developer. A bubble prevents developer from reaching the underlying film, and so leaves a small, clear circular spot on the radiograph.

f. *Streaking* is caused by a variety of technical errors and is one of the most troublesome types of film defects. It usually results from: (1) failure to agitate the films in the developer; (2) failure to rinse the films adequately; (3) failure to agitate the films when first immersed in the fixer; and (4) failure to stir the processing solutions thoroughly after replenishment.

AUTOMATED PROCESSING

A major improvement in radiography has been the perfection of automated processing of films. Successful marketing of automated processors has brought about a reduction in cost, thereby making such equipment available even to the department with a limited budget.

There are many outstanding advantages of automated processing. First, it *shortens total processing time to as little as one and one-half minutes,* in contrast to about one hour for the manual method. (Even with hypo neutralizer and wetting agent, manual processing can be reduced, at best, to about forty minutes.) Second, it *improves quality control* by more precise temperature regulation and replenishment in the automated processor. Third, it increases the capacity of the Radiology Department or office by expediting work flow. One-minute processing is now available.

The manufacturers of the first automated processors were beset by four main problems: transport mechanism, processing chemicals, temperature control, and film characteristics. A discussion of these problems and their solution is essential to the understanding of the principles of automated processing, especially as they depart from manual processing.

1. **Transport Mechanism.** The basic mechanism of the automated processor is a series of rollers which transport the films from the loader through the various sections—developing, fixing, washing,

and drying. The speed of transport must be constant to assure correct sojourn of the films in each section. Spacing between the rollers must be accurate to an extremely small tolerance to avoid slipping or jamming of the films. Besides these operations, the transport system provides brisk *agitation* of the solutions over the film surfaces to assure uniform action of the developer. The squeezing action of the last few rollers in each section removes a large amount of solution from the films before they enter the next section, thereby lessening contamination of the solutions. Finally, the terminal rollers in the wash section squeeze out most of the water and so hasten drying of the films. Figure 17.5 shows diagrammatically a typical automated processor; its apparent simplicity belies the difficulties encountered during its early development.

2. **Processing Chemicals.** Although the basic principles of photographic chemistry apply to both manual and automated processing, conventional solutions were found to cause insurmountable difficulties in the first automated experimental models. Roller spacing was so critical that the swelling of the emulsion in the developer, and its subsequent shrinkage in the fixer, caused the films either to jam between the rollers or to wrap around them. Further complications arose when an attempt was made to speed up processing through the use of stronger solutions.

Figure 17.6 shows a variation of about eleven units of *gel swell* of the emulsion during manual processing. This renders conventional solutions completely unsuited to automated processing; in fact, it turns out that the problem is largely *physical.*

To adapt solutions to automation with its decreased processing time, certain changes had to be made. These may be summarized as follows:

a. *Filters in Water Supply Line* to remove small-particle impurities. There is a separate porous filter for the hot and cold water supply. The filters must be replaced at regular intervals, depending on the purity of the water. Should the cold water filter become clogged and the hot water continue to enter the processor, the solutions would become overheated, resulting in overdeveloped, dark radiographs. Furthermore, the emulsion could soften, causing slippage or wraparound of the films as they proceed through the rollers. On the other hand, should the hot water line become clogged, the thermostat would cycle on and off more frequently to maintain

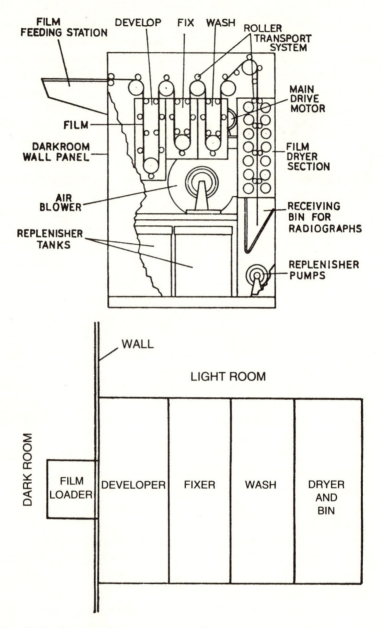

Figure 17.5. Automated processing. In the upper part is shown a cutaway diagram of the processor (side view). In the lower part is shown a diagrammatic top view. (*Courtesy of Eastman Kodak Company.*)

processing temperature, increasing the likelihood of thermostat failure. Some processors operate on only a cold water supply requiring a single filter; in these units the thermostat and heater are designed to temper (warm) the incoming water without the need of a separate hot water supply.

b. *Increased Concentration of Solutions* to shorten greatly the processing times in the various sections. Hydroquinone and pheni-done are the developing agents, and cycon is added as an antioxidant (retards oxidation). A fresh developer bath consists of replenisher, properly mixed, to which has been added a *starter solution* to prevent excess alkalinity. Thereafter, a pump automatically feeds properly mixed replenisher solution from a storage tank into the development section. The fixer also has a higher concentration than in manual processing; here, too, a pump automatically feeds fixer replenisher solution from a storage tank into the fixer section.

c. *Increased Processing Temperatures* to help speed up processing; for example, 35 C (95 F) in developer and fixer, 31 C (88 F) in wash, and 57 C (135 F) in dryer are typical of ninety-second processors. Drying time is reduced to 20 seconds. To combat chemical fog at these high temperatures, antifogging agents such as aldehydes are added to the developer.

d. *Hardening of Emulsion* to prevent softening by the solutions, as well as sticking to the rollers. For this purpose, special harden-ing agents such as glutaraldehyde or sodium metaborate are included in both the developer and fixer.

e. *Control of Emulsion Thickness* to keep it as constant as pos-sible throughout processing, to permit the use of constant spacing between the rollers. Compounds such as sulfates are added to the developer to minimize swelling of the emulsion.

f. *Precise Replenishment* of the developer and fixer to maintain the proper alkalinity of the developer, acidity of the fixer, and the chemical strength of both solutions. Replenishment must be adjusted to a constant rate for each film processed, the fixer being replenished at twice the rate of the developer.

3. **Temperature Control.** As just noted, *high temperatures* prevail in automatic processing. This has resulted in marked shortening of processing time, but now extremely accurate temperature control has to be maintained in each section of the processor. This is accomplished by a pump, which circulates water at the proper temperature around the developer and fixer sections; the water finally enters the wash section. Incoming water temperature is

usually lower than that of the processing solutions so that it can be heated to the correct temperature under thermostatic control. Temperature is even more critical than in manual processing because the technologist can no longer adjust development or fixing time for differences in temperature. Moreover, the very short developing time at high temperature makes the films much more susceptible to serious under- or overdevelopment when these factors differ even slightly from the normal.

4. **Film Characteristics.** The manufacture of films requires even more precise control than before to maintain constant thickness of the base and emulsion. Curling tendency must be eliminated to prevent wrapping around rollers or misdirection to the wrong rollers, causing jamming of the films. The emulsion characteristics must conform to the new types of processing solutions. Finally, stickiness of the emulsion has to be minimized to prevent adherence to the rollers and wrap-around.

The various modifications just described have made automatic processing so dependable that it has become practicable even in the private office and small hospital. Figure 17.6 shows how constant the thickness of the emulsion remains. The overall variation in gel swell from start to finish in automatic processing is about four units, but after the initial relatively slight swelling of the emulsion in the developer, the thickness remains virtually constant well into the drying phase.

Table 17.1 contrasts the important steps in manual and automatic processing. It bears out the assertion at the beginning of this section that automation has removed one of the main bottlenecks in the Radiology Department.

The question naturally arises as to the frequency of breakdown of automatic processing equipment. Experience has shown that the greatest cause of breakdown is *failure to keep the rollers scrupulously clean,* as recommended by the manufacturer. Another cause is improper rate of automatic replenishment which should be carefully checked at regular intervals.

Sensitometry is coming into wider use to check the correctness of developer replenishment. It is long overdue as a method of quality control. Sensitometric testing of the developer should be done periodically, in addition to the usual test for replenishment, to assure uniformity of development during the useful life of the solution.

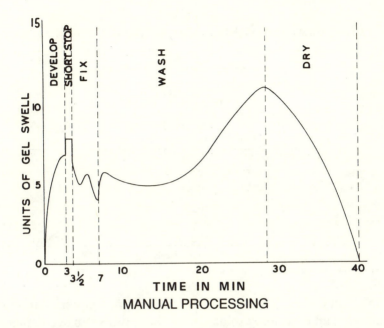

MANUAL PROCESSING

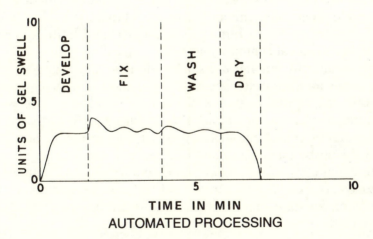

TIME IN MIN
AUTOMATED PROCESSING

Figure 17.6. Comparison of manual and automatic processing on the basis of relative swelling of film emulsion during the various phases. The units of gel swell are arbitrary. Note the much smaller variation in the degree of emulsion swelling with automatic processing; this results from special additives in the solutions. (*Courtesy of R.E. Humphries, GAF.*)

TABLE 17.1
COMPARISON OF MANUAL AND AUTOMATIC PROCESSING

	Manual	Automated*
		1½-minute
Developing Temperature	20 C (68 F)	35 C (95 F)
Fixing Temperature	20 C (68 F)	35 C (95 F)
Washing Temperature	20 C (68 F)	35 C (95 F)
Drying Temperature	43 C (110 F)	57 C (135 F)
Developing Time	3–5 min	25 sec
Fixing Time	2–10	21
Washing Time	15–30	9
Drying Time	15–20	20
Surface Change	Manual Agitation	Transport Rollers
Replenishment	Manual	Automatic

*Data vary with type of equipment, films, and chemicals.
**Sixty-minute processors are now available.

One and one-half minute processing time is now standard; occupying an extremely small space, these units can accommodate about 300 films of assorted sizes per hour. Newer models offer 1-minute processing with almost no loss of quality.

To assure optimum film speed, contrast, and latitude, proper film and chemical combinations must be used.

As a rule, two smaller units are preferable to one larger unit of the same total capacity because (1) in the event of equipment failure, the second unit is still available; and (2) there is less waiting time in loading two processors during peak hours.

Several brands of automated processors are available and selection should be made carefully because they vary in performance. One should especially *avoid* those requiring more than ninety seconds because they lack important features and also demand an annoying increase in processing time. Furthermore, one should have ready access to competent repair service.

QUESTIONS

1. Define latent image. How is it produced?
2. Of what do the black areas on a roentgenogram consist? What factors determine the various densities in a radiograph?
3. What is the purpose and theory of development? Fixation?

4. List the ingredients of a manual developing solution and describe the function of each one.
5. List the ingredients of a manual fixing solution and describe the function of each one.
6. What is meant by time-temperature development?
7. Discuss replenishment of developer and fixer in manual and automated systems.
8. What is included in the term "film processing"?
9. Name five causes of film fogging and state how they can be avoided.
10. Describe four causes of streaking of radiographs.
11. Why should overfixing be avoided?
12. Describe the important differences in the composition of processing solutions in automated and manual systems.
13. What problems would arise if manual processing solutions were used in an automatic processor?
14. List the important advantages of automated processing over manual processing.
15. State and explain briefly the three functions of the transport system.
16. Why are automated processors operated at high temperature? How critical is the developer temperature in automated processing in comparison with manual processing?
17. How can one decrease the frequency of automated processor breakdown?
18. Why are filters used in the water supply of automated processors? What is the result of clogging of these filters?
19. Compare manual and automatic processing times.

Chapter 18

RADIOGRAPHIC QUALITY

Medical radiography seeks to provide the maximum diagnostic information about a particular anatomic area by recording the radiographic image, usually on film. Its success depends on the production of radiographs of superb quality. A radiograph of poor quality may cause serious error in diagnosis through inadequate recording of information, which in this instance is a collection of images.

What do we mean by radiographic quality? This refers to the visible sharpness of images of structural details such as bone trabeculae and small pulmonary vessels. We use the expression *recorded detail* to denote radiographic quality. Other terms include definition, sharpness, detail, and visibility of detail, but an all-inclusive term like recorded detail is preferred so as to avoid confusion.

Recorded detail refers to the distinctness of radiographic image margins. When such margins are blurred (ie, fuzzy), recorded detail is poor. The sharpness of image margins can be measured objectively as *acutance* — the abruptness of the boundary between a detail and its surroundings. Another concept is *resolution*, which measures the ability of an image receptor (screen-film, fluoroscopic screen, image intensifier, etc) to produce separate images of closely spaced small objects. Although acutance and resolution are related, they are affected differently by such factors as contrast and density.

Five principal factors influence recorded detail: blur, density, contrast, distortion, and noise. These will now be considered separately and as they relate to one another.

BLUR

There are four causes of image blurring or unsharpness. One is geometric in origin, depending primarily on the measurable size of

318

the focal spot. The other three have to do with motion of the object, the nature of the image receptor (eg, film-screen system), and the shape of the object being radiographed.

Geometric or Focal Spot Blur

The arrangement in space relating the x-ray source (focal spot), anatomic part, and film, controls the degree of geometrically produced blurring. We shall simply call this *geometric* or *focal spot blur.* The terms *penumbra* and *edge gradient* may be used as synonyms for geometric blur.

Geometric blur depends on three factors: (1) effective focal spot size, (2) source-to-image receptor distance (SID) or focus-film distance (FFD), and (3) object-to-image receptor distance (OID) or object-film distance (OFD) (see Figure 18.2.) All these factors affect recorded detail (image sharpness), which improves as geometric blur decreases. The term focus-film distance has been replaced by *source-to-image receptor distance (SID);* this refers to any medium, like a fluorescent screen or x-ray film, that converts incident x-ray photons either into visible light or into some other form that can, in turn, be changed to visible light.

1. **Size of Effective Focus.** According to elementary geometry, the effective or apparent size of the focal spot (see pages 209–211) has a marked influence on recorded detail. Figure 18.1 shows an anatomic object situated between an x-ray tube and a film. If the focal spot were assumed to be a point (having no dimension), all x rays passing the edge of the object would produce a point-for-point sharp image of the object margin on the film. But in practice the focal spot has a finite width, usually ranging from 0.3 to 2.0 mm. Therefore, as shown in Figure 18.2, x rays originate over innumerable points on the focal spot, spread as they pass the edge of the object, and proceed toward the film. This produces a blurred margin, the width of the blur being proportional to the effective focal spot size (compare Figure 18.2A with Figure 18.2B). In other words, to obtain minimal blur we should use the smallest practicable focal spot. Since blur impairs recorded detail, we may conclude that *tubes with smaller focal spots provide better recorded detail, with improved image quality.*

2. **Source-to-Image Receptor Distance (SID).** The degree of blurring (ie, blur width) also depends on the distance between the

**POINT
FOCAL SPOT**

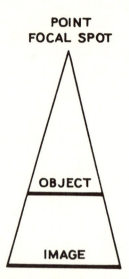

Figure 18.1. Image production by a hypothetical (imaginary) *point* focal spot. All the rays passing the edge of the object would produce a point-for-point sharp image of the object margin. The result would be nearly perfect geometrically recorded detail (no blur or penumbra).

source (focal spot) and the image receptor. As the SID (FFD) is increased while the OID (OFD) remains unchanged, the x rays arising from innumerable points on the focal spot undergo less spread after passing the edge of an object, on their way to the film. As shown in Figure 18.3A, at a shorter SID, the image margin is blurred more than in Figure 18.3B where the SID is longer. Thus, blurring decreases and recorded detail improves with increasing SID.

3. **Object-to-Image Receptor Distance (OID).** The third geometric factor in blurring is the object-to-image receptor distance (OID) or (OFD). If the focal spot size and the source-to-image receptor distance remain constant, a decrease in the OID decreases blur (improves recorded detail). This is readily seen by comparison of Figures 18.4A and 18.4B.

The above three factors in geometric blurring (penumbra) may be conveniently summarized by a simple equation derived from Figure 18.3, where F is the *effective* focal spot size, a the source-to-object distance (SOD), b the object-to-image receptor distance (OID), and B the blur width, all in cm. According to the rule for similar triangles (shaded areas on right),

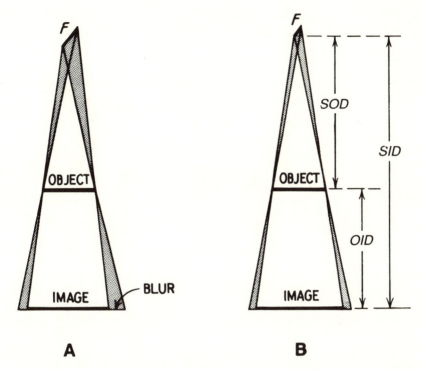

A   **B**

Figure 18.2. Influence of focal spot size on recorded detail. In *A*, the larger effective focal spot produces more blurring (wider penumbra) than the smaller effective focal spot in *B*. Focal spot blurring is caused by x rays, which are emitted from the entire focal spot area, converging toward the edge of the object and then diverging toward the edge of the image at the film or other image receptor. Thus, *a smaller focal spot produces better geometrically recorded detail.*

$$B/F_e = b/a$$
$$B = F_e \, b/a$$
$$B = \frac{\text{effective focal spot size} \times \text{OID}}{\text{SOD}} \text{ cm} \tag{1}$$

As will be shown later (see pages 347, 351), recorded detail deteriorates with increased magnification, which results from an increase in OID and a decrease in SID (see Figure 18.4).

In summary, then, recorded detail (image sharpness) is enhanced by any factor that decreases geometric blur (penumbra): namely, (1) small focal spot, (2) long source-to-image receptor distance (SID), and short object-to-image receptor distance (OID). Therefore,

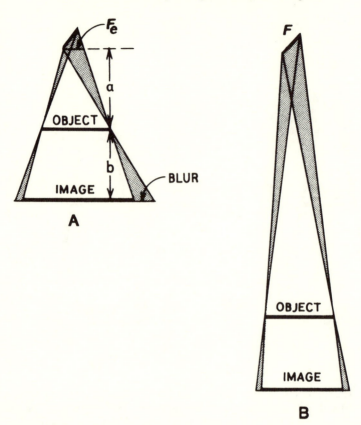

Figure 18.3. Effect of source-image receptor distance (SID) on recorded detail. The object-image receptor distance (OID) is the same in *A* and *B*, but the SID (FFD) is longer in *B* than in *A*. Therefore, there is less blur of the image in *B*. Thus, *increasing the SID (FFD) improves geometrically recorded detail.*

you should place the part to be radiographed as close to the film as possible and use an appropriately long SID with the smallest applicable focal spot.

Focal Spot Evaluation

As we have seen, the size of the focal spot strongly influences recorded detail. Unfortunately, there is no simple method of evaluating the focal spot, nor is there universal agreement, even among physicists, as to how it should be done. At the present time there are three general ways of designating focal spot characteristics:

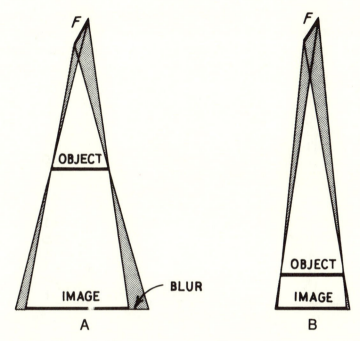

Figure 18.4. Effect of object-image receptor distance (OID) on recorded detail. The SID (FFD) is the same in A and B, but the OID is shorter in B than in A. Therefore, image blur is less in B than in A. Thus, *decreasing the OID (part-film distance) improves geometrically recorded detail.*

1. **Nominal Size.** As quoted by the manufacturer, this is based on *computation* or *measurement* at very low mA (pinhole camera). However, the nominal size often understates the *effective* or *projected size* of the focal spot by a significant margin. Based on the ±50 percent tolerance customarily allowed in manufacture, a 0.3 mm focal spot may be as large as 0.45 mm, a 0.6 mm focal spot 0.9 mm, and a 1.0 mm focal spot 1.5 mm.

We shall now turn to the methods of practical evaluation of focal spot size.

2. **Pinhole Camera.** The *projected (effective) focal spot* dimensions can be measured by a *pinhole camera* as described in detail in *ICRU Report No. 10f.* A pinhole camera consists of a tiny hole drilled in a plate made of a gold-platinum alloy. For example, hole diameter is 0.03 mm for focal spots smaller than 1 mm, and 0.075 mm for focal spots 1 to 2.5 mm. These very small holes minimize blur (penumbra), making for more accurate measurement of the

focal spot image. At the same time, the small pinhole requires very long exposures, so that care must be taken not to damage the target of the x-ray tube. It is extremely important that a pinhole camera conform to rigid specifications and be purchased from a reputable dealer.

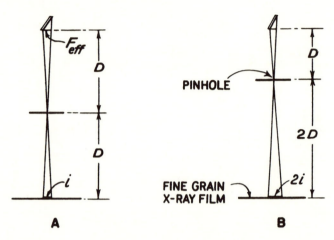

Figure 18.5. Principle of the pinhole camera for measurement of focal spot size. In A the pinhole is located midway between the tube focus and the film, so the image i of the focal spot has the same size as the effective or apparent focal spot F_{eff}. In B, with the pinhole located twice as far from the film as from the focus, the image $2i$ is twice the size of the focal spot. Thus, proper correction must be made for the position of the pinhole in measuring focal spot size. In addition, the vertical axis of the pinhole must be perfectly aligned with the central ray.

In actual use, the pinhole is placed between the focal spot and a fine grain x-ray film, perfectly aligned with the central ray, perpendicular to it and parallel to the film. Figure 18.5 shows the arrangement. For focal spots up to 2.5 mm the pinhole-film distance is twice the focus-pinhole distance to give a magnification factor of two. Measurement of the pinhole image is made with a calibrated lens and corrected for magnification.

As shown in Figure 18.6, the x rays emitted by the focal spot have a nonuniform intensity distribution (ID), being concentrated along two edge bands. In fact, such an ID makes the focal spot behave as though it consisted of two narrow focal spots, the effect being to impair image resolution when compared with an ID that has a central peak of intensity. Doi and Rossman have shown that with a computerized simulated system in small vessel angiography, focal spot size may be more important than ID.

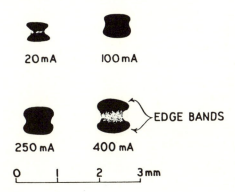

Figure 18.6. Effect of milliamperage (mA) on focal spot size. Note that x-ray emission is most intense along the edges of the focal spot—the so-called *edge* or *band effect.* Also, the size of the focal spot increases significantly with mA, a condition known as *blooming.* (*Adapted from Bernstein, Bergeron, and Klein, courtesy of Radiology.*)

Also note that focal spot size increases with increasing mA, a condition known as *blooming* (but decreases with increasing kV). Since the focal spot is smaller when hot than when cold, blooming does not depend on temperature; rather it results from space charge effects associated with higher mA. It is important to know that in angiography, for example, the projected focal spot size increases about 50 percent in going from 200 to 1000 mA. Because focal spots are measured at the factory at mA values much below those prevailing in radiography, their sizes are significantly understated. Thus, the pinhole camera permits the evaluation of focal spots as to size, shape, and intensity distribution under actual operating conditions in the Radiology Department.

3. **Resolution Test Patterns.** Inasmuch as the pinhole camera does not show directly the resolution capability of a focal spot, another method has had to be devised—the *star pattern resolution object.* Resolution designates the smallest distance between two objects such that they appear as two separate images. Figure 18.7A shows a star pattern object. It provides a continuous change in spacing between thin radiopaque lines (usually lead), the limit of resolution being indicated by the blending of the line images.

The theory underlying the use of resolution test objects is too complex to be considered here. However, it should be noted that there may be a ringed zone of apparently good resolution within the blurred zone, but this represents spurious resolution.

Another type of resolution test object consists of a series of grouped *parallel opaque lines* that differ in thickness and spacing (see Figure 18.7B). With this test object, resolution is related to the thinnest and most closely spaced line pairs resolved by the particular focal spot and is expressed as the maximum number of line pairs per mm. A line pair consists of a line and one adjacent space.

In summary, then, focal spot measurement is necessary to assure satisfactory recorded detail, since the true effective focal spot size may be appreciably larger than the nominal size. However, such measurements require experienced personnel using precision instruments. The two most widely used devices are the standard pinhole camera and some type of resolution test object; the former measures the physical size of the projected (effective) focal spot (also, shape and intensity distribution), whereas the latter measures its resolution capability. It behooves the purchaser of an x-ray tube to demand a suitably narrow tolerance range, especially for fractional focal spots such as 0.3 mm, and for angiography and magnification radiography. But in any case, measurement should be verified by one or both of the described methods.

A   **B**

Figure 18.7. Examples of resolution test patterns. *A*. Star type, useful in measuring focal spot resolution. *B*. Parallel line type for measuring resolution of intensifying screens. (*Courtesy of Nuclear Associates, Inc.*).

Motion Blur

Motion of the part being radiographed may be regarded as the greatest enemy of recorded detail because it produces a blurred image and cannot be completely avoided. Motion can be minimized in three ways: (1) by careful immobilization of the part with sand bags or compression band, (2) by suspension of respiration when examining parts other than the limbs, and (3) by using exposures that are as short as possible, generally with intensifying screens of adequate speed, provided quantum mottle is not excessive (see pages 328–329). Although screens contribute blur, this detrimental effect is usually more than compensated for by the reduction of motion blur.

Screen Blur

Recorded detail is better in radiography with cardboard holders than with intensifying screens, provided *immobilization is adequate.* However, when fast exposures are needed to minimize the effects of uncontrolled motion, screens must be used; under these circumstances, the gain in sharpness with screens and short exposures usually more than compensates for the blur that would result from the increased effect of motion accompanying the use of cardboard holders or detail (slow) screens with long exposures. In general, screen blur is greater than that caused by a 1 mm focal spot. On the other hand, extremely short exposures using high speed screens with very low mAs can impair recorded detail as a result of excessive quantum mottle (see pages 328–329). Insofar as *film-screen combination* is concerned, better recorded detail results from a fast film and medium speed screen combination than from the reverse. This is explained by the following data on calcium tungstate screens:

Screen blur is generally of the order of about six times inherent film blur.

Blur with medium-speed *screens* is about one-half that with high speed screens.

Blur with high and medium speed *films* is virtually negligible in radiography.

Why is recorded detail impaired by intensifying screens even when the patient is adequately immobilized? We may summarize the reasons as follows:

1. **Crystal Size.** Recall from Chapter 15 that intensifying screens produce a radiographic image because of *fluorescence of crystals* in the active layer. Since each crystal, despite its very small size, produces a spot of light of similar size on the film, image lines are broader than when they are formed directly by x-ray photons. Image blur increases with increasing crystal size, but this is of relatively minor importance in practice because crystals do not vary widely among various types of screens.

2. **Active Layer Thickness.** The diffusion (spreading) of light in the active layer due to its measurable thickness increases image blur (see page 282).

3. **Film-screen Contact.** Cassettes are provided with metal clamps to secure close contact between the film and screens. Even very slight separation between them allows the light from any point in the screen to spread on its way to the film, thereby contributing to blur, with impaired recorded detail (see pages 285–286).

Because of these factors, slow screen-film systems (detail screens) should be used in the radiography of small parts (hands, feet, wrists), especially in infants and small children, to obtain adequate recorded detail. However, in the radiography of parts measuring 8 to 10 cm thick, medium or high speed screens should be used routinely; and above 10 cm, a grid also.

4. **Noise.** In examining, with a magnifying glass, a radiograph that has been exposed by means of *intensifying screens,* you will find that it has a mottled or grainy appearance. Only a small part of this is contributed by radiographic mottle, which comprises the granular structure of the screens and the clumps of silver in the radiographic image.

More important is *quantum mottle,* produced by the nonuniform intensity over the cross section of an x-ray beam as it leaves the tube port. Recall that the beam consists of photons or quanta having a random distribution in space. Thus, in a film that has been exposed to such a beam, different areas have actually received different numbers of photons. Quantum mottle is a form of *noise* which may be defined as any random audible or visible disturbances that obscure information. Other examples of visible noise on a radiograph are *fogging* of film by light or by scattered x rays.

With *slow image receptors* such as slow screen-film combinations, or direct exposure film, a large number of photons are needed for a particular degree of film density (darkening). Under these conditions the photons striking the image receptor are closely crowded, the variation in the number of photons from area to area is small, and the image is relatively uniform. On the other hand, with *fast image receptors* such as high-speed screen-film combinations, a smaller number of photons can provide the same overall density, but now there is a larger variation in the number of photons from area to area (statistical fluctuation) and the image appears mottled. This is a simplified version of quantum mottle.

Quantum mottle increases with high contrast films because density differences are exaggerated. But increased diffusion of light in intensifying screens, such as those with thicker active layers, increases image blurring and at the same time makes quantum mottle less apparent. Quantum mottle, as a form of *noise,* impairs recorded detail.

Aside from the image receptors themselves, there is another factor in the production of quantum mottle. This is the *kV.* As kV is increased, each x-ray photon gives rise to more light photons, so the screen intensifying factor increases. Thus, fewer information-carrying x-ray photons are ultimately required for a given degree of film darkening with a particular film-screen combination. This explains why high kV increases quantum mottle, especially with very low mA.

In summary, then, quantum mottle increases with fast image recording systems and with high energy photons. In any given instance, the decision has to be made as to whether high speed or minimal mottle is more important.

Object Blur

Most anatomic structures, which we here refer to as objects, have rounded borders. These introduce a blur factor, demonstrated in Figure 18.8. Note that in *A* the object has a shape conforming to that of the x-ray beam, so all the rays pass through the full thickness of tissue. As a result, the radiographic image presents a sharp boundary or *density gradient* as seen from the density trace at the bottom of the figure. In *B*, the rays pass through progressively thinner portions of the object toward its periphery so that image density fades off gradually at the border. In *C*, the round object also gives rise to an image with fading density at its boundary. Thus, in both *B* and *C* the image border has a long density gradient with increased blurring, its degree depending on the distance over which the density falls off. The effect of object blur is greater than is often appreciated, and it may even exceed geometric blur.

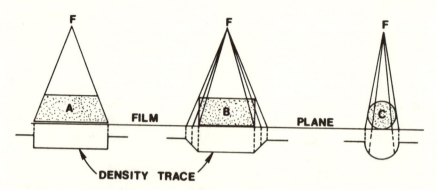

Figure 18.8. Object blur caused by object shape. In *A* the shape of the object conforms to that of the beam, so the density fall off sharply at the edge of the image which therefore has an abrupt or sharp border. In *B* a progressively smaller thickness of the object intercepts the beam toward the edge, so the density falls off gradually at the image boundary, which is therefore less sharp than in *A*. In *C* a rounded object also produces a less sharp image than in *A*. (*Adapted from Seemann, H. E.: Physical and Photographic Principles of Medical Radiography;* by permission of John Wiley & Sons, Inc.)

DENSITY

The amount of darkening of an x-ray film, or of a certain area on the film, is called **radiographic density**. It should be recalled that any region of the film emulsion exposed to light from intensifying screens, or directly to x rays, becomes susceptible to the action of the developer. The silver halides that have been so affected are changed by the developing agent into tiny particles of **metallic silver** which appear black because of their finely divided state. The greater the total amount of radiation that reaches the film, the greater will be the final degree of blackening. Areas receiving only a small amount of radiation undergo little or no subsequent action by the developer, such underexposed areas appearing gray or translucent in the finished radiograph. Thus, in the final analysis, *radiographic density* (ie, degree of blackening or opacity) *depends on the amount of radiation reaching a particular area of the film and the resulting mass of metallic silver deposited per unit area during development.* Standardization of development should produce optimal film density for a correctly exposed film.

Density is measured by an instrument known as a *densitometer,* which indicates the relationship between the intensity of light falling upon one side of a given area of a radiograph (as from an illuminator) and the intensity of the light passing through (transmitted). This relationship is given by the following equation:

$$\text{density} = \log \frac{\text{incident light intensity}}{\text{transmitted light intensity}} \qquad (2)$$

The equation can best be explained by numerical examples. If the incident light intensity is ten times the transmitted intensity, then the density is $\log 10 = \log 10^1 = 1$; if the incident light intensity is one hundred ·times the transmitted intensity, then the density is $\log 100 = \log 10^2 = 2$; etc (see pages 23–24). One need not be conversant with logarithms to use a densitometer, since it is calibrated to read density directly. With this method, clear film base has a density of 0.06 to 0.2, depending largely on the amount and shade of blue dye present. A diagnostic radiograph usually has densities varying from about 0.4 in the lightest areas to 3.0 in the darkest. Obviously, excellent radiographic quality requires optimal density for any particular anatomic structure; the correct exposure is supposed to be incorporated in the technic chart.

Density is an extremely important factor in radiographic quality because it carries *information.* Without density there is no image and therefore no recorded detail. However, density must be optimal because, in excess, it may conceal information through loss of visibility of recorded detail.

Five factors govern the radiation exposure and resulting density of a radiograph: (1) kilovoltage, (2) milliamperage, (3) time, (4) distance, and (5) thickness and nature of part being radiographed.

1. **Kilovoltage.** An increase in kV applied to the x-ray tube increases both the exposure rate and the percentage of higher energy (short wave length) photons. These more penetrating photons are not so readily absorbed by the structures being radiographed,

and therefore a larger fraction of the primary radiation eventually reaches the intensifying screens. Thus, an increase in kV increases the exposure rate at the film and the resulting radiographic density. An *increase of 15 percent in kV approximately doubles the exposure.* For example, we can double the exposure at 40 kV by adding 6 kV, and at 70 kV by adding 10 kV. Note that although radiographic density increases as exposure increases, this relationship is not strictly proportional. At the same time, subject contrast decreases (see pages 339–340).

Under ordinary conditions, with grids having a ratio of 8:1 or less, it is unwise to exceed 85 kV because this produces an excess of *scattered radiation,* which fogs the radiograph and impairs contrast. On the other hand, the efficiency of x-ray equipment improves at higher kV because of the considerably smaller heating load on the anode (see page 220) and the greater certainty of penetration. It has been found that *high voltage radiography* at 100 to 130 kV and a high ratio grid (12:1 or 6:1) offer advantages over conventional radiography, including greater latitude and a smaller dose to the patient.

2. **Milliamperage (mA).** This is a measure of electron flow per second from cathode to anode of the x-ray tube. As this flow rate is increased, more photons per sec are produced at the target. In fact, x-ray exposure rate (R/sec) is proportional to mA: doubling mA doubles x-ray exposure rate, tripling mA triples exposure rate, etc. Note that mA determines *exposure rate only;* it has nothing to do with the penetrating ability of an x-ray beam, which is governed by kV.

3. **Time.** An increase in exposure *time* causes a proportional increase in the *number* of photons emitted by the target. Thus, doubling exposure time doubles total exposure, and tripling exposure time triples total exposure—a longer exposure time allows the radiation at a given exposure rate to act longer, thereby affecting more silver halide crystals in the film emulsion and increasing radiographic density. In practice, mA and exposure time are multiplied: *mA* × *sec* = *mAs,* called *milliampere-seconds.* This is a measure of the *charge* transferred from cathode to anode during an exposure. For example, if the selected technique calls for 100 mA and 0.1 sec, then multiplying $100 \times 0.1 = 10$ mAs. If the same exposure is to be given in a shorter time, so that radiographic density remains unchanged, mA can be increased to 400 and

exposure time to 0.025 (= 1/40 sec): 400 mA × 0.025 sec = 10 mAs as before. We can generalize this relation by the simple equation

$$mA_2s_2 = mA_1s_1 \qquad (3)$$

where, if mA_1s_1 is known and we select a particular time s_2, we can determine mA_2 from equation (3). This relation, known as the *reciprocity law*, holds true provided the x-ray generating equipment has been *properly designed* and *calibrated.* You should cultivate the habit of thinking in terms of mAs because it greatly facilitates the modification of established technics, thereby increasing the flexibility of the technic chart.

4. **Distance.** The effect of distance on exposure rate is not so simple as some of the other factors, but it is easily understood by keeping in mind the following elementary geometric rules:

a. X-ray photons actually originate from innumerable points on the focal spot of the x-ray tube, spreading in all directions from their points of origin.

b. In considering the effect of distance on exposure rate, we shall assume that the focal spot acts as a point source from which x rays spread in the form of a cone after leaving the circular port of the tube housing. We normally modify the beam by a collimator (beam-limiting device), which gives it a rectangular or square cross section.

Since the x-ray photons diverge (spread) *the width of the beam increases as the distance from the target increases.* This means that the same amount of radiation is distributed over a larger area the farther this area is from the tube focus. Obviously, the same radiation spread over a larger area must spread itself "thinner." If, at a certain distance from the tube focus, the beam were to cover completely a film of a certain size, then at a greater distance it would cover a larger film. However, the radiographic density of the latter would be less in a given exposure time because each square centimeter of this film would have received less radiation than each square centimeter of the first film. We may therefore conclude that *radiographic exposure rate decreases as the focus-film distance increases.*

We can determine by simple geometry *how much* the exposure rate decreases as the focus-film distance increases. In Figure 18.9 the slanting lines represent the edges of a beam emerging from the focus at the target, *T,* and restricted by a square collimator. Two planes, *ABCD* and *EFGH,* are chosen perpendicular to the direction of the central ray of the beam (represented by the dotted line). Both planes are assumed to be squares. Plane *EFGH* is located twice as far from the target as plane *ABCD.* Therefore each side of the lower plane, such as *HE,* is twice as long as a side of the upper plane, such as *DA,* because triangles *FHE* and *FDA* are similar and their corresponding sides are proportional. To simply the discussion, let X equal a side of the upper plane. Then $2X$ must equal a side of the lower plane. The area of the upper plane will then be $X \times X = X^2$, and the area of the lower plane will be $2X \times 2X = 4X^2$. Thus, the lower surface, *EFGH* has four times the area of the upper surface, *ABCD.*

It is evident that **when the distance is doubled, the same radiation is spread over an area four times as great.** Therefore, the brightness or illumination must be ¼ as great. The **inverse square law of**

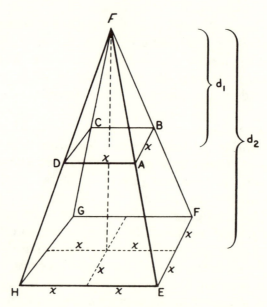

Figure 18.9. Inverse square law of radiation. The lower plane surface (*EFGH*) is selected at twice the distance from the point source of radiation (tube focus) than is the upper plane (*ABCD*). Each side of the lower plane ($x + x$) is twice as long as each side of the upper plane (x). It is evident from the diagram that the lower surface area is four times as large as the upper surface area, which means that at twice the distance from the target the x-ray beam covers four times the area and therefore the brightness of illumination must be only one-fourth as great.

radiation applies to both x rays and light, and may be stated as follows: *the intensity or exposure rate of radiation at a given distance from a point source is inversely proportional to the square of the distance.* For example, if the exposure rate of an x-ray beam at 20 in. from the focus is 100 R/min, what will it be at 40 in.? Let us set up the inverse square law in equation form:

$$I/i = d^2/D^2 \qquad (4)$$

where

I is the exposure rate at 40 in. = ?
i is the exposure rate at 20 in. = 100 R/min
$d = 20$ in.
$D = 40$ in.

Note again the inverse square proportion. Substituting the above values in equation (4):

$$I/100 = (20)^2/(40)^2$$

$$I = \frac{\cancel{20} \times \cancel{20} \times 100}{\underset{2 \times \ \ 2}{\cancel{40} \times \cancel{40}}}$$

$$I = 25 \text{ R/min}$$

Thus, at twice the distance the exposure rate is one-fourth the initial value (25 as compared with 100). This means that in radiography, doubling the distance while the kV and mAs are held constant reduces the exposure to one-fourth. Therefore, to keep the exposure constant the mAs has to be multiplied by four.

Let us take another example. If the exposure rate of radiation at 60 in. is 10 R/min, what will the exposure rate be at 20 in.? Again using equation (4), let i represent the unknown exposure rate at point d located 20 in. from the focus. Then, using the data given in the above problem,

$$10/i = (20)^2/(60)^2$$

$$1/i = \frac{\cancel{20} \times \cancel{20}}{\underset{3 \quad\ \ 3}{10 \times \cancel{60} \times \cancel{60}}}$$

$$1/i = 1/90$$

$$i = 90 \text{ R/min}$$

In other words, the distance has been reduced to 20 ÷ 60 = ⅓ and the exposure rate has increased by 90 ÷ 10 = 9 times.

There is a ***much simpler method*** of applying this law. Rearranging equation (4),

$$ID^2 = id^2 \qquad (5)$$

This means that for a given set of factors (kV, mA, filter, etc) the exposure rate at a particular point in the beam, times the square of the distance of the point from the focal spot, is a ***constant.*** Thus, in equation (5), id^2 is a constant. In the problem on page 335, it is stated that $i = 100$ R/min at distance d, 20 in. Therefore,

$$id^2 = 100 \times 20 \times 20$$

The problem is now solved by substituting this constant and the ***new*** distance, 40 in., in equation (5).

$$I \times 40 \times 40 = 100 \times 20 \times 20$$

$$I = \frac{100 \times \cancel{20} \times \cancel{20}}{\underset{2}{\cancel{40}} \times \underset{2}{\cancel{40}}}$$

$$I = 25 \text{ R/min}$$

In actual practice, a therapy machine must be calibrated for each treatment distance because of deviation from the inverse square law due to the large focus. However, the law is sufficiently accurate for approximation in diagnostic radiology and in protection problems.

In radiography we use a ***direct square law*** when changing focus-film distances because we are ***compensating*** for the reduced or increased exposure as the distance is increased or decreased, respectively. Assuming that the kV remains constant, we may obtain the mAs required to compensate for a change in distance by using the following equation:

$$\textit{new mAs/old mAs} = \textit{(new distance)}^2/\textit{(old distance)}^2 \qquad (6)$$

For example, if the factors 100 mAs, 100 kV, and 40-in. (100 cm) distance produce a radiograph having the proper density, what will be the mAs needed to maintain the same radiographic density at a focus-film distance of 60 in. (150 cm)?

$$\text{new mAs}/100 \text{ mAs} = (60)^2/(40)^2$$

$$\text{new mAs} = 100 \times 3600/1600 = 225 \text{ mAs}$$

5. **Radiographic Object.** The human body consists of tissues and organs that differ in thickness and density with resulting differences in radiolucency — the ability to transmit x rays. Thicker

and denser anatomic parts, whether normal or pathological, attenuate (ie, absorb or scatter) x rays to a greater degree, leaving less exit or remnant radiation to reach the film. Ranging from the greatest to the least density or radiopacity are: (1) dental enamel, (2) bone, (3) tissues of "water density" such as muscle, glands, and solid nonfatty organs, (4) fat, and (5) gas. Pathologic processes such as pneumonia increase the density of lung tissue. On the other hand, destructive bone diseases reduce the density of bone. Exposure factors—kV and mAs at a particular distance—are optimally selected for the thickness and density of the part being radiographed. Finally, opaque media rank high on the list of dense materials.

CONTRAST

A radiograph consists of light and dark areas; that is, it shows variations in density. The range of density variation among the light and dark areas is called *radiographic contrast.* While density represents the amount of silver deposited in a given area, contrast represents the relative distribution of silver in various areas of a radiograph. To be perceptible to the average human eye, the difference in density of adjacent areas must be at least 2 percent. *Optimum contrast enhances recorded detail.*

There are three kinds of contrast. *Radiographic contrast*—the overall contrast of a radiograph—depends on *subject contrast* and *film contrast.* These will be discussed separately and interrelated. However, the unqualified term contrast usually means radiographic contrast.

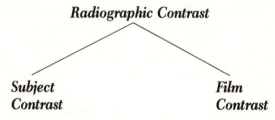

Radiographic Contrast

Subject Contrast   *Film Contrast*

Radiographic Contrast

We usually designate radiographic contrast as *long scale* (low) or *short scale* (high). This concept may be explained as follows: suppose a radiograph were cut into small squares representing each of the different densities present, from the lightest to the darkest, and these squares were then arranged in order of increasing density. If there should be many such squares with very little difference in the density of successive squares, then the original radiograph must have had *long scale* or *low contrast*. On the other hand, if there should be relatively few squares with a large difference in density between successive ones, then the original radiograph must have been of *short scale* or *high contrast*. In other words, a long scale or low contrast radiograph has a long range of tonal gradation from white, through many shades of gray, to black; whereas a short scale or high contrast radiograph has a short range between white and black (see Figure 18.10).

Contrast improves recorded detail and is optimal when it produces sufficient difference in density among the various details to make them distinctly visible in all areas of the radiograph. This is usually achieved by *medium scale contrast*. Excessively short scale contrast tends to impair detail; an example is the so-called "chalky" radiograph which may lead to error in interpretation, especially in the detection of fine fracture lines. However, short scale contrast is desirable under special circumstances such as angiography.

Subject Contrast

As indicated earlier, *subject contrast* is one of the factors in radiographic contrast, the other being film contrast. Before defining subject contrast, we must first introduce appropriate basic concepts.

An x-ray beam undergoes *attenuation* or loss of photons while passing through a patient, because of absorption, scatter, and inverse square law. But attenuation is not uniform throughout the cross section of the *exit beam*, that is, the beam leaving the patient. If we could somehow "see" the exit beam in cross section from the standpoint of the film, we would observe a variation in the number of photons per cm^2 at different locations in the beam. This results from unequal degrees of attenuation of x rays by various tissues. It is the spatial distribution of photons in the cross section of the exit beam that constitutes the *aerial image* (see Figure 17.3), which will eventually be recorded on the film (or some other image receptor) as the *radiographic image*. This distinction must be kept in mind.

Thus, *subject contrast may be defined as the contrast in the aerial*

Figure 18.10. Effect of kilovoltage on contrast, using an aluminum stepped wedge. The short range (high) contrast with 40 kV, as compared with long range (low) contrast with 100 kV, is obvious.

image. It is actually the ratio of the number of photons in two or more equal small zones of the aerial image.

Note that the aerial image contains information derived from the patient. But in addition to the wanted information, the aerial image becomes contaminated with *noise,* this being **unwanted** radiation that impairs the quality of the information and ultimately degrades the quality of the resulting radiograph. Such noise factors, which include scattered radiation, quantum mottle, and fogging, obscure recorded detail.

We shall now explore the factors that affect subject contrast, namely, radiation quality, radiographic object, scattered radiation noise, and fog noise.

1. **Radiation Quality** — determines penetrating ability (transmission) of an x-ray beam. This is the inverse of attenuation. We found in Chapter 12 that an increase in the kV across the x-ray tube increases the penetrating ability of the resulting x-ray beam. A beam of high penetrating ability goes through tissues of various

densities more uniformly than does a beam of low penetrating ability. For example, in radiography of a leg, the bony structures absorb a much greater fraction of the primary x rays than do the soft tissues. As the kV and consequent penetrating ability of the primary photons increase, the difference in the absorption of the radiation, as between bone and soft tissues, becomes smaller. Consequently, the transmission of x rays through bone is more nearly like that through soft tissue and there is less difference in radiographic density between them; in other words, there is a reduction in subject contrast. Figure 18.10 shows the effect of various kilovoltages on contrast. You can see that an ***increase in kV produces longer scale or lower contrast*** in the radiograph of an aluminum stepped wedge (a bar made up of a series of increasing thicknesses of aluminum).

An additional factor at higher kV is the loss of contrast resulting from the increase in the fraction of scattered radiation, but this is largely removed by the use of high ratio grids.

2. **Radiographic Object** —atomic number, density, and thickness of the object being radiographed.

Atomic Number. As noted before, photoelectric absorption increases dramatically with increasing atomic number. Therefore, attenuation is higher per cm of bone (high atomic number of calcium and phosphorus) than per cm of soft tissue.

Density. An increase in density (g/cm³), when expressed in terms of electrons/cm³), increases x-ray attenuation per cm of tissue because with higher density, more electrons are present per cm of tissue to engage in Compton scatter.

Thickness. With increased total thickness of tissue traversed by the x-ray beam, there are more atoms available for interactions (absorption and scatter).

The degree of attenuation in a tissue is called its ***radiopacity;*** conversely, the degree of transmission is its ***radiolucency.*** Greater attenuation means that less radiation is transmitted through the patient; that is, less exit radiation is available for radiography.

As noted above (pages 336–337), the body contains tissues of various densities. For example, a radiograph of the abdomen has a light zone in the center representing the vertebral column which attenuates more radiation than an equal mass of soft tissue. Various soft tissues are represented as shades of gray, depending on differential attenuation. The fat around the kidneys, along the psoas muscles, and in the abdominal wall, appears as dark gray or nearly black lines.

The differential attenuation of x-ray photons produces subject contrast in the aerial image, ultimately contributing radiographic contrast, which improves recorded detail. A number of organs and anatomic regions do not have sufficient inherent contrast for plain radiography. In such cases adequate contrast enhancement becomes possible by the use of *contrast media.* For example, dense media such as barium sulfate in the gastrointestinal tract, and iodinated compounds in the gall bladder, kidneys, vascular system, and bronchial tree. Note that the photoelectric effect produces a sharp peak in absorption of x-ray photons, approximately equivalent to a 65-kV_p beam, because the binding energy of the K-shell electrons of iodine is 32 keV. Optimal contrast would therefore be obtained at about 65 kV_p. However, we find in practice that adequate contrast between iodine and soft tissue occurs with x-ray beams generated at 65 to 80 kV_p. Sometimes a less dense medium can be used, such as air in the ventricles of the brain. At this point you should review the physical reasons for the difference in attenuation of x rays by various tissues, discussed on page 181.

3. **Scattered Radiation** — a noise factor, obscuring information, or detail, in the radiographic image. Scattered radiation impairs contrast by a fogging effect mainly on the **lighter areas** of the radiograph, and may be controlled by means of stationary or moving grids and by limiting the size of the x-ray beam with a collimator, cone, or aperture diaphragm (see Chapter 19).

4. **Fogging** from any cause contributes to noise, imparting an overall gray appearance to the radiograph with reduction in contrast and visibility of recorded detail.

Film Contrast

As we have already pointed out, film contrast and subject contrast make up radiographic contrast. However, in considering film contrast we must include the type of image receptor (in this instance, film), its use with or without screens, and the processing system. It is therefore more appropriate to speak of the *film imaging system,* a designation which includes film and processing, with or without screens.

Films themselves vary in their *inherent contrast,* depending on their emulsion characteristics. Thus, films are specifically designed for long, medium, or short scale contrast.

The *development process* also affects film contrast. For example, a developer containing sodium hydroxide as an accelerator produces greater contrast than one containing sodium carbonate. Within the normal range of time-temperature development, contrast is not appreciably influenced by the temperature. However, at excessively high temperature or unduly prolonged developing time chemical fog may occur, reducing contrast. Gentle agitation during development improves contrast. More important, a decrease in radiographic exposure with appropriate increase in developing time, within reasonable limits, enhances contrast. For optimum contrast, strict adherence to the correct time-temperature curve is essential. With automated processing, optimum contrast is readily achieved by precise control of time and temperature.

Finally, you should recall that *intensifying screens* convert 98 percent or more of the radiologic image to light. In other words, the aerial image is changed almost completely to a light image which is then recorded by the film. This conversion process enhances contrast because screen-type film has *more inherent contrast* for the light emitted by screens than for x rays directly. In general, contrast increases and latitude decreases with faster film-screen systems.

There is no simple definition of *film* contrast. It can be understood only by studying the so-called *characteristic* or *sensitometric* curve of a particular film recording system, shown in Figure 18.11. This type of curve was first used in photography by Hurter and Driffield and is often referred to as an H & D curve. Film manufacturers use such curves to monitor film quality and consistency. H & D curves show the density response of film to various exposures under experimental conditions. Radiographic density is then plotted as a function of exposure (actually, density versus log relative exposure). Referring to Figure 18.11, we note that at a value of 0.5 on the horizontal axis the beginning of the curve already lies above this axis because of *inherent* or *base film density of 0.2.* As the exposure is increased, film density at first increases gradually with an upward curve in the *toe portion,* then more steeply along a *straight line,* and finally along a *shoulder portion.* The slope of the straight line joining the point on the curve near the toe corresponding to base density 0.2 + 0.25 = 0.45, and the point on the curve near the shoulder corresponding to base density 0.2 + 2.0 = 2.2, has been defined arbitrarily as the *average gradient* (see dashed line in Figure 18.11). The average gradient includes the useful exposure-density range of a film and varies with the type of film. *Those films with greater average gradients show greater contrast for a given subject contrast.* A film with an average gradient greater than 1 will heighten subject contrast.

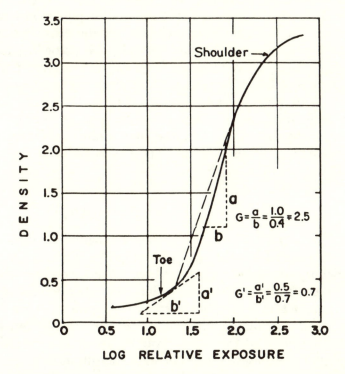

Figure 18.11. Characteristic (H & D) curve of a typical screen-type x-ray film. $G = a/b$, the slope (steepness) of the straight portion of the curve, represents the region of maximum film contrast. $G' = a'/b'$, the slope of one extremely short segment of the toe portion; note that here the slope varies along the curve and is less than that of the straight portion so that contrast is low. In the shoulder portion contrast is also low. Furthermore, the toe portion is a region of under-exposure, and the shoulder a region of overexposure. The dashed line represents the *average gradient* (average slope) and encompasses that portion of the curve corresponding to the useful exposure-density range of the film in question.

The characteristic curve also shows that at low densities (toe portion) and at high densities (shoulder portion) contrast is poor.

Maximum contrast occurs in the straight portion. An average radiograph displays poor contrast in both the lightest (toe) and darkest (shoulder) regions. Furthermore, underexposure exists in the toe region, and overexposure in the shoulder region. It is precisely in these regions where recorded detail is poor.

Most radiographic *technic charts* are based on the different densities of various parts of the body, and on different thicknesses of the same part of the body in different individuals. The thicker or denser the part, the greater is the required exposure in terms of kV, mA, and time. Some technic charts keep the mAs constant for a given region of the body and vary the kV according to part thickness. Others show *optimum kilovoltage* for a given part and vary the mAs. If high kV equipment is available, we can use 100 to 140 kV with grids having a ratio of 12 or 16. High voltage radiography produces radiographs with good definition, wide latitude, and moderate contrast, provided suitable filters and grids are employed; besides, patient exposure is reduced with high kilovoltage exposure.

Soft Tissue Radiography — Mammography

In soft tissue radiography, of which mammography is a specialty, we are dealing with essentially two types of structures — those of *water density* (eg, muscle, glands, fibrous tissue, blood vessels) and *fat*. With conventional exposure factors, these tissues exhibit poor subject contrast. However, by selection of suitable exposure factors we can enhance radiographic contrast between fat and any tissue of water density. But there is no way to distinguish, by ordinary radiography, the various tissues of water density unless they happen to be surrounded by fat, short of using contrast agents. This important principle applies directly in mammography in that lesions such as tumors (water density) can be separated radiographically from the surrounding fat which is often present in sufficient amount to act as a natural contrast agent, especially in older or obese women. Furthermore, since about one third of breast cancers contain microcalcifications (0.05 mm or less) that require viewing with a hand lens, superb recorded detail is essential.

Film Mammography. Since conventional radiography, especially with ordinary films, does not provide the high soft-tissue contrast and excellent recorded detail needed in mammography, we must use special exposure factors and films: (1) low kV, (2) minimum filtration of the beam, (3) small focal spot, and (4) special, fine-grain mammographic film and a single intensifying screen, with increasing use of a 5:1 moving grid.

An extremely important consideration is the *dose* to the breast during mammography. Based on studies of breast cancer incidence in survivors of the atom bomb attack on the Japanese cities (Hiroshima and Nagasaki) by the U.S.A. at the end of World War II, experts *estimate* that approximately 6 excess breast cancers appear per million women per year per cGy (rad) exposure. This incidence is extrapolated (extended) from large doses, downward to vanishingly small doses (ie, non threshold response, according to the Beir Report of 1986). However, when one considers the benefit/risk ratio in terms of lives saved at the extremely small doses used today in mammography, Feig (1977) estimates 2900 lives saved per million women screened at a midplane breast dose of 0.15 cGy (rad) for two views, culminating in a benefit/risk ratio of about 145 to 1 (that is, 145 lives saved at the risk of inducing one cancer).

Detecting the subtle changes in the soft tissues or the tell-tale cancer calcifications, while minimizing radiation exposure, became possible when, in 1966, a new type of radiographic tube was invented in France by J. R. Bens and J.-C. Delarue. Knowing that optimum contrast is achieved in soft tissue radiography with x rays ranging in wavelength from 0.6 to 0.9 Å or average energy of about 17.5 kV, they designed an x-ray tube with a *molybdenum (Mo) target.* They selected Mo because at about 30 kV_p it emits characteristic radiation with energies of 17.14 and 19.5 kV (average 19 kV). Note that the energy of this radiation matches fairly closely the optimum for soft tissue radiography (in fact, it is slightly higher than the optimum 17.5 kV, and this is desirable). Furthermore, they introduced a 0.03 mm *Mo primary filter* into the beam to remove much of the radiation with energy above and below the desired range, thereby producing a *nearly monoenergetic beam.* Using the *same* metal—Mo—for *both* the target and the filter utilizes the principle of the *spectral window;* that is, a filter readily transmits the characteristic radiation produced initially by the same element, while absorbing much of the general radiation (continuous spectrum). Note that a tungsten target emits only a continuous spectrum (ie, no characteristic radiation) at 30 kV;

although it is nearly monoenergetic, it is of such low intensity that a large mAs must be used.

Mammographic tubes are generally available with rotating anodes having dual 0.1 mm, and 0.3 or 0.4 mm focal spots to provide exquisite recorded detail. These tubes have molybdenum filters and beryllium windows, as explained above. Some tubes have hooded anodes to minimize off-focus radiation (see Figure 18.12).

Molybdenum targets are now standard in low-dose film mamography. Although tungsten targets are preferred for xeromammography, some experts accept molybdenum-target tubes for this purpose. It is now mandatory for all mammography to be performed with dedicated units. In fact, one should also use a *special film processor* to maintain optimal radiographic quality and minimize film artefacts.

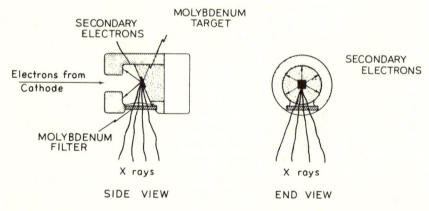

Figure 18.12. Stationary anode tube with a hooded molybdenum target, which helps minimize off-focus radiation by trapping scattered and secondary radiation. (*Courtesy of CGR Medical Corporation.*)

Optimal technique is essential in low-dose mammography. Not only does it demand the use of dedicated x-ray units with molybdenum targets and specifically designed mammographic film-screen combinations, but equally important, it requires *correct positioning* of the breast together with firmly applied *compression* to reduce its thickness. The latter entails two highly beneficial effects: (1) a diminution in scattered radiation, a significant noise factor that impairs the already low soft-tissue contrast, and (2) a reduction in breast glandular dose due to the lower exposure factors. All mammographic technologists should have special training in this field before assuming the responsibilities of a mammographer.

Before moving on, you should turn to pages 257–260 for a discussion of high-frequency generators, and pages 435–438 for xeroradiography.

DISTORTION

A radiographic image does not faithfully represent the anatomic part, but differs from it in varying degrees of size and shape. Such *misrepresentation of the true size and shape of an object is called distortion.* The amount of distortion depends on several factors, to be discussed in this section. While distortion generally has a degrading effect on radiographic quality, we cannot completely eliminate it because the image lies in a single plane and is two dimensional, whereas the object, being solid, is three dimensional. In fact, we often use distortion deliberately to bring out a structure that would otherwise be hidden, for example, in oblique radiography of the gall bladder to separate it from the vertebral column.

The two kinds of distortion—*size* and *shape*—will now be described.

1. **Size Distortion—Magnification.** X rays emerge from the tube focus in all directions, but the tube port delimits them to form a beam that is circular in cross section. As the beam advances it broadens, becoming cone shaped. Finally as it passes through the collimator, the x-ray beam acquires a rectangular cross section.

Geometric rules prevail in the realm of x-ray image formation just as they do in photography with light. We can easily show that when an object is held between a source of light and a white surface, the size of the shadow enlarges as the object is moved nearer the light, and shrinks as it is moved closer to the white surface. This is due to the *divergence of image-forming light rays in* the beam. In Figure 18.13 the divergent beam of light *magnifies* the shadow more in *A* than in *B;* that is, the shorter the distance between the object and the source of light the greater the magnification. *The law of image magnification states that the width of the image is to the width of the object, as the distance of the image from the light source is to the distance of the object from the light source.*

$$\frac{\text{image width}}{\text{object width}} = \frac{\text{image distance}}{\text{object distance}} \qquad (6)$$

Precisely the same law applies to radiographic image formation. For example, in cardiac radiography the standard procedure is to use a 2-meter (6-ft) focus-film distance to minimize size distortion or magnification of the heart, so that measurement of the transverse cardiac diameter on the radiograph gives virtually the true diameter. On the other hand, contrary to what is often taught, the cardiothoracic ratio, or the ratio of the transverse cardiac diameter to the transverse thoracic diameter, undergoes no significant change as the SID is varied between 100 cm (40 in.) and 1.8 m (72 in.), if the patient's position remains constant. This results from the fact that as the SID is changed, these two diameters change at virtually equal rates, as shown in Figure 18.14.

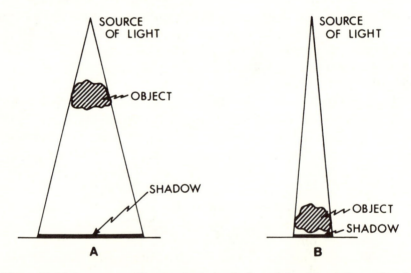

Figure 18.13. Image magnification. In A the shadow of the object is larger than in B, because in A, the object is nearer the source of light. The same principle applies to the formation of x-ray images.

Thus, magnification is unavoidable in radiography because of the geometric nature of image formation. As a rule, *magnification or size distortion can be decreased either by reducing the OID or by increasing the SID.*

Radiographic magnification is easily derived from the geometric relationship of similar triangles (see pages 17–19). For example, suppose that an object measures 15 in. (37.5 cm) in diameter and lies 10 in. (25 cm) above the level of the film. If the SID is 40 in. (100 cm), what is the magnification of the radiographic image? We must first determine the width of the image by constructing a diagram as shown in Figure 18.15 and applying proportion (6):

$$\frac{\text{image width}}{\text{object width}} = \frac{\text{SID}}{\text{SOD}}$$

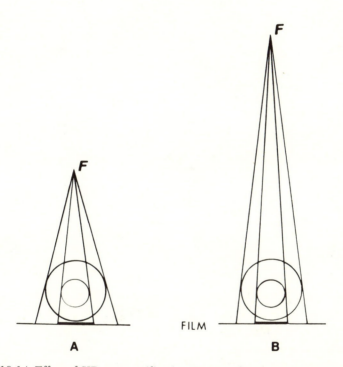

Figure 18.14. Effect of SID on magnification. In *A*, with a shorter SID, magnification is greater than in *B*. However, in this example, the *ratio* of the diameter of the *image* of the inner circle to that of the outer circle is essentially the same in both *A* and *B*.

$$\frac{\text{image width}}{15} = \frac{40}{30}$$

$$\text{image width} = \frac{40 \times 15}{30} = 20 \text{ in.}$$

a. The *linear magnification of the image,* or the enlargement of its width relative to that of the object, may now be expressed in one of two ways:

a. *Magnification Factor.* This is defined as the ratio of image width to object width:

$$\text{magnification factor} = \frac{\text{image width}}{\text{object width}} \qquad (7)$$

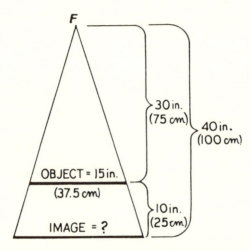

Figure 18.15. Magnification of the image in radiography. Image width/object width = 40/30.x/15 = 40/30.x = 20 in.

In the above example,

$$\text{magnification factor} = \frac{20}{15} = 1\frac{1}{3}$$

b. *Percentage Magnification.* We define this as the difference between image width and object width, relative to object width, expressed in percent.

$$\% \text{ magnification} = \frac{\text{image width} - \text{object width}}{\text{object width}} \times 100 \qquad (8)$$

In the above example,

$$\text{percentage magnification} = \frac{20 - 15}{15} \times 100$$

$$= \frac{1}{3} \times 100$$

$$= 33\frac{1}{3}\%$$

The same geometric law of image formation gives the *diameter of the object* being radiographed when the image diameter, the SID and the OID are known—draw a diagram as in Figure 18.15, insert the known values, and find the object diameter by using the proportion in equation (6).

2. **Shape Distortion.** This is caused by misalignment of the central ray, the anatomic part, and the film. In Figure 18.16, an oval piece of lead placed in the x-ray beam nonparallel to the film, casts a circular shadow. A rectangular piece of lead can be placed in a beam so that it casts a square shadow. With these objects placed perpendicular to the central ray and parallel to the film, the images will, of course, show no appreciable shape distortion and appear oval and rectangular, respectively. Under certain conditions, deliberate distortion may bring out parts of the body that are obscured by overlying structures; for example, in radiography of the gall bladder the patient is often rotated to displace the gall bladder shadow away from the vertebral column.

We can summarize the data on distortion as follows: *there are two types of distortion—size distortion and shape distortion. Size distortion* is magnification caused by progressive divergence (spread) of the image-forming x rays in the beam: the shorter the OID and the longer the SOD, the less the degree of size distortion.

Shape distortion results from improper alignment of the central ray, object, and film. It is, of course, possible to have shape and size distortion occurring together if the causative factors of both conditions are present.

Since distortion (both size and shape) and recorded detail are influenced by the same factors—SID and OID—the greater the distortion of the radiographic image the poorer the recorded detail (see Figures 18.4 and 18.16). You can readily demonstrate this principle by observing the shadow cast by an object such as a pencil on a white surface. As the pencil is moved toward the light (away from the white surface), the shadow not only becomes larger (magnification distortion) but also becomes more blurred (poorer recorded detail). As the pencil is moved nearer the white surface, the shadow becomes smaller and sharper.

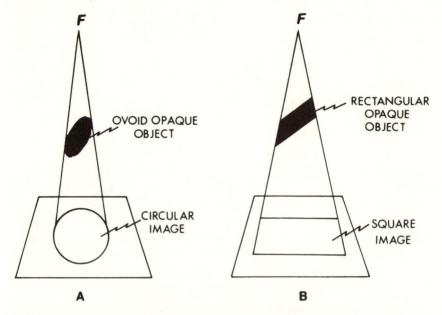

Figure 18.16. Shape distortion. In *A* the projected image of an oval opaque object is circular because the object is not parallel to the film. In *B* the image of a rectangular opaque object is almost square.

Direct Magnification or Enlargement
Radiography (Macroradiography)

We have already learned that as we move an object away from the film, toward the x-ray tube, image magnification increases if the focus-film distance remains unchanged (see Figure 18.15). At the same time, blurring increases and recorded detail worsens (see Figure 18.4). If we could preserve recorded detail while achieving magnification, we should have a satisfactory method of *direct magnification radiography*, also known as *macroradiography.*

The introduction of tubes with *fractional focal spots*, that is, *microfocus tubes*, confirmed the feasibility of producing directly enlarged radiographic images. Although the 0.3 mm focal spot initially seemed to give reasonably good results insofar as image quality was concerned, subsequent work indicated that even smaller focal spots were needed to *minimize geometric blur and enhance recorded detail* during magnification to produce images with truly superb resolution. In fact, with two-diameter magnification a focal spot larger than an anatomic object (eg, tiny blood vessel) will generate blur wide enough to wipe out the image of the detail as shown in Figure 18.17. This explains why a focal spot no larger than 0.3 mm should be used in macroradiography.

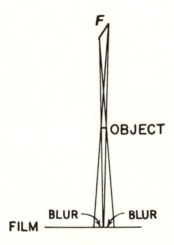

Figure 18.17. When the object is smaller than the effective focal spot, blur *B* dominates the image, so the true image or umbra disappears. This explains why, in two-diameter macroradiography, vessels smaller than the focal spot disappear in an angiogram and also emphasizes the importance of using the smallest practicable focal spot when utmost recorded detail is sought.

Tubes have been designed with focal spots as small as 0.1 mm and, in fact, some experts believe that really high quality radiographic enlargement cannot be achieved with focal spots larger than 0.2 mm (Rao and Clark). However, their use in radiography is still severely restricted by the very limited heat loading capacities of these tubes.

In practice, two times linear (four times area) magnification is most often used. This requires placing the *anatomic part or object midway between the tube focus and the film* (see Figure 18.18). There we see that the magnification factor depends on the following relation:

$$\text{magnification factor} = \frac{\text{source-to-image recptor distance}}{\text{source-to-object distance}} \quad (9)$$

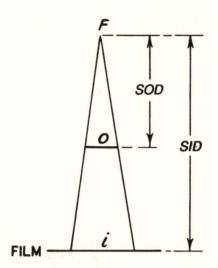

Figure 18.18. Principle of radiographic magnification. *Image size/object size = SID/SOD*; or, magnification factor $m = i/o = SID/SOD$.

or

$$m = \frac{SID}{SOD} \tag{10}$$

Since, in Figure 18.19, SOD is equal to ½ SID, substituting the value ½ SID for SOD,

$$m = \frac{SID}{\frac{1}{2}SID} = 2$$

At the same time, Figure 18.19 shows that the geometric blur B (penumbra) at two times magnification equals the effective focal spot size.

It should be noted that the discussion thus far has dealt with *transverse magnification;* that is, the magnification of an object lying in a plane parallel to the film. On the other hand, the concept of *longitudinal magnification* (Doi and Rossman, 1975) refers to magnification of an object lying in a plane perpendicular to the film, relative to the size of its image in conventional radiography. Important for us is the fact that longitudinal magnification increases the resolution of vertically oriented objects by increasing their separation in the magnified image. Thus, two objects lying above

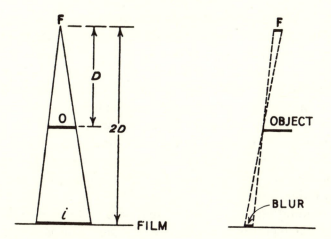

Figure 18.19. In the diagram on the left, the object is located midway between the tube focus and the film. Therefore, the magnification factor $m = 2$, because $m = 2D/D = 2$. In the diagram on the right, blur B equals the effective focal spot size when m = 2. o = object. i = image.

each other would show greater image separation (resolution) during magnification than they would in ordinary radiography (see Figure 18.20).

It is obvious that an appreciable *air gap* must exist between the

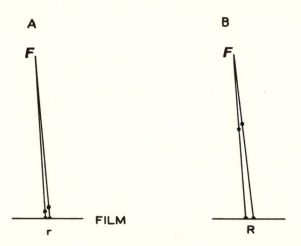

Figure 18.20. In A, transverse magnification is less than in B. At the same time, longitudinal magnification, or the resolution of two points lying at different distances above the film, is greater in B. (R in B is larger than r in A.)

object and the image receptor (film-screen system) in magnification radiography (see pages 352–354). For example, the SID is normally 40 in. (100 cm), so that with a magnification factor of two, an air gap of 20 in. (50 cm) will be present between the object and the film (neglecting the thickness of the object itself). In fact, a 20 in. (50 cm) air gap provides cleanup of scattered radiation as well as a 15 ratio grid. Therefore, we can omit the grid in magnification radiography. At the same time, close beam limitation and precise alignment to the anatomic area of interest are essential to high quality magnification radiography, just as in ordinary radiography.

Bear in mind that bringing the anatomic part closer to the tube increases the exposure at the level of the part. This is governed by the inverse square law; for example, with a magnification factor of two, the FOD is decreased by one-half relative to the usual SID in ordinary radiography. Reduction of the distance by one-half increases exposure four times = (magnification factor)^2 = 2^2. However, this applies only to the center of the object; the tube side of the patient will receive an even greater increase in exposure. But the increased dosage is partly offset by dosage reduction through elimination of the radiographic grid in the presence of the air gap.

Insofar as *screen* and *film blur* are concerned, the latter is not significant. Much more important is screen blur resulting from diffusion of light in the active layer. Fortunately the *intensifying screen becomes more efficient when the same information in the enlarged image is spread over a larger area,* thereby enhancing recorded detail beyond that in conventional radiography.

Although the above discussion of macroradiography would seem to indicate that the last word has been written on the subject, you should know that experts differ as to its value in routine radiography. For example, convincing data by Rao and Clark favor the use of detail screens or nonscreen technic in the radiography of *small* parts that can be well immobilized, the image being viewed with the aid of an excellent convex hand lens. This gives a degree of resolution approaching that in macroradiography. At any rate, there is general consensus that tubes with focal spots larger than 0.3 mm should not be used in magnification radiography. Full advantage of the fractional focus tubes prevails only in close collimation to the area of interest, close apposition of the beam-limiting device to the tube aperture, and accurate centering of the beam.

As noted above, excellent recorded detail occurs with nonscreen or detail screen exposures for small parts, using focal spots 0.3 mm or smaller and optical magnification with a hand lens.

In general, we can say that the advantages of magnification radiography, where applicable, depend on three factors:

1. *Sharpness factor* — the magnified image is spread over a larger area of the intensifying screens and therefore exhibits better recorded detail than the corresponding nonmagnified image.

2. *Noise factor* — while the image is magnified, screen mottle is not. Thus, recorded detail improves (greater resolution) but the noise factor (screen mottle) remains unchanged, so we say that the *signal-to-noise ratio is increased.*

3. *Visual factor* — visual recognition of images improves with magnification.

We may therefore conclude that the advantages of radiographic magnification in certain kinds of examinations outweigh the disadvantages, such as the limited heat-loading capacity of microfocus tubes, the slightly increased geometric blur, and the limited field size. Macroradiography will therefore continue to play an important part, especially in such procedures as angiography. In fact, the perfection of tubes with focal spots as small as 0.1 mm having an adequate tube rating and the introduction of faster screens should result in even wider application of direct magnification radiography in the future.

MODULATION TRANSFER FUNCTION

The term *modulation transfer function* (*MTF*) often appears in textbooks and literature dealing with radiographic quality. The technologist should therefore be aware of its significance, at least in a general way. This has prompted a short, simplified treatment of the subject.

As noted earlier, the recording of a radiologic image by any type of image receptor (radiography, image intensification, TV, etc) entails the transfer of information from the object to the image. In the process there is inevitably some loss of information, such as recorded detail. The MTF represents a mathematical method of expressing the efficiency of any imaging system, that is, the part of the information put into the system, which becomes recognizable as the image. In a practical sense, then, *MTF may be regarded as a measure of imaging system quality.*

It should be pointed out that we can obtain MTFs for the individual components of an imaging system; for example, in radiography we can derive the MTF separately for the focal spot, geometry, film-screen system, motion of the object, etc. Multiplication of the individual MTFs gives the total MTF of the system:

$$MTF_{total} = MTF_1 \times MTF_2 \times MTF_3 \times \ldots \ldots \times MTF_n$$

How is MTF obtained? It turns out that while this involves a complex mathematical procedure, we can still gain sufficient practical knowledge about it without necessarily understanding all of the involved mathematics. Basically, MTF is derived from what is known as *line spread function (LSF)*. If we were to expose *directly* (without screens) a film to an extremely narrow beam of x rays collimated by a slit about 10 microns wide, we would obtain a sharp line image. However, a film-screen combination exposed to the same beam would record a line having blurred edges due to diffusion of light in the active layer of the screen. The LSF represents the width of this line as measured by a microdensitometer, with relative density plotted as a function of distance from the center of the line (see Figure 18.21).

By a complex mathematical process known as Fourier analysis, LSF may be converted to MTF which is then plotted (as in Figure 18.22). Here MTF on the vertical axis receives a maximum value of one, since the ratio of the amount of information recorded in the image to the amount put into the system can never exceed unity, and is usually much less. In the examples shown in the figures, we are considering two different film-screen systems. You can see from Figures 18.21 and 18.22 that the superior film-screen system insofar as image quality (sharpness, resolution, and contrast) is concerned, has a narrower LSF, and an MTF that is higher at larger "spatial frequencies" (sine functions analogous to the number of line pairs resolved per mm). The MTF for other components of the imaging system can be obtained by mathematical transformation of the appropriate LSF.

At present, MTF is the best available means of evaluating the quality performance of an imaging system, derived from the MTFs

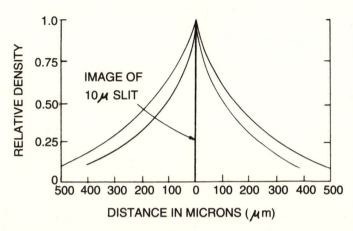

Figure 18.21. Hypothetical example of line spread function for high speed (upper curve) and medium or par speed (lower curve) screens. The image of the slit with high speed screens is wider, giving poorer resolution, than with par speed screens. *Adapted from K. Rossman, courtesy of Radiology.*

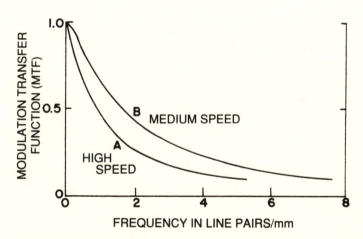

Figure 18.22. Hypothetical modulation transfer function (*MTF*) for high speed (*A*) and medium or par speed (*B*) screens. Par speed screens have a larger *MTF* for a particular value of resolution (spatial frequency in terms of line pairs/mm) and are able to resolve more line pairs/mm than high speed screens (*MTF* curve *B* extends to higher values of resolution).

of its constituent parts. As time goes on, we anticipate a growing use of this concept in the specification of radiographic equipment and image receptors, both as a whole and with regard to their individual components.

QUESTIONS AND PROBLEMS

1. Define radiographic quality. How does it relate to recorded detail?
2. What is meant by sharpness; blur; acutance; resolving power; image receptor? How are they related?
3. List the five factors that affect recorded detail.
4. What are the three factors in geometric blur? Explain each one.
5. State the equation that relates the factors in geometric blur.
6. Describe briefly the two main methods of evaluating focal spot size.
7. Define focal spot blooming. What exposure factor influences the degree of blooming?
8. How important is motion of the part being radiographed insofar as recorded detail is concerned? How can motion be controlled?
9. Discuss the factors of crystal size, active layer thickness, film-screen contact, and mottle as they affect blur.
10. How does object shape influence blur?
11. Define radiographic density. Explain the five factors that determine it.
12. State the inverse square law in your own words, and the two forms of the applicable equation. How is it modified in radiographic practice?
13. At a point 50 cm from the tube focus the exposure rate of an x-ray beam is 20 R per min. What will the exposure rate be at 25 cm? 100 cm?
14. A certain technic calls for 50 mA and 0.1 sec at 65 kV. If we have an uncooperative patient and wish to reduce motion by using an exposure time of 1/60 sec, what will the new mA value be?
15. What is meant by the penetrating ability of an x-ray beam? How can the penetrating ability of a beam be changed?
16. A certain technique calls for an exposure of 70 kV and 180 mAs at a distance of 60 in. (150 cm). If we wish to decrease

the distance to 20 in. (50 cm), what new mAs value will be required to maintain the same radiographic density?

17. How do intensifying screens affect density? contrast? recorded detail?

18. List the main materials in the human body in decreasing order of their radiographic density.

19. Define radiographic contrast. On what two kinds of contrast does it depend?

20. What is the function of radiographic contrast? Can there be too much contrast?

21. How are the following terms related: low contrast; high contrast; long scale contrast; short scale contrast?

22. Define subject contrast. How is it related to the aerial image? remnant (exit) radiation?

23. What is film contrast? Draw and explain the characteristic curve of a film. Discuss average slope (average gradient).

24. Name the factors that influence subject contrast.

25. Discuss the advantages of a molybdenum anode in mammography. With which type of image receptor should it be used?

26. Describe the xerographic process. What advantages does it offer in mammography?

27. How do xerographic images differ from film images?

28. Define distortion. Name and explain the two kinds.

29. What effect does distortion have on recorded detail?

30. How can size and shape distortion be minimized?

31. An anatomic part is located 4 in. (10 cm) above the image receptor (film-screen system). The SID is 40 in. (100 cm), and the diameter of the anatomic part is 9 in. (23 cm). Find the size of the image, percent magnification, and magnification factor.

32. What is quantum mottle? State its importance in radiography. Under what conditions is it accentuated?

33. Discuss the principle of direct magnification radiography (macroradiography). How do we compensate for the increased blurring inherent in this system?

34. How does magnification radiography affect patient exposure?

35. Why can the radiographic grid be eliminated in magnification radiography? Is this always desirable?

36. On what is modulation transfer function based? What does it measure?

Chapter 19

DEVICES FOR IMPROVING
RADIOGRAPHIC QUALITY

I N THE PRECEDING chapter we described the important factors affecting radiographic quality. We shall now show how the quality of a radiograph can be improved by the use of various special devices. Recall that recorded detail is the ultimate criterion of radiographic quality.

As already noted, the rotating anode tube with its small focus provides excellent recorded detail. Furthermore, this tube has made possible the use of high mA with very short exposures, thereby minimizing motion, the greatest enemy of recorded detail. Thus, while the inherent construction of modern equipment should provide radiographs of superb quality, certain auxiliary devices are needed to reduce *scattered radiation,* a form of noise that impairs image quality by reducing contrast and obscuring recorded detail.

SCATTERED RADIATION

You should recall from Chapter 12 that the primary x-ray beam leaving the tube focus is polyenergetic in that it contains photons of various energies. The primary beam consists of *brems radiation* resulting from the conversion of the energy of the electrons as they are stopped by the target, and *characteristic radiation* emitted by the target metal due to excitation of its atoms. As the primary beam passes through the patient, some of the radiation is absorbed, while the rest is scattered in many directions. In the diagnostic range of 30 to 140 kV the scattered radiation generated in the body consists mainly of scattered photons produced by Compton interaction, but also includes characteristic radiation resulting from photoelectric interaction. Recall that the energy of the Compton (scattered) photons increases with increasing kV. Characteristic photons have extremely low energy and are absorbed locally in the tissues, but *many of the Compton (scattered) photons have enough energy to pass through the body and approach the film from many directions.*

This multidirectional scattered radiation is a *noise factor* which seriously impairs radiographic quality by its fogging effect, diffusing x rays over the surface of the film and thereby *lessening contrast* (see Figure 19.1). You can see that the larger the percentage of scattered radiation relative to the primary radiation, the greater will be the loss of contrast of a detail such as *P*. In a radiograph of good quality, less than one-fourth the density should result from scattered radiation.

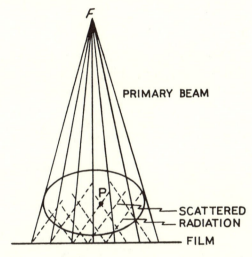

Figure 19.1. Deleterious effect of multidirectional scattered radiation on radiographic quality. This causes *loss of radiographic contrast* of a detail such as *P*. *F* is the focal spot.

The ratio of scattered radiation to primary radiation is not uniform over the surface of the radiograph. In fact, with relatively opaque objects, the ratio increases toward the image border, so that scattered radiation tends to *degrade edge contrast* (Jaffe and Webster).

Of the total radiation reaching the film, the *ratio of scattered radiation relative to primary radiation increases with the following factors:*

a. An *increase in the area of the radiation field and the thickness of the part* traversed by the beam. As the beam increases in cross section it irradiates a larger volume of tissue and generates more scattered radiation.

b. An *increase in tube potential.* As the kV is increased, there is an increase in the probability of Compton interactions *relative to* photoelectric interactions (see page 183); recall that the Compton interaction produces scattered radiation. Furthermore, at higher kV the 45° scattered photons, because of their greater energy and penetrating ability, are more likely to reach the film. In general, the energy of the scattered photons is only slightly less than that of the primary photons, as shown in Table 19.1. Finally, an increase in the applied kV causes more and more of the radiation to scatter in a forward direction, that is, at an angle of 45° or less with relation to the direction of the primary beam. Thus, increasing kV increases not only the energy of the scattered photons but also the number directed toward the film, so that a larger fraction of the scattered radiation can reach the film and reduce contrast.

Table 19.1
RELATIONSHIP BETWEEN PRIMARY PHOTON ENERGY
AND ENERGY OF PHOTONS SCATTERED AT 45°

Primary Photon	45° Scattered Photon
keV	keV
40	39
60	58
80	76
100	94
120	112

c. An *increase in the density of the tissues* traversed by the primary beam; for example, as between air and soft tissue. In nongrid exposures of the chest and abdomen, the following are approximate relative values of the scattered and primary radiation reaching the film:

chest   50% scattered   50% primary
abdomen   90% scattered   10% primary

REMOVAL OF SCATTERED RADIATION BY A GRID

Since scattered radiation cannot be entirely eliminated because of the very nature of x rays and their interaction with matter, we can only attempt to remove as much of it as possible once it has been produced, so as to improve radiographic contrast. This is especially important in the radiography of large anatomic areas such as the abdominal viscera, because of the greater amount of scattering by such large volumes of tissue. Most effective in removing scattered radiation are the *stationary grid* and the *moving grid.* Although the latter is by far the more frequently used of the two, the stationary grid will be discussed first because of its historical priority and relative simplicity.

Stationary Grid

In 1913 Gustav Bucky introduced the *stationary radiographic grid,* a device placed between the patient and the cassette for the purpose of reducing the amount of scattered radiation reaching the film and so *improving radiographic contrast.* As shown in Figure 19.2, Bucky's original grid was of the *cross hatch type,* consisting of wide strips of lead arranged in two parallel series, which are mutually perpendicular. Despite the coarseness of the early grids, with their 2 cm × 2 cm spacing, they removed a significant amount of scattered radiation, as shown by the frontal skull radiograph at the Smithsonian Institution.

A *modern stationary grid* has a very different appearance (see Figure 19.3). It consists of extremely thin, closely spaced lead strips measuring about 0.05 mm in width. They are separated by radiolucent material, usually plastic or aluminum, measuring about 0.33 mm wide. *Aluminum* is preferred for improved durability of the grid, and to absorb scattered radiation from the lead strips. A grid is manufactured by stacking and bonding together alternate sheets of lead and spacer material of appropriate thickness like the strips just mentioned. The stack is then sliced crosswise, the thickness of each slice representing the height of the grid. Thus, each slice is a grid with alternating thin strips of lead and aluminum (or plastic).

Figure 19.4 shows an enlarged diagram of a grid. Only those x-ray photons passing directly through the narrow aluminum spacers

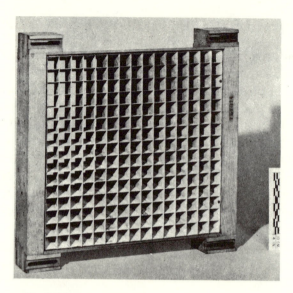

Figure 19.2. Photographic reproduction of Bucky's original radiographic grid. Note the cross-hatch pattern and the very coarse spacing. Yet, this grid achieved significant cleanup of scattered radiation, although a pattern of the lead strips appeared on the radiograph. (*Photograph [No. 47022] furnished by courtesy of Smithsonian Institution.*)

will reach the film-screen system. Since the scattered photons proceed at various angles, most of them will strike the lead strips and undergo absorption without reaching the film. Such a grid may absorb as much as 90 percent of the scattered radiation and entail a notable improvement in contrast and recorded detail.

The stationary grid has one main disadvantage—as the lead strips absorb radiation, they cast shadows on the radiograph as thin white lines; but at the usual radiographic viewing distance, and especially with high quality grids, these lines are almost invisible. Besides, their presence is more than compensated for by the improved quality of the radiographic image. Because of the interposition of the grid in the x-ray beam, an appreciable part of *both primary and scattered radiation is absorbed,* and therefore the exposure must be increased according to the type of grid being used.

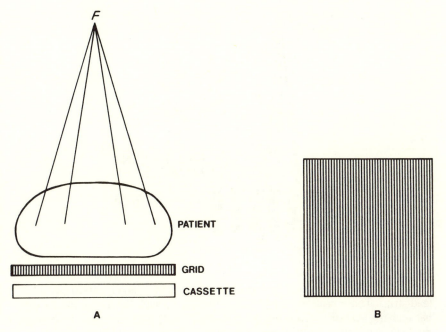

Figure 19.3. Stationary parallel or nonfocused grid, seen in cross section in *A* and top view in *B*. *F* is focal spot of x-ray tube.

Efficiency of Radiographic Grids

Certain criteria have been established to designate the quality or efficiency of radiographic grids; that is, *their ability to remove scattered radiation,* or *"cleanup."* These apply to both stationary and moving grids (to be described later), and include *physical* and *functional* aspects.

Physical Factors in Grid Efficiency. The three physical factors mainly responsible for grid efficiency are *grid ratio, grid frequency,* and *lead (Pb)* content.

1. *Grid Ratio* is defined as *the ratio of the height of the lead strips to the distance between the strips,* as expressed in the following equation:

$$r = \frac{h}{D} \qquad (1)$$

where *r* is the grid ratio, *h* is the height of the lead strips, and *D* is

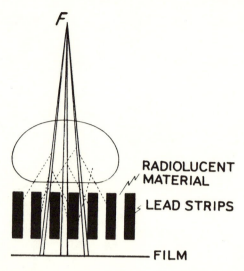

Figure 19.4. Principle of the radiographic grid. Scattered radiation, represented by the dotted lines, proceeds in various directions and is largely absorbed by the lead strips. Those primary rays that pass directly through the aluminum spacers between the lead strips reach the image receptor (ie, film-screen system). Although the lead strips are very thin, any primary rays striking them are absorbed—an effect known as *cutoff.*

the width of the spaces between the strips. These factors appear diagrammatically in Figure 19.5. As an example, if the lead strips are 2 mm high and the space separating them is 0.4 mm, the grid ratio is 2/0.4 = 5:1.

Notice in Figure 19.5 that grids A and B have the same ratio, despite the difference in their heights. Grid D has the same height as B, but D's ratio is greater because its lead strips are closer. Further comparison of the grids in Figure 19.5 reveals that the paths of the most divergent scattered photons in grids A and B make the same angle with the lead strip, whereas they make half that angle in grid D. Thus, the higher the grid ratio, the "straighter" the rays have to be to get through the grid interspaces. It is like a line of cars passing through a tunnel; any car that strays too far off the road will hit the side of the tunnel. All other factors being equal, the higher the grid ratio, the better will be the "cleanup" of scattered radiation and the resulting radiographic contrast, with improved recorded detail.

2. **Grid Frequency,** defined as the number of lead strips per cm

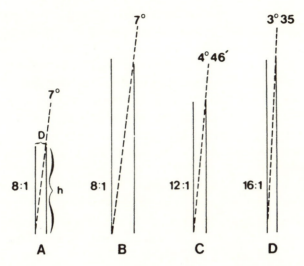

Figure 19.5. Effect of grid ratio on maximum angle of obliquity of a scattered ray that can get through the grid. In *A* and *B*, it is obvious that the ratio may be the same in grids of different thickness, and 7° is the maximum angle that a ray can make with the vertical and still pass between the lead strips. In *D*, with the grid ratio twice that in *A*, the maximum obliquity of a transmitted ray, 3° 35′ is half as large as that in *A*. Note the smaller frequency of grid B relative to grid A despite their identical grid ratio.

or per inch, has an important bearing on grid efficiency. Available grids range in frequency from 20 to 40 lines per cm (50 to 100 lines per in.). In general, the greater the grid frequency, the thinner the lead strips have to be, and the greater will be the likelihood of scattered photons passing *laterally* through the strips, especially at high kV. With grids of equal ratio, the one having fewer and thicker strips per cm possesses superior quality but at the same time the visibility of the grid lines becomes more objectionable (except in moving grids). Conversely, *as the grid frequency increases, the grid ratio must also be increased to maintain the same efficiency.*

Closely related to grid ratio and frequency is the **lead content** in terms of mass per unit area—g/cm² —of grid surface. For grids of equal ratio, the one with thicker strips (lower frequency) will be more efficient in removing scattered radiation and improving contrast, by virtue of its greater lead content. But, at the same time, it must be kept in mind that it will also remove more primary radiation and so require a larger exposure for the same film density. In general, lead content serves as an indicator of performance if the grid has an adequate ratio and frequency.

Functional Factors in Grid Efficiency. Knowing that a grid has a particular ratio and frequency (as well as lead content), how can we judge its ability to minimize scattered radiation? In other words, is there a quantitative measure of grid efficiency based on its actual function? Fortunately, there are several criteria of grid performance, the best ones being *selectivity* and *contrast improving ability,* as defined by the ICRU in *Handbook 89.*

1. *Selectivity.* Since the grid lies in the path of the x-ray beam, its lead strips absorb a fraction of *both* primary and scattered radiation. The absorption of primary radiation by a grid is known as *grid cutoff.* Obviously, a grid should transmit as large a fraction of the primary radiation as possible, to keep the exposure of the patient small. At the same time, it should absorb a maximum of scattered radiation to provide the best possible improvement in radiographic contrast. As specified by the ICRU, if Greek letter Σ (sigma) represents selectivity, T_p the fraction of *primary* radiation transmitted through the grid, and T_s the fraction of *scattered* radiation transmitted through the grid, then

$$\Sigma = \frac{T_p}{T_s} \tag{2}$$

Thus, the larger the ratio of the transmitted fraction of the primary radiation, to the transmitted fraction of the scattered radiation reaching the film, the greater will be the selectivity of the grid. At the same time, it must be remembered that the larger the fraction of the *total radiation* (primary plus scattered) that is transmitted and reaches the film, the less will be the exposure required for a given radiographic density.

2. *Contrast Improvement Factor.* The ultimate criterion of the efficiency of a grid is its actual performance in improving radiographic contrast, as explained earlier in this chapter. The contrast improvement factor, *K,* is defined as the ratio of the radiographic contrast with grid divided by the contrast without grid:

$$K = \frac{\text{radiographic contrast with grid}}{\text{radiographic contrast without grid}} \tag{3}$$

Unfortunately, this factor is so dependent on kV, field size, and thickness of part that it cannot be specified for a given grid under all conditions. (The ICRU has given special recommendations for measuring K in a 20-cm thick water phantom at 100 kV.)

We must emphasize that selectivity and contrast improvement factor have certain limitations because they are measured in a water phantom, which is inherently different from the human body. In making these measurements, certain conditions of standardization have to be laid down, and these may or may not hold during practical radiography. However, selectivity and contrast improvement serve as useful criteria for the intercomparison of grids.

TYPES OF RADIOGRAPHIC GRIDS

Stationary Grids. These thin *wafer grids* come in various sizes and have to be taped to the front of the cassette. Because they are easily damaged—a slight bend will impair their quality—these grids are now built into the cassette front. Moreover, because grid technique requires increased exposure factors, grids must be used with intensifying screens. The thinness of the grid does not appreciably increase OID, so it results in negligible blurring of the image. The cassette with a built-in grid is called a *grid cassette.*

A special *crossed grid* is available, consisting of two sets of grid lines arranged perpendicular to each other. They are not often used in general radiography but may be advantageous in special procedures such as cerebral angiography with biplane equipment. A more versatile arrangement is the superimposition of two lower-ratio grids, one of which is turned 90° relative to the other; the grid ratio of such a combination of crossed grids equals the sum of the individual grid ratios. For example, crossing grids with ratios of 5:1 and 8:1 give a cross-grid combination with a ratio of 13:1, although centering of the x-ray beam need not be any more perfect than with an 8:1 grid (Cullinan). On the other hand, the grids can be used separately when a lower grid ratio suffices. Crossed grids do not lend themselves to angled beams because of excessive cutoff.

Grids are used with intensifying screen cassettes because of the relatively heavy exposures that would otherwise be required. Because the grid is very thin and does not appreciably increase the object-film distance, it causes little or no distortion of the radiographic image. The **grid cassette,** whose front has a built-in grid, is much more convenient to use than an ordinary cassette with a separate grid.

1. **Parallel or Nonfocused Grid.** In this type of grid, seldom used now except in fluoroscopy, the lead strips are all oriented parallel to one another. When such a grid is used at too-short SIDs (see Figure 19.6) the more slanting rays toward the periphery of the beam strike the sides of the more peripheral lead strips. This causes progressively greater absorption of primary radiation toward the edges of the grid, with corresponding underexposure of a radiograph toward its sides relative to its center. What has just been described is an example of **peripheral cutoff,** caused by **distance decentering** of the grid. To minimize this problem, parallel grids should be used at a distance of 48 in. (125 cm) or more with films larger than 8 in. × 10 in., because the rays are practically parallel at distances beyond 48 in.

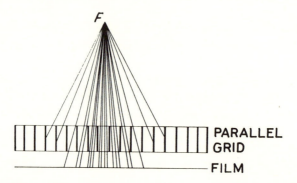

Figure 19.6. The effect of *distance decentering* of a *parallel* grid that was caused by excessively short focus-film distances (less than 48 in. or 125 cm). Because of the divergence of the rays in the beam, the more slanting rays toward the edges of the beam will undergo greater absorption by the grid than will the central rays. Therefore, the lateral portions of the radiograph will be progressively more underexposed relative to its center. This is *symmetrical grid cutoff.*

If the beam is not perpendicular to the grid, with a tilt across the long axis of the lead strips, progressively more radiation is absorbed by the grid in the direction of the beam as shown in Figure 19.7. This is another example of **peripheral cutoff,** but it is now asymmetrical in that the side of the radiograph toward which the beam was angled is underexposed. Tilting of the beam across the lead strips as just described may be called *angulation decentering*. However, with the beam tilted parallel to the long axis of the strips, density will not vary across the radiograph, except for the heel effect (see pages 394–397).

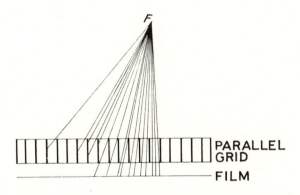

Figure 19.7. The effect of *angulation decentering* of a *parallel* grid, which was caused by angulation of the beam across the long axis of the lead strips. In this case, the radiograph will be underexposed on the side toward which the beam is directed. This is *asymmetrical peripheral cutoff.* Angulation decentering is also called *off-level decentering.*

2. *Focused Grid.* This most frequently used type of grid comprises parallel strips that slant more and more toward the lateral edges of the grid as shown in Figure 19.8. If one extends the planes of these tilted strips upward, these imaginary planes intersect along a line called the *convergence line*, which is parallel to the grid surface. Naturally, when seen end-on as in Figure 19.9, the convergence line appears as a point. The vertical distance between the convergence line C and the center of the grid is called the *focusing distance* f_o, a term that must *not* be confused with focus-film distance.

Moving Grid

To eliminate the grid lines produced in a radiograph by a stationary grid, Dr. Hollis Potter, in 1920, conceived the idea of *moving the grid* between the patient and the film during x-ray exposure, in a direction perpendicular to the lead strips. The moving grid has come to be called, loosely, a Potter-Bucky diaphragm, but

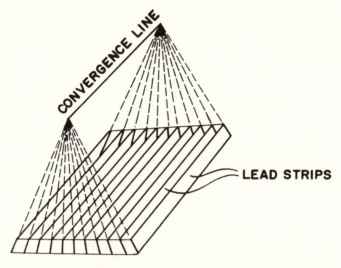

Figure 19.8. Three-dimensional representation of a focused grid, showing the inclined lead strips whose imaginary extensions upward intersect along the *convergence line*.

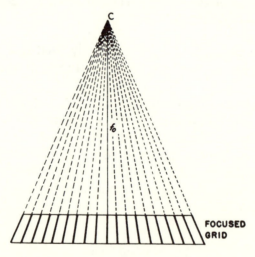

Figure 19.9. Focused grid. The lead strips are inclined so that their extensions, indicated by the broken lines, meet along a specified imaginary line, *C*, above the grid, called the *convergence line* (seen end-on). The vertical distance from the center of the grid surface to the convergence line is called the *grid focusing distance*, f_o. See also Figure 19.8.

the ICRU recommends instead the term *moving grid,* reserving the name Potter-Bucky for the mechanism that activates the grid. However, the term Bucky for the grid and mechanism together has become so standardized in diagnostic radiology that it would seem futile to try to discard it.

The direction of grid travel is very important. While moving parallel to its surface, the grid must move perpendicular to the long axis of its lead strips. Thus, in Figure 19.10, during exposure the grid would move back and forth across the table.

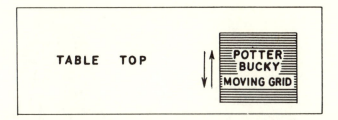

Figure 19.10. Top view of a radiographic table equipped with a Potter-Bucky diaphragm. Note that the lead strips are parallel to the long axis of the table.

The moving grid is of the focused type, but the lead strips are thicker than those in a stationary grid because its motion during the exposure blurs out the lead shadows ("grid lines") on the radiograph.

Precautions in the Use of Focused Grids

Care must be taken to use *focused* grids properly to take advantage of their maximum efficiency; that is, to achieve the least absorption of primary radiation and the greatest absorption of scattered radiation in the lead strips. In the following sections we shall deal with the factors governing *grid cutoff,* which may be defined as the increased absorption of primary radiation in the lead strips resulting from an incorrect SID or improper alignment or centering of the beam to the grid.

1. *Source-Image Receptor Distance.* A grid with a ratio of 8:1 or less tolerates, without appreciable loss of efficiency, a range of equal to the focusing distance ±25 percent. Thus, with a focusing distance of 40 in. (100 cm), SIDs of 30 to 50 in. (about 75 to 125 cm) give satisfactory results with an 8:1 grid. Below or above these

limits, ***peripheral cutoff*** becomes significant as shown in Figure 19.11. With a grid ratio of 12:1, and even more so with a ratio of 16:1 tolerance is extremely small, so the SID must more nearly coincide with the focusing distance to minimize peripheral cutoff.

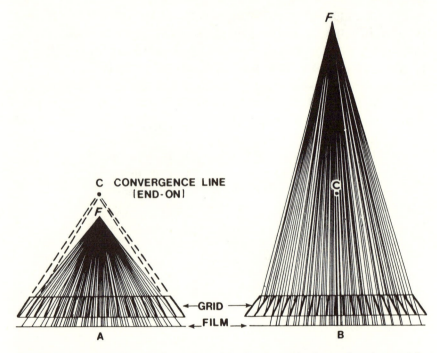

Figure 19.11. The effect of the SID on the efficiency of a focused grid. In *A*, the SID is less than the grid focusing distance minus 25 percent, and more and more rays are absorbed by the lead strips near the edge of the film. In *B*, the SID is longer than the grid focusing distance plus 25 percent, and again there is relatively greater absorption of the beam at the edges of the film. In either case, the periphery of the radiograph will show reduced density — *grid cutoff*.

2. ***Angulation of the Beam.*** With a *focused* grid, angulation of the beam across the long dimension of the lead strips causes uniform decrease in density of the radiograph. (Recall that with nonfocused or parallel grids, such angulation causes asymmetrical peripheral cutoff.) This resembles the decentering of the beam relative to the center of the grid, as will be shown below. The tube, however, may be angled in a direction parallel to the strips without causing grid cutoff, provided the vertical SID remains within the correct limits of grid focusing distance. In practice, a

moving grid is located under the table with the lead strips oriented parallel to the long axis of the table, as shown in Figure 19.10. Consequently, the tube may be angled only in the direction of the long axis of the table without entailing grid cutoff.

3. *Centering of the Tube.* The tube must be centered to the center of the grid to avoid off-axis decentering. With the tube decentered, that is, off-centered across the long axis of the lead strips, *the effect on radiographic density depends on the SID relative to the grid focusing distance.* With the SID equal to the grid focusing distance, and the tube decentered across the lead strips, there will be uniform reduction in the density of the radiograph, that is, uniform cutoff, as in Figure 19.12A. Indeed, this resembles an underexposed or underdeveloped radiograph. Uniform cutoff intensifies with increasing grid ratio and lateral decentering distance, and diminishes with increasing grid focusing distance.

If the tube focus is at an appreciable distance *above* the conver-

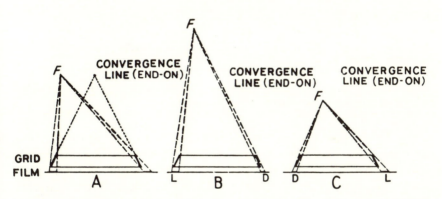

Figure 19.12. Effects of focus-film distance and lateral decentering on efficiency of a focused grid.

In *A*, the tube focus is decentered but level with the grid convergence line. The projections of the lead strips on the film are equal in size, but are broader than they would be if the tube were not off-center. Therefore, the radiographic density is uniformly decreased over the entire radiograph.

In *B*, the tube focus is decentered but is *above* the level of the grid convergence line. The projections of the lead strips lying more directly under the tube are broader than those of the corresponding lead strips on the opposite side (*L* is broader than *D*). Therefore, the radiograph will show less density on the side toward which the tube is shifted.

In *C*, the tube focus is decentered but at a level below the grid convergence line. The result is opposite that in *B*. *L* = lighter. *D* = darker.

gence line, and is at the same time decentered, the area of the film directly beneath the focus will be lighter than the remainder of the film. If the tube focus is *below* the convergence line and decentered, then the area of the film directly below the focus will be darker than the rest of the film. These relationships are illustrated in Figure 19.12.

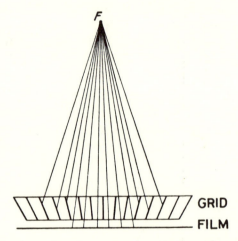

Figure 19.13. The focused grid is inverted, with the "tube side" away from the tube focus. Therefore, the outer regions of the radiograph will show decreased density or *peripheral grid cutoff.*

4. *Tube Side versus Film Side.* The focused grid has a "tube side" and a "film side." In all the above figures, the grid is shown correctly with the tube side toward the tube target. If the grid is inserted reversed, as in Figure 19.13, the tilted strips near the edges of the grid will absorb progressively more radiation, causing severe peripheral cutoff. However, with parallel grids either side may face the tube or film.

PRACTICAL APPLICATION OF RADIOGRAPHIC GRIDS

Bucky or Grid Factor. Absorption of both primary and scattered radiation by a grid demands an exposure increase above nongrid technique. Moving grids with a 12:1 ratio are standard in conventional radiography at or below 100 kV; here the *Bucky* or *grid factor* ranges from about four at 70 kV, to about 5 at 100 kV

(Liebel-Flarsheim Company). The Bucky or grid factor is defined as follows:

TABLE 19.2
GRID OR BUCKY mAs FACTORS AT TYPICAL kV VALUES
FOR VARIOUS GRID RATIOS*

Grid Ratio	70 kV	95 kV	120 kV
No Grid	1	1	1
5:1	3	3	3
8:1	3.5	3.75	4
12:1	4	4.25	5
16:1 crossed	4.5	5	6
5:1 crossed	4.5	5	5.5
8:1 crossed	5	6	7

*Experimental data from Liebel-Flarsheim Company. In practice, these may have to be increased by 25 to 30 percent for grids with ratios of 12:1 or greater.

mAs with grid = grid factor × mAs without grid

for equal density at a particular kV. Table 19.2 summarizes the grid factors over a practical kV range. One may, instead, prefer to compensate for the grid by changing kV instead of mAs. Table 19.3, adapted from Cullinan, gives the approximate change in kV relative to nongrid technic for various grid ratios. In any case, because of uncertainties in the percentage of radiation absorbed by any particular grid, the quoted grid factors should be confirmed by trial exposures with a phantom. Since grids must be used in the radiography of parts measuring 11 cm or more in thickness, intensifying screens are required to reduce the patient's exposure and minimize the effect of motion.

At this point we should state the relative merits of grids with

TABLE 19.3
APPROXIMATE CHANGE IN kV (IN THE USUAL kV RANGE) RELATIVE
TO NONGRID TECHNIC FOR VARIOUS GRID RATIOS

Grid ratio	kV increase
5:1	+ 8
8:1	+15
12:1	+20–25
16:1	+20–25

*Adapted from J. E. Cullinan: *Illustrated Guide to X-ray Technics*, Philadelphia, Lippincott, 1980. (*By permission of the publisher.*)

ratios of 12: or 16:1 in *routine radiography below 100 kV* a 12:1 ratio *is preferred* for the following reasons:

1. Its efficiency in cleaning up scattered radiation (contrast improvement) is not much less than that of a grid with a 16:1 ratio in routine radiography below 100 kV.

2. Centering is less critical, allowing a larger permissible margin of error.

3. Less exposure of the patient is required for a particular film exposure.

4. Lead strips can be thicker (more lead content), thereby absorbing more efficiently the higher energy scattered radiation.

Above 100 kV a 16:1 grid must be used to obtain radiographs of optimum quality. Table 19.4 compares the practical characteristics of various types of grids. In general, positioning latitude — permissible error in centering, angulation, and distance — is worse with parallel than with focused grids of like ratio at SIDs under 40 in. (100 cm).

TABLE 19.4
CHARACTERISTICS OF VARIOUS TYPES
OF GRIDS, STATIONARY AND MOVING

Ratio	Type	Recommended kV	Positioning Latitude	Cleanup
16:1	linear	above 100 kV	very poor	superlative
6:1	crossed	above 100 kV	good	
12:1	linear	up to 100 kV	poor	excellent
5:1	crossed	up to 100 kV	excellent	
8:1	linear	up to 90 kV	fair	good
6:1	linear	up to 80 kV	excellent	moderate
5:1				

Grid Specifications. In 1963 the ICRU, in *Handbook 89*, recommended that manufacturers include certain data to specify the important characteristics of focused grids:

1. *Grid ratio* — the height of the lead strips divided by the distance between successive pairs.

2. **Grid frequency** — the number of lead strips (lines) per in. or per cm; for example, 100 lines/in.

3. **Focusing distance** — the distance between the convergence line and the grid surface.

4. **Source-grid distance limits** — the maximum percent variation in the tube focus-to-grid surface distance within which no excess grid cutoff (ie, absorption of primary radiation) occurs. This is usually equal to the focusing distance ± 25 percent.

5. **Contrast improvement factor** — under standard conditions.

Although the first four items are to be included in grid labeling, the fifth is also included in grid specification.

Patient Dose with Grids. Various combinations of kV and grid ratio affect the radiation exposure of the patient. Table 19.5 compares these factors obtained at the University of Rochester School of Medicine, in the radiography of the lumbar spine in the lateral projection. All the radiographs were diagnostically equivalent. Evidently, in radiography at 70 kV, the radiation exposure of the patient doubles as the grid ratio increases from 8:1 to 16:1. Therefore, **grids with a ratio of 16:1 should not be used in conventional radiography** — below about 100 kV. On the other hand, **above 100 kV** a 16:1 grid produces far better cleanup of scattered radiation than lower ratio grids without a significant increase in patient exposure. For general radiography **up to 100 kV** a 12:1 grid serves as a satisfactory compromise both from the standpoint of grid efficiency and patient exposure.

Potter-Bucky Mechanisms. As mentioned before, the Potter-Bucky grid moves during the x-ray exposure. In the obsolete **single-stroke** type, one first "cocked the Bucky"; that is, pulled it to one side of the table by a lever, putting a spring under tension. When the grid was released (either by a string or by an electromagnetic tripping device), the spring pulled it across the table, its motion being smoothly cushioned by a piston acting against oil in a cylinder.

The **reciprocating** Potter-Bucky mechanism has no separate timer, since it oscillates continuously during the exposure. It is moved rapidly forward (0.3 sec) by a solenoid, while tension is placed on a spring, and returns slowly (1.7 sec) by action of this spring against oil in a chamber as in the single-stroke Bucky. Minimum exposure is 1/20 sec.

TABLE 19.5
EFFECT OF GRID RATIO AND CORRESPONDING kV AND mAs
ON ENTRANCE SURFACE EXPOSURE OF PATIENT
IN LATERAL PROJECTION OF LUMBAR SPINE.
ALL RADIOGRAPHS OF COMPARABLE DIAGNOSTIC QUALITY*

Grid Ratio	kV	mAs	Entrance Exposure	Entrance Exposure as % of Maximum
			R	%
16:1	70	1,000	18.9	100
8:1	70	500	9.5	50
16:1	100	160	5.9	31
8:1	100	120	4.5	25

*University of Rochester School of Medicine, Department of Radiology.

The *recipromatic* Potter-Bucky is activated entirely by an electric motor, the speed of both strokes being identical; exposure times may be as short as 1/60 sec.

The *oscillating* or *trill* Potter-Bucky mechanism is the most advanced type available today and is relatively simple. The grid is supported on four blade springs near its corners (see Figure 19.14). Just before the start of an exposure, the grid is pulled an appropriate distance across the table by a series of relays and then suddenly released. Thereupon, it oscillates or vibrates over a rapidly decreasing distance until after the end of the exposure. The trill Bucky allows very short exposures without grid lines because of the continually changing position of the grid with relation to the bursts of x-ray photons in the beam throughout the exposure.

Moving grids have their greatest application in the radiography of thick parts, measuring 11 cm or more.

There are several causes of *grid lines* in a radiograph made with older types of moving grids. Since they have relatively thick lead strips, the appearance of these grid lines in the radiograph as alternating light and dark stripes—the so-called corduroy effect—is especially objectionable because it impairs recorded detail. The causes of grid lines with moving grids of various kinds include:

1. Synchronism. We found in Chapter 12 that an x-ray beam is not generated continuously, but rather in intermittent showers or bursts of photons corresponding to the peaks in the voltage applied to the tube. If the travel of the grid is such that a different lead

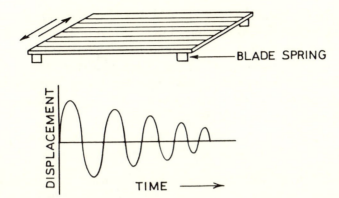

Figure 19.14. Oscillating "trill" mechanism for moving grids. The upper figure shows the grid mounted on four blade springs. A system of relays, synchronized with the exposure switch, starts the grid vibrating in the direction of the arrows just before the beginning of the exposure and stops the exposure just before the grid stops vibrating. The lower figure shows the "damped" (decreasing) oscillation or vibration of the grid. Very short exposures are possible with this type of moving grid mechanism, without the possibility of synchronism.

strip always happens to be above a given point on the film at the same instant that a kV peak is reached, the images of different lead strips will be superimposed on the same point on the film, even though the grid is moving (see Figure 19.15). The net effect is the same as though the grid were stationary. This series of events cannot occur with a trill Bucky under normal operating conditions.

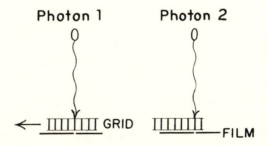

Figure 19.15. Grid synchronism. If the grid moves to the left at certain speeds, a different photon is intercepted by a different lead strip over the *same* point on the film. Hence, a shadow of the lead strip appears on the film just as though the grid were stationary.

2. Exposure starting before grid has reached full speed, or continuing after grid travel has slowed down or stopped. Correct relationship between grid travel and exposure time prevents the appearance of grid lines. An electric contactor starts the exposure only when grid travel has reached the correct speed, and stops the exposure before the grid stops moving. Grid lines appear whenever exposure occurs with the grid stationary.

3. Grid moving too slowly may produce either clearly visible grid lines or density variations in the radiograph.

In recent years, **stationary grids** have been so perfected that the grid lines are virtually invisible at the usual viewing distance. Such ultrafine line grids may often be used in place of moving grids. Stationary grids are considerably less expensive because the Bucky mechanism is eliminated. Furthermore, there is less exposure of the patient with a stationary grid for the same radiographic density because no decentering occurs during the exposure. On the other hand, with a moving grid, as much as 3 cm of decentering occurs with reduction in density of almost 20 percent (Curry, Dowdey, and Murry).

REMOVAL OF SCATTERED RADIATION BY AN AIR GAP

Although the radiographic grid is the most effective and widely used device for removing scattered radiation arising in the patient, one can do this nearly as well by increasing the object-film distance. The space between the patient and the image receptor (eg, film-screen system) is commonly referred to as an **air gap**. Referring back to Figure 19.1, you can see that when an x-ray beam enters the body, some of the radiation is scattered in all directions. That which reaches the film, being non-image forming, causes loss of image contrast and impairs recorded detail. Increasing the thickness of the air gap allows more and more of the scattered photons to move laterally outside the film area (see Figure 19.16). This is especially true of photons that have been scattered at large angles (ie, more oblique to the central ray). In other words, less scattered radiation reaches the film as we increase the object-film distance. This is **entirely a geometric effect;** there is virtually no absorption of scattered radiation in the air gap. Because of the loss of scattered photons, air gap technic requires a compensatory increase in exposure factors to keep radiographic density unchanged, provided the focus-film distance remains constant.

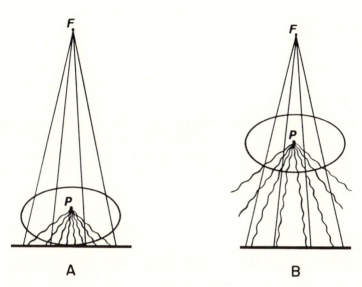

A   **B**

Figure 19.16. Effect of air gap thickness on reduction of the amount of scattered radiation that reaches a film or other image receptor. In *B*, the object is farther from the film, (nearer the tube) and more of the scattered radiation generated in the patient at a point, such as *P*, misses the film than is the case in *A* where the object is nearer the film.

Comparison of an air gap with a grid insofar as contrast improvement in a water phantom is concerned has been reported (Gould and Hale). The practical consequences of this study are that a 10-in. air gap gives about the same degree of cleanup as a 15:1 grid for a thin patient measuring 10 cm, but the same grid is a little more efficient than the 10 in. air gap for a 20-cm patient.

The effectiveness of an air gap is not influenced by kV. In the radiographic range of 40 to 120 kV, scattering occurs almost equally in all directions.

Air gap technic has a place in chest radiography, magnification radiography, and in cerebral and renal angiography. Further improvement in recorded detail can be achieved, especially in cerebral and renal angiography, by the use of a low-ratio grid in addition to the air gap.

In chest radiography, a 10-in. (25-cm) air gap effectively reduces the amount of scattered radiation. The accompanying image magnification must be compensated by an increase in focus-film distance from the customary 6 ft (2 m) to 10 ft (3.3 m). At the same time, this helps reduce the blurring associated with the increased OID. However, with nongrid technique at both distances, an increase in SID from 6 ft to 10 ft requires a much larger exposure by a factor of about $10^2/6^2 = 100/36 = 2.8$ times (according to inverse square law). Air gap technic in chest radiography therefore calls for 110 to 120 kV.

Magnification technic (see pages 352–357) can aid in the detection of suspected hairline fractures. An air gap is inherent in this procedure. The increase in patient exposure that would occur with the part midway between the tube and image receptor is partly offset by elimination of the radiographic grid. Similar conditions prevail in cerebral and renal magnification angiography. In each instance, the loss of recorded detail associated with the increased object film-distance (air gap) must be compensated by use of a fractional focal spot.

REDUCTION OF SCATTERED RADIATION BY LIMITATION OF THE PRIMARY BEAM

By modifying the primary x-ray beam (ie, the beam coming from the focal spot), we can decrease the fraction of the photons that might be scattered in the body. This depends on two well-known facts. First, *an increase in kV increases the fraction of radiation undergoing scatter.* Therefore, the kV for a given anatomic part should be just enough to penetrate it. Excessively high kV should be avoided, unless a grid with higher grid ratio is used. Under ordinary conditions, with a 12:1 grid, 100 kV should not be exceeded. Above this level, one should use a 16:1 grid.

Second, *the fraction of radiation scattered increases as the area (and volume) of irradiated tissue increase.* This results in impairment of radiographic quality through loss of contrast and, also, an increase in patient exposure. In fact, halving the width of a square beam reduces the exposed *area* to 1/4, and the *volume* to 1/8. Thus, the cross-sectional area of the beam should be limited to encompass the smallest area that includes the selected anatomic region. We accomplish this by placing various *beam limiting devices* (*BLDs*) or *beam restrictors* in the path of the x-ray beam as close to the tube port as the housing permits.

Beam limitation is most effective when the field is less than 6 in. (15 cm) in diameter, especially in nongrid radiography. However, with larger areas such as the abdomen, the proper *grid* excels as a means of reducing scattered radiation; beam limitation further improves radiographic contrast when large areas are to be covered, but this is most effective if the collar of the beam-limiting device reaches and surrounds the tube port to remove as much off-focus radiation as possible. Such off-focus radiation arises when unfocused electrons strike parts of the anode other than the target area itself and may amount to as much as 25 percent of the on-target radiation (Ter-Pogossian).

About 50 to 90 percent of the density of a radiograph may result from scattered radiation, so that restriction of the primary beam requires an increase in exposure to compensate for the loss of density. The increase in exposure must be found by trial or in published tables. In setting up a technic chart, we must state the dimensions of the beam (or film size) for each technic. Any further increase or decrease in beam area may entail a compensatory decrease or increase in exposure, respectively.

1. *Aperture Diaphragms.* Figure 19.17 shows how an aperture diaphragm decreases scattered radiation, thereby enhancing contrast. The aperture diaphragm would be very efficient if the x rays originated at a true point source. However, since the tube focal spot has a finite (measurable) area, x rays originate from it at innumerable points and proceed in all directions. Some of these photons produce a blur or penumbra of the edges of the collimator opening; they appear to "undercut" these edges and pass outside the desired confines of the beam, showing up on the radiograph as a gradually fading rather than a sharp border. Thus, a diaphragm fails to provide adequate beam limitation (see Figure 19.18), although it may be better than nothing under unusual circumstances.

2. *Cones.* A radiographic cone is a metal tube that flares outward, away from the tube aperture (see Figure 19.19), and is supposed to limit the beam. However, a cone deserves the same criticism as an aperture diaphragm in that it fails to provide adequate beam limitation, as shown in Figure 19.19A. A cone limits a beam by virtue of its upper opening only, which actually acts as an aperture diaphragm. A modification of the cone, the extension cylinder, has

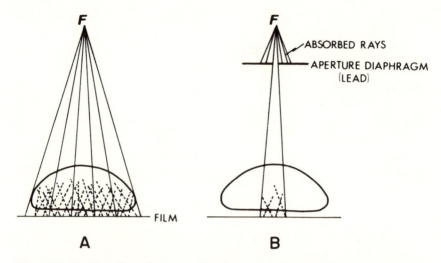

Figure 19.17. The effect of aperture diaphragm on the volume of tissue irradiated and the resulting scattered radiation. In *A* there is no restriction of the x-ray beam; a large volume of tissue is irradiated, with resulting abundance of scattered radiation. In *B* the aperture diaphragm narrows the beam, a small volume of tissue is irradiated, and there is less scattered radiation.

openings of equal size at both ends, so the lower opening acts as an aperture diaphragm. Because of the longer distance of the lower aperture from the focal spot, beam limitation is better than that with a cone (see Figure 19.19B).

Another serious shortcoming of the flared cone is the excessive area of the circular field required to cover a rectangular film of any given size (see Figure 19.20). This is objectionable because not only does it lead to the production of more scattered radiation, but it also exposes the patient to radiation outside the area of interest. A partial solution might be to have cones with rectangular openings to cover each film size at various focus-film distances, but this would be impracticable. The variable aperture device does this much more efficiently, as described in the next section.

3. *Variable Aperture Beam-limiting Devices.* These—also known as *collimators*—combine the best features of the cone and aperture diaphragm. The term collimator, although generally used, is inaccurate because, strictly speaking, it is a device that is supposed to line up rays so as to make them parallel to one another. In radiographic usage, the collimator simply limits or shapes the beam to

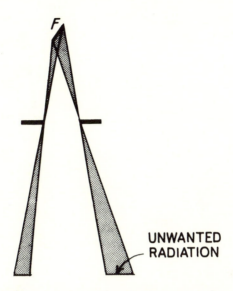

Figure 19.18. Inefficiency of an aperture diaphragm as a beam-limiting device or collimator. Since, in practice, a focal spot has a finite (measurable) size, the edge of the beam is not sharply defined—penumbral radiation produced at the margins of the diaphragm, traverses the patient, and reaches the film.

the desired area; the radiation continues to diverge after leaving the collimator. However, the term collimator has become so ingrained in radiology and is so much simpler than the term variable aperture beam-limiting device that we shall continue to use it. It should be mentioned that some are using the term **BLD** (for beam-limiting device) to designate a collimator.

A typical collimator comprises a box-like apparatus (see Figure 19.21) equipped with two or more sets of adjustable diaphragms, stacked one above the other, which can be opened or closed to delimit accurately a beam of any desired rectangular configuration. In addition, the multiple diaphragms more effectively reduce the amount of radiation passing outside the edge of the beam, thereby giving it a sharper margin as shown in Figure 19.21B, provided they are properly aligned. The best collimators have a collar extending to and surrounding the tube port for the purpose of further improving recorded detail by removing off-focus radiation. As stated above, circular beams shaped by aperture diaphragms or cones allow radiation to pass beyond the zone of interest (see Figure 19.20), increasing the fraction of scattered radiation and

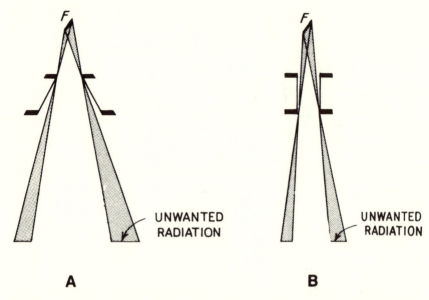

Figure 19.19. Comparison of the effectiveness of a flared cone and a cylinder on beam limitation. In *A*, the upper aperture limits the beam, similar to an aperture diaphragm. In *B*, the lower aperture of a cylinder limits the beam. The width of the interval between the inner and outer rays at the film represents radiation outside the desired beam. Thus, the cylinder in *B* provides better beam limitation than the flared cone in *A*.

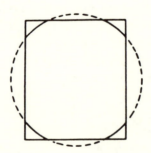

Figure 19.20. The x-ray beam cuts the corners of the film when limited by an aperture diaphragm or cone, but the dashed circle indicates the excessive volume of irradiated tissue.

unnecessarily exposing the patient. Thus, they do not adequately limit the beam (see Figure 19.21A).

When used with an extension cone, the collimator should be narrowed to the smallest square that encloses the circular field to

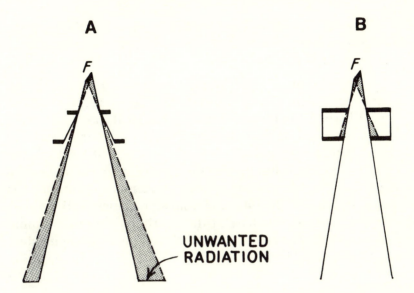

Figure 19.21. Effect of a well-designed collimator (variable aperture beam-limiting device) on beam restriction as compared with a cone. In *A*, a cone reduces the beam area, but, in effect, the upper opening of the cone serves as an aperture diaphragm so that the beam is not sharply defined (see Figure 19.18). In *B*, a double-diaphragm collimator markedly improves beam restriction by absorbing penumbral radiation in the lower diaphragm. Further improvement can be achieved with more than two levels of variable diaphragms.

improve further the efficiency of beam limitation. However, not all collimators can accept an extension cone.

Collimators are equipped with an ***illuminated field and beam-centering device.*** They also have calibrated scales to allow aperture adjustment for films of various sizes at different SIDs. According to federal regulations, collimators for all new equipment must be electrically interlocked with the clamps in the Bucky tray so that the apertures adjust automatically to the size of the cassette. Such a collimator is known as a ***positive beam-limiting device (PBLD).*** A manual override allows beam-shaping for special field sizes.

Film Coverage by Aperture Diaphragms and Cones. As just noted, collimators are calibrated, thus obviating the need of calculations. However, with aperture diaphragms and cones, we can compute film coverage by apertures of various sizes at desired focus-film distances. Since film sizes are still generally stated in inches, we shall use these units in our example. Computation of film coverage requires the following information:

1. Distance from focus to aperture.
2. Distance from focus to film.
3. Diameter of aperture.

The computation of film coverage by a beam limited by an aperture diaphragm is not difficult. In Figure 19.22, with an opening 1 in. in diameter and 4 in. from the focal spot, and an SID of 40 in., what will be the diameter of the beam at the surface of the film, *disregarding geometric blur?* Note that the x-ray beam continues to broaden after passing through the aperture. In longitudinal section, this forms a larger triangle with its apex at the tube focus and its base on the film, and a smaller triangle with its apex at the tube focus and its base at the aperture. These are similar triangles (see pages 17–19), so by the application of elementary geometry we obtain the proportion:

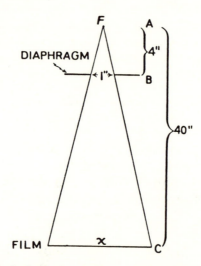

Figure 19.22. Method of calculating film coverage with an aperture diaphragm. In this example, x/1 = 40/4. Therefore, 4x = 40 and x = 10 in., disregarding geometric blur.

$$AB/AC = d/x$$

where AB = focus-to-aperture distance
    AC = SID
    d = diameter of aperture
    x = diameter of beam at film plane

Substituting the numerical values in the proportion,

$$4/40 = 1/x$$
$$4x = 40 \text{ in.}$$
$$x = 10 \text{ in.}$$

An x-ray beam 10 in. in diameter is obviously too small to cover a 10 in. × 12 in. film but will cover adequately an 8 in. × 10 in. film. Since the diagonal of this film is 12.8 in., its corners will be clipped (see Figure 19.20).

The same principle applies to the *flared cone* whose **upper** opening limits the beam (see Figure 19.19A). Calculations must be made as though an aperture diaphragm were located at that level, the lower opening being ignored.

With a *cylinder* the *lower* aperture always limits the size of the beam (see Figure 19.23), and therefore the calculations are made as though an aperture diaphragm were present at the level of the lower opening. The following proportion may be used to find the film coverage diameter, that is, the diameter of the beam at the film:

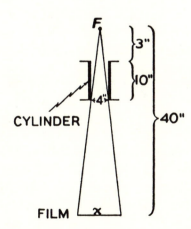

Figure 19.23. The effect of a cylinder in limiting the size of an x-ray beam.

$$\frac{\text{film coverage diameter}}{\text{cylinder diameter}} = \frac{\text{SID}}{\text{source-to-lower aperture distance}}$$

where the "source-to-lower aperture distance" is obtained simply by adding the distance between the focus and the top of the cylinder, to the length of the cylinder. Substituting the corresponding values from Figure 19.23 in the above equation.

$$\frac{film\ coverage\ diameter}{4} = \frac{40}{3 + 10}$$

$$film\ coverage\ diameter = \frac{4 \times 40}{13} = \frac{160}{13} = 12.3\ in.$$

Thus, an 8 × 10 film will have about 1/4 in. cut off all four corners under these conditions, provided the beam has been centered exactly to the film. Bringing the x-ray tube and cylinder close to the patient greatly improves beam limitation, but increases dose.

OTHER METHODS OF ENHANCING RADIOGRAPHIC QUALITY

Moving Slit Radiography

Introduced by Jaffe and Webster as a method of improving radiographic contrast, moving slit radiography extends the principle that a very narrow (slit-limited) x-ray beam engenders less scattered radiation than does a circular beam of the same area. A suggested prototype model (see Figure 19.24) consists of two lead sheets of appropriate thickness, in which have been cut a series of twelve parallel slits 3 mm in width spaced 2.5 cm apart. One lead sheet would be placed above, and the other below, the patient. Corresponding upper and lower slit pairs must be coaxially aligned in the beam. Movement of the assembly during exposure wipes out the images of the slits in the same manner as a moving grid. With this system, radiographic contrast should be enhanced by a factor of two to three relative to that with a grid. However, this would require a tenfold increase in x-ray output by the tube. The ultimate practicability of this device must await further development.

The Anode Heel Effect—Anode Cutoff

As it leaves the focal spot, the primary beam displays as nonuniform distribution of radiation intensities (see Figure 19.25). The exposure rate at any point in the beam depends not only on the distance from the focus (inverse square law) but also on the *angle the rays make with the face of the target*. Figure 19.25 shows

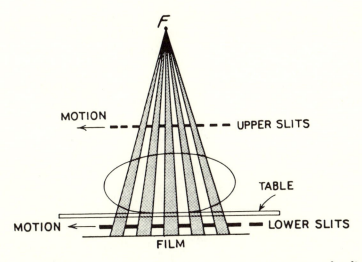

Figure 19.24. Principle of moving slit radiography to remove scattered radiation. The upper and lower lead plates, one above and the other below the patient, each has a series of twelve parallel slits. These are aligned coaxially with the beam as shown. The lead plates move together across the patient in the direction of the arrows. Movement of the lead wipes out the images of the slits on the radiograph in the same way as those of the lead strips in a moving grid system. Image contrast is said to be enhanced two or three times relative to that with a grid but requires about a tenfold increase in exposure.

a 20° anode with the angular distribution of the relative exposure rates in an x-ray beam, based on the assumption that the exposure rate on the central axis of the beam (at a given distance) is 100 percent. Thus, if the beam falls on a 14 in. × 17 in. (35 cm × 43 cm) film placed lengthwise with relation to the long axis of the tube, at an SID of 40 in. (100 cm), the exposure rate at the end of the film below the cathode will be about 105 percent, and at the end of the film below the anode about 80 percent of the exposure rate at the center of the film. This discrepancy becomes less pronounced as the SID is increased and the size of the film is decreased. The variation in exposure rate according to the angle of emission of the radiation at the focal spot is called *anode cutoff,* or *anode heel effect.* The smallest exposure rate occurs at the anode end of the beam. This is useful in certain procedures; for example, in radiographing a region of the body, which has a wide range of thicknesses, we can take advantage of the heel effect by placing the thickest part toward the cathode end of the beam (where the exposure rate is greatest) and the thinnest part toward the anode (where the exposure rate is least). Accordingly, a more uniform radiographic density is recorded despite varying thicknesses of the object. For example, in radiography of the thoracic spine in the anteroposterior projection the patient should be placed with the head toward the anode and the feet toward the cathode. Figure 19.26 explains the physical basis of the heel effect. Note that it becomes more pronounced with a smaller anode angle.

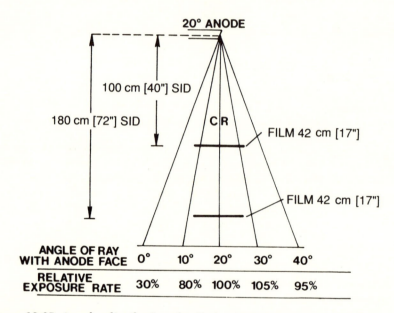

ANGLE OF RAY WITH ANODE FACE	0°	10°	20°	30°	40°
RELATIVE EXPOSURE RATE	30%	80%	100%	105%	95%

Figure 19.25. Angular distribution of radiation intensities (exposure rates) in an x-ray beam. The exposure rate is maximal in the central portion of the beam, falling off more toward the anode side of the beam; this is *anode cutoff,* known also as the *heel effect.* Note that anode cutoff becomes less pronounced as the SID increases and as the film size decreases.

COMPENSATION FILTERS

Various types of filters can be used in radiographing parts of the body that differ greatly in thickness or density so as to record them with more uniform density on a single film. Such compensation filters have traditionally been made of an aluminum or barium-plastic compound. For example, in radiography of the foot in the

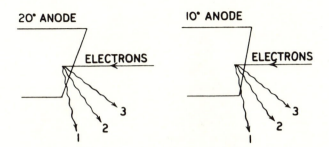

Figure 19.26. Explanation of anode cutoff (heel effect). X-ray photons such as *1* undergo greater attenuation than photons such as *3* because they pass through a larger thickness of anode "heel." This effect becomes more pronounced as the anode angle decreases.

anteroposterior projection, an aluminum wedge can be placed with its thickest part under the toes so as to even out the image density of the toes and heel. However, this increases the part-film distance and impairs recorded detail. One solution has been to place a wedge filter in the filter slot of the tube housing, but this is impossible with modern equipment.

There is now a type of transparent filter material composed of clear plastic incorporating 30 percent lead by weight. It is the Clear-Pb™ compensation filter (Victoreen Nuclear Associates), which can be mounted on the face of the collimator by a specially designed magnetic system. These filters are thin and lightweight. The simplicity of this system will undoubtedly find wide application in general radiography in such examinations as combined mediastinal-pulmonary tomography, aortic arch angiography, and anteroposterior projections of the foot.

Some dedicated chest radiographic units provide for a ***trough filter*** —one that is thinner at the center than at the edges—to permit heavier exposure of the mediastinum than of the lungs (see Figure 19.27).

A simple method of compensation for differences in tissue opacity calls for the insertion of a sheet of black paper between one screen and the film in that part of the cassette which is to lie behind the more radiolucent tissues. Sufficiently heavy exposure can then be given to penetrate the denser structures without overexposing the less dense ones, provided the difference in opacity is not excessive. For example, in radiography of the chest when

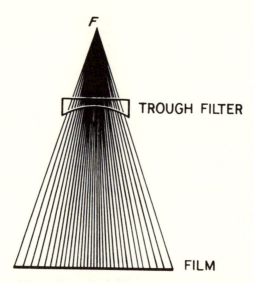

Figure 19.27. Principle of the trough filter. There is more attenuation of the beam as it passes through the lateral portions of the filter than in the center. Such a filter provides greater exposure of the mediastinal structures than of the aerated lungs, thereby compensating for inherent density differences.

one lung has become partly opacified by disease, we can use a sheet of black paper to block the half of one screen behind the normal lung. We can then increase the exposure to penetrate the diseased lung without overexposing the normal lung, provided there is only a moderate density difference between them.

SUMMARY OF RADIOGRAPHIC EXPOSURE

In Figure 19.28 a typical arrangement of x-ray equipment and patient in radiography is shown to summarize the important items in radiographic exposure.

QUESTIONS AND PROBLEMS

1. Discuss the three factors affecting the percentage of scattered radiation relative to primary radiation at the film surface.
2. Describe the principle and construction of a radiographic grid.
3. Explain the physical factors in grid efficiency, including grid ratio and grid frequency.

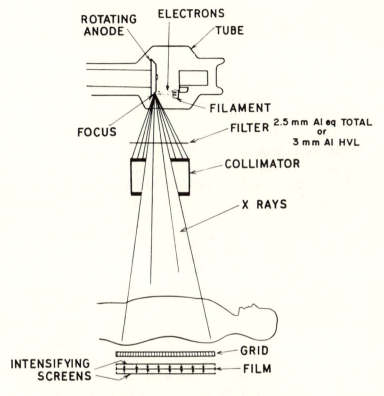

Figure 19.28. Summary of radiographic exposure.

4. Discuss the functional characteristics of grid efficiency, including selectivity and contrast improvement factor.
5. What is meant by the convergence line of a grid? The focusing distance of a grid?
6. Discuss four precautions in the use of a grid.
7. What is the preferred grid ratio in conventional radiography? In high voltage radiography?
8. Explain the decreased density at the edges of a radiograph, which may occur when a grid is used. What is it called?
9. Describe briefly the main types of Potter-Bucky mechanisms.
10. Name three causes of grid lines with a moving grid and state how each can be corrected.
11. Under what conditions may the x-ray beam be angled with respect to a radiographic stationary or moving grid?
12. Describe four causes of grid cutoff.

13. Define grid ratio. What is the relationship between grid ratio and grid efficiency?

14. Explain how air gap technic reduces the amount of scattered radiation reaching an image receptor?

15. How does the efficiency of an air gap compare with that of a grid in improving contrast (ie, "cleanup")?

16. Describe three examples in which air gap technic may be advantageous. How does one minimize blurring associated with an air gap, and what effect does it have on exposure?

17. How does an aperture diaphragm diminish scattered radiation?

18. An aperture diaphragm measuring 2 in. (5 cm) in diameter is located 5 in. (12.7 cm) below the tube focus. What will be the diameter of the beam at 40 in. (100 cm)?

19. In problem 18, if the effective focal spot diameter is 1 mm, find the width of the penumbra (blur) of the image of the aperture edge on the radiograph.

20. Describe a typical radiographic collimator (variable beam-limiting device). Why is it the most desirable type of beam limiter?

21. A cylinder measuring 12 in. (30.5 cm) in length is used to limit an x-ray beam. The lower opening is 15 in. (38 cm) from the tube focus and measures 4 in. (10 cm) in diameter. What is the largest standard film that will be covered (except for the corners) at a focus-film distance of 40 in. (100 cm)?

22. Summarize the methods of minimizing scattered radiation in radiography.

23. Define and explain the cause of the anode heel effect. How is it influenced by the focus-film distance? By the anode angle? By the size of the film?

24. Draw a simple diagram summarizing radiography.

25. What is the purpose of a compensation filter? Discuss the available materials for compensation filters and give their relative advantages.

26. Describe the form, function, and application of a trough filter.

Chapter 20

SPECIAL EQUIPMENT

A NUMBER OF USEFUL modifications of radiographic equipment are available for special purposes. Of these we shall describe, in particular, image intensification, stereoscopic radiography, tomography, computed tomography, photofluorography, xeroradiography, and digital fluorography.

Bright Fluoroscopy (Image Intensification)

Fluoroscopy, an essential procedure in diagnostic radiology, makes possible the visualization of organs in motion, positioning of the patient for spotfilming, instillation of opaque media into hollow organs, insertion of catheters into arteries, and a variety of other procedures. Unfortunately, **direct** or **dark** fluoroscopy with a zinc cadmium sulfide screen is an extremely inefficient process, both physically and visually. This results from the very low brightness level of the direct fluoroscopic image, about $1/30,000$ that of a radiograph on a fluorescent illuminator, but also to the nature of human vision which we shall explore briefly in introducing the subject. Therefore, direct fluoroscopy has become virtually obsolete.

The human eye is a living camera. Light reflected from an object enters the eye and passes through the lens which focuses it on the *retina*, a membrane lining the inside surface of the back of the eye (see Figure 20.1). Within the retina there are two types of light-sensitive cells, the *cones* and the *rods*. Present in other areas, but concentrated in a small spot at the center of the retina—the *fovea centralis*—the cones respond to light at ordinary brightness levels, and not to the dim light from a zinc cadmium sulfide fluoroscopic screen.

The rods, microscopic rod-shaped cells distributed throughout the periphery of the retina, respond to light at the low **brightness levels** prevailing in direct fluoroscopy, but they exhibit very poor visual acuity—the ability to discriminate small images. In fact, we

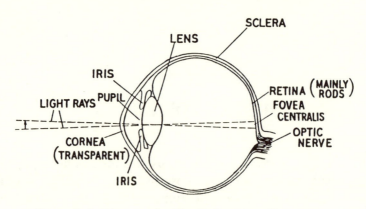

Figure 20.1. Diagram of midsagittal section of human eye.

know that the contrast (difference in brightness) of two adjacent images must approximate 50 percent to be perceived by rod vision, but cones can discriminate images with a contrast of about 2 percent in high level radiographic illumination. Since fluoroscopic image contrast averages one-fourth to one-half that in a radiograph, the visibility of details is poor on the dim fluoroscopic screen. Another factor in rod vision is the prolonged dark adaptation, 20 to 30 min, required for the rods to attain adequate sensitivity, which then has to be preserved during fluoroscopy in an almost completely darkened room. Fortunately, the rods are particularly sensitive to the green light emitted by the zinc cadmium sulfide screen.

In summary, then, direct fluoroscopy is an exceedingly inefficient process because of the low brightness level, the low contrast, the dependence on rod vision with its poor visual acuity, and the need of prolonged dark adaptation.

In 1942 Dr. W. E. Chamberlain, in a classic paper, predicted that if the brightness of the fluoroscopic image could be increased about 1,000 times, cone vision would be brought into play and visual acuity enhanced by a factor of ten (ie, 10 times greater) relative to rod vision. Besides, dark adaptation and a completely darkened room would become unnecessary.

After extensive research by various manufacturers, marked brightening of the fluoroscopic image was realized through the development of the *x-ray image intensifier.* Its main component is the image tube, an electronic device that promptly converts, in several steps, an x-ray image pattern into a corresponding visible light pattern of significantly higher energy per square cm of viewing screen. Figure 20.2 shows the basic construction of a typical image intensifier. After passing through the patient, the x-ray beam enters the *highly evacuated glass envelope* of the image tube and produces a fluorescent image on the cesium iodide *input screen.* The light from this screen falls on the closely applied *photocathode,* liberating electrons in the form of an *electronic image* that duplicates the light image but consists of various densities of electrons instead of photons. Sharpness of the electronic image is enhanced by having the photocathode curved as shown in Figure 20.2. The electrons forming the image undergo acceleration toward the cathode by an applied high voltage—about 25 kV—and are focused, by a series of electrostatic "lenses," onto the *output screen* whose phosphor is *silver-activated zinc cadmium sulfide.* This phosphor layer has a thin coating of aluminum on the side toward the photocathode to permit the passage of electrons and at the same time prevent the flow of light back to the photocathode. Furthermore, the aluminum is grounded to prevent the accumulation of electrons.

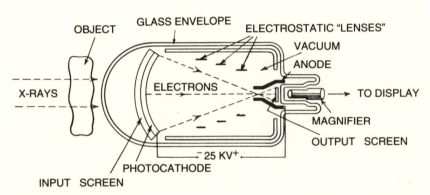

Figure 20.2. Image intensifier. X rays produce an image on the *input screen,* whose light liberates *electrons* from the *photocathode.* The concentration of the electrons is proportional to the light intensity at any given point on the photocathode, so they form an *electron image* in space. The electrons are then accelerated at 25 kV and electrostatically focused on the *output screen,* where the final intensified image is formed. This image may be viewed by a system of lenses and a mirror, which has been largely replaced by closed circuit TV; it may also be recorded on cine film, video tape, or video disc.

In the output screen the electrons' energy is converted to a corresponding but inverted visible light pattern—an image whose brightness is about 4,000 to 7,000 times that of the light image on the input screen, a comparison designated as **brightness gain.** The image finally undergoes processing through an appropriate system for one or more of the following options: direct viewing with a mirror, closed circuit television, cine (motion picture) recording, videotape, or videodisc.

Brightness gain depends on two main properties of the image intensifier. In the first place, in an image tube with a 9-in. input screen and a 1-in. output screen, the ratio of the output screen area to the input screen area is $(1/9)^2$ or $1/81$. Since the light photons have been "crowded" into an area about $1/80$ as large, the brightness will increase eighty times simply through reduction in size or **minification** of the image; this is called **minification gain.** In the second place, the acceleration of the electrons results in a multiplication of photons at the output screen, a process known as *flux gain* — the ratio of the number of photons at the output screen to the number at the input screen for a given exposure. In general,

brightness gain = minification gain $\times$ flux gain (1)

With a minification gain of 80 and a flux gain of 50,

$$brightness\ gain = 80 \times 50 = 4,000$$

which means that the brightness of the image at the output screen is 4,000 times that at the input screen.

Further increase in brightness can be achieved by application of higher accelerating voltage, but this is not unlimited. On passing through the optical lens and mirror system (see Figure 20.2), the final image will have regained about 80 to 90 percent of its original size without appreciable loss of brightness.

The ICRU has proposed a standard for specifying the intensifying ability of an x-ray image intensifier, based on strictly defined conditions. This is the **conversion factor,** defined as the **ratio of the brightness (ie, luminance) of the output screen to the x-ray exposure rate at the input screen.** The conversion factor is a more precise concept than brightness gain and may eventually replace it.

The main advantage of the image intensifier lies in its brightness gain, with activation of cone vision (also called photopic vision) and improved visual acuity—ability to see fine detail. Besides, it has done away with the inconvenience of dark adaptation. However, contrast in the intensified image may be lessened by the "fogging" effect (noise factor) of some x-ray photons passing directly from the input screen to the output screen, bypassing intermediate conversion to light and electron images. Moreover, some light photons pass in the reverse direction, from the output screen to the input screen, eventually contributing to image fog and loss of contrast. Contrast can be significantly improved by a television system.

The introduction of cesium iodide (CsI) as the input screen phosphor has resulted in notable improvement in recorded detail by virtue of the increased image resolution and contrast. This depends on the fact that CsI crystals can be crowded much more closely and in a thinner layer than zinc cadmium sulfide crystals onto the input screen. CsI is also a more efficient absorber of x-ray photons, thereby improving recorded detail and contrast. As a result, CsI has become the preferred phosphor in image intensifiers.

Recorded detail is actually a little poorer with image intensification than with direct fluoroscopy because of slight loss of information at each stage of image formation (ie, x rays, light, electrons, light, final viewing system). Also, there is poorer focusing of electrons toward the outer portions of the output screen, especially with large image intensifiers. This causes not only some loss of recorded detail but also distortion of lines (curving inward) toward the edges of the screen ("pincushion effect").

In summary, the advantages of image intensification far outweigh the disadvantages and have led to the obsolescence of direct or dark fluoroscopy. Image-intensified fluoroscopy:

1. Increases brightness level and image contrast. Cone vision is brought into play by the higher brightness level, resulting in improved visual acuity.

2. Eliminates dark adaptation and an almost totally darkened fluoroscopic room.

3. Decreases radiation exposure rate to patient, although this is less than was first anticipated because a certain irreducible minimal brightness is needed at the input screen to reduce quantum mottle and provide good recorded detail in the final intensified image.

4. Can be used to localize certain structures such as the optic foramina for spot filming.

5. Can be used with a television system for more convenient

viewing and for enhancement of contrast (see Figure 20.3). Note that the image intensifier by itself cannot increase the inherent contrast of the image presented to it on the input screen.

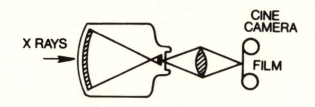

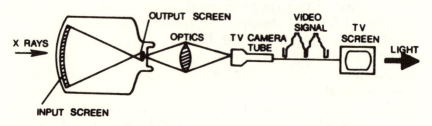

Figure 20.3. Two common auxiliary methods of recording the image in an image intensifier. A movie camera is used in the upper diagram, and closed circuit television in the lower diagram. (*Adapted from ICRU Handbook 89.*) In modern TV systems, optical coupling is no longer used, the camera vidicon tube being directly coupled to the output screen of the image intensifier.

6. Can be used in cineradiography (see Figure 20.3) or in a video tape or video disc system.

The basic x-ray image intensifier has been so improved in its maneuverability, phosphor (cesium iodide), optical system, and adaptability to routine fluoroscopic procedures, that it has become an indispensable part of the equipment, not only in the hospital Radiology Department, but also in the private radiology office. The improvement in the auxiliary systems, such as television, cine, video tape, videodisc, and digital fluorography, has immeasurably widened its field of usefulness.

STEREOSCOPIC RADIOGRAPHY

Stereoscopy means "seeing solid," that is, seeing in three dimensions. An ordinary photograph appears flat, lacking depth, because the camera has one "eye" and projects the image of an object on the film in only two dimensions. But when a pair of human eyes sees the same object, the brain perceives it as a three-dimensional solid body.

How is this brought about? Each eye sees the object from a slightly different angle, depending on the distance of the object, its size and shape, and the distance between the pupils of the eyes (see Figure 20.4). As a result, slightly different images are formed on the retinas of the respective eyes and are carried separately over the optic nerves to the brain. Here the two images are fused and perceived as one having three dimensions—height, width, and depth.

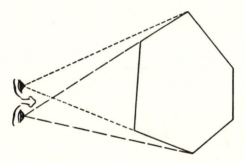

Figure 20.4. Stereoscopic vision. Each eye sees a slightly different view of the solid object. These two images are blended in the brain to give the impression of one image having a solid appearance; that is, height, width, and depth.

Stereoscopic vision also permits us to judge the relative distances of various objects. You can easily verify this by trying to touch something with one eye closed, and then with both eyes open. Still, it is possible for a one-eyed person to learn to judge distance fairly well, although stereoscopy is impossible.

We are not born with stereoscopic vision but develop it during infancy. Many years ago Dr. O. V. Batson showed that we differ in our ability to fuse stereoscopic images, and that in many instances this ability can be improved by corrective glasses, exercise, and practice.

Principle of Stereoradiography. We can produce a stereoscopic image radiographically by applying the principle of stereoscopic vision. First, we make *two separate exposures of the same object from different points of view;* the x-ray tube takes the place of the eyes, exposing two films from slightly different positions (see Figure 20.5). The part being examined is immobilized while the tube is shifted to obtain the two stereoradiographs. Second, we must *view properly* the pair of stereoradiographs if we are to achieve stereoscopic or depth perception.

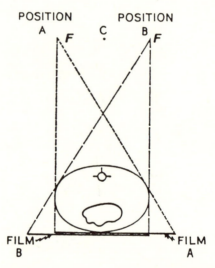

Figure 20.5. Stereoradiography. Two films are exposed separately, one with the tube focus at *A,* and the other with the focus at *B.* Note that the object is not moved. The focus positions are at equal distances from the center *C.*

Tube Shift in Stereoradiography. The question arises, "How far must the tube be shifted between exposures to obtain a satisfactory pair of stereoradiographs?" The interpupillary distance, that is, the distance between the pupils of the eyes, averages 2.5 in. (6 cm), whereas the average viewing distance is 25 in. (64 cm). These two distances are in the ratio of $2.5/25 = 1/10$, and the same ratio governs the relationship between tube shift and SID:

$$\frac{\text{tube shift}}{\text{SID}} = \frac{1}{10} \qquad (2)$$

According to this equation, stereoradiography requires the following typical shifts:

SID 40 in. (100 cm)   tube shift 4 in. (10 cm)
SID 72 in. (180 cm)   tube shift 7 in. (18 cm)

We may conclude that the **tube shift is theoretically one-tenth the SID for a viewing distance of 25 in.** (64 cm). While, in most instances, this ratio lays down the conditions for excellent stereoradiographs, some observers find it difficult to fuse the images unless the ratio is reduced to $1/13$ or even $1/16$. However, the resulting apparent depth of the stereoscopic image will be less than that with the larger ratio of $1/10$.

The **direction** of the tube shift relative to the anatomic area should be perpendicular to the dominant lines; thus, chest stereoradiography requires a tube shift across the ribs. With low-ratio grids, tube shift perpendicular to the grid strips should entail no significant difference in the density of the two radiographs if the tube has been correctly centered to the grid. But with a tube shift parallel to the grid strips, there may be a difference in the densities of the radiographs due to **anode cutoff** or **heel effect.** Also, recorded detail will be different in the two radiographs because the size of the effective focus will differ in the two positions. This is shown in Figure 20.6. But with a **high ratio Bucky grid such as 12:1 or 16:1** because of the criticality of centering, the shift **must be parallel to the long axis of the table** to avoid off-center cutoff.

We must be careful to open the collimator so as to cover the films in both positions of the tube. If the beam is collimated excessively, the tube will have to be tilted to cover both films.

The steps in stereoradiography may now be summarized:

1. Center accurately the part to be radiographed.
2. Determine the total shift of the tube for the given focus-film distance. Suppose this shift is to be 4 in. (that is, $1/10$ of 40 in.).
3. Determine the correct direction of shift.
4. Now shift the tube one-half the total required distance, away from the center. In this case, $1/2 \times 4 = 2$ in. Make the first exposure.
5. With the patient immobilized in the same position, change the cassette. (For chest stereoradiography, patient should suspend respiration.)
6. Now shift the tube the total distance, in this case 4 in., in a

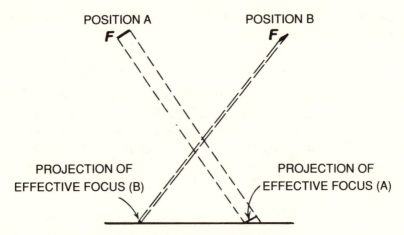

Figure 20.6. Difference in size of the projected or effective focus when the tube is shifted parallel to the long axis of the tube. Note that in position *B* the effective focus is smaller and therefore recorded detail is better than in position *A*. But the density of the radiograph will be less in position *A* because of anode cutoff (heel effect).

direction exactly opposite the first shift, and make the second exposure. (This is the same as shifting the tube one-half the total distance to the opposite side from the center point.) The exact method of shifting is shown in Figure 20.7.

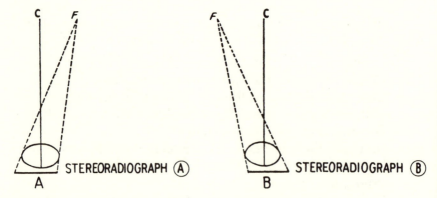

Figure 20.7. Method of shifting the tube in stereoradiography. In *A*, the tube has been shifted one-half the total distance to one side of the center, *C*, for the first exposure. In *B*, it has been shifted the full distance in the opposite direction for the second exposure.

7. Process the films.
8. The stereoradiographs are now ready for viewing.

Viewing Stereoradiographs. There are various methods of viewing a pair of stereoradiographs so that they are perceived three-dimensionally. These will now be described.

A number of optical devices facilitate stereoscopic viewing. They all have the capability of shifting the image of one of the stereoradiographs onto its mate. The simplest device, a 30° or 40° *glass prism,* is available at optical supply houses. Figure 20.8 shows how a prism bends the light rays coming from one radiograph so that they seem to come from the other one, thereby superimposing the images. To be viewed properly, the stereoradiographic pair is held up with the tube side facing the observer; in other words, the eyes see the radiograph just as the tube did. At the same time, the stereoradiographs are positioned so as to represent a *horizontal shift of the tube.* With a vertical tube shift, as in radiography of the chest, the radiographs would have to be turned 90°, the chest then appearing to lie on its side. On one radiograph, the image appears *farther from the left hand edge;* this radiograph was made with the tube in the left position, and is therefore placed in the *left illuminator.* The other radiograph, exposed with the tube in the right position, is placed in the right illuminator. The observer now holds the prism between the thumb and index finger of the right hand in front of the right eye, with the *thinnest portion near the nose,* as shown in Figure 20.8. (If preferred, the prism can be held with the left hand in front of the left eye, again with the thinnest edge near the nose.) By slightly adjusting the position of the prism and stepping forward and backward, the observer soon succeeds in fusing or blending the images. This method requires very little practice, is inexpensive, and provides a convenient *monocular stereoscope.* If preferred, two 20° prisms may be used, one being held before each eye, again with the thinnest part near the nose; this is the simplest type of *binocular stereoscope.*

Elaborate stereoscopes (Wheatstone; Stanford) were modifications of those just described, making use of prisms or mirrors to displace the images. Because of their bulk and high cost, they have been largely replaced by various types of small *hand-held binocular stereoscopes.*

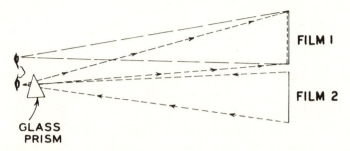

Figure 20.8. Basic principle of stereoradiographic viewing. The glass prism bends the light rays from stereoradiographic 2 in such a way as to superimpose its image on stereoradiograph *1*. This facilitates the fusion of the images to provide depth perception.

One must be careful to place the stereoradiographs in the illuminators correctly to achieve a genuine stereoscopic effect. Only then will an anteroposterior radiographic projection appear to be anteroposterior, and conversely. For example, with a chest stereoradiographed in a posteroanterior position, the anterior ribs should appear to curve away from the observer. Such verification, which is important in the relative localization of lesions or foreign bodies, can be simplified by taping an identifying marker — "stereo" — to the table top or cassette holder, to indicate the part that was farthest from the tube during radiography.

There is wide variation in the frequency with which stereoradiography is performed, depending on the preference of the individual radiologist. In fact, there has been a decline in the use of stereoradiography in recent years. However, stereoradiography helps not only in localizing a lesion or a foreign body but also in improving the perception of its shape, structure, and spatial relationship to other anatomic parts.

TOMOGRAPHY

In ordinary radiography a three-dimensional object — the body — is represented on a two-dimensional film. In effect, the object has been compressed virtually to zero thickness in the direction of the x-ray beam so that the images of a number of structures are superimposed on each other in the radiograph. Not infrequently, this makes radiographic interpretation difficult, if not impossible.

We have three methods of separating such superimposed images through modification of ordinary radiography:

1. *Increased Distortion.* We can separate confusing images by simply rotating the patient or by changing the direction of the x-ray beam. Thus, we use an oblique projection of the gall bladder to separate its shadow from that of the vertebral column or gas-filled bowel.

2. *Decreased Source-Image Receptor Distance.* When structures are relatively far apart along the axis of the x-ray beam, we can use a very short SID. This produces more geometric blurring (see page 319) of the images of structures farther from the film than of structures nearer the film, so the latter remain reasonably sharp. An example is radiography of the mandible in the posteroanterior view with a very short SID, blurring the image of the cervical vertebrae.

3. *Motion of a Part.* When two structures are directly aligned in the x-ray beam and the undesired one happens to be movable, we can set it in motion and blur its image while the desired image remains sharply defined. For example, the dens (odontoid process) of the second cervical vertebra can be well shown in the anteroposterior projection by having the patient move his/her jaw during the exposure.

While the above methods may succeed, they often do not. In fact, they fail completely when the desired structure is buried within a denser one, as exemplified by a cavity inside a densely consolidated or scarred part of the lung. Here we make use of a principle that departs from ordinary radiography and is called *tomography.*

Tomography provides a reasonably sharp image of a particular structure, while images of structures lying in front of or behind it are blurred beyond recognition. Ideally, the desired image stands out sharply in a hazy sea of blurred shadows. *Tomography simply consists of motion of the tube and film in opposite directions in a prescribed manner during the exposure.*

It is of interest that the principles of tomography, one of the outstanding innovations in radiology, were discovered independently in France by Bocage (1921), and Portes and Chasse (1922); in the Netherlands by Ziedses des Plantes (1921); and in the United States in 1929 by Kieffer, a radiographer. The device is called a *tomograph,* and the product a *tomogram.*

Although various systems of tomography have been developed—

planigraphy, stratigraphy, laminagraphy, tomography—the ICRU, in *Report 10f,* proposed the inclusive term **tomography** for all these systems. The device is called a **tomograph,** and the radiographs **tomograms.** We shall adhere to this terminology.

Principles of Tomography. We can most readily understand this by first considering how a **moving** source of light casts a shadow of an opaque object. The model consists of a pen flashlight, a pencil, and a sheet of paper. On arranging these items in a darkened room as in Figure 20.9A and moving the light to the right as in Figure 20.9B, we see that the shadow of the pencil moves to the left across the paper. Suppose we next replace the paper with a sheet of *film* and repeat the experiment. Now, as the shadow of the pencil moves across the film, its image on the film will be blurred. But if we move the film to the left at the same speed as the shadow, the position of the shadow on the film remains unchanged and its margin will be sharp.

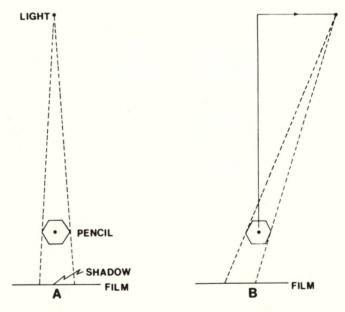

Figure 20.9. In *A* the shadow of the pencil falls directly below. In *B* the light has moved to the right, displacing the shadow to the left. If the shadow were to fall on a moving film which moved to the left at the same speed as the *shadow,* the latter would have a sharp (unblurred) image.

Suppose, now, we line up three pencils one above the other in the path of light from a stationary source as in Figure 20.10. They will cast a single shadow representing the superimposed shadows of all three pencils. But if we move the light source to the right, as shown in Figure 20.10, while moving the film to the left at the same speed as the shadow of one of the pencils such as *1*, its shadow image will remain sharply defined on the film. At the same time, the shadow of pencil *2*, moving faster than the film, and the shadow image of pencil *3*, moving slower than the film, will both be blurred. This, basically, is the tomographic principle.

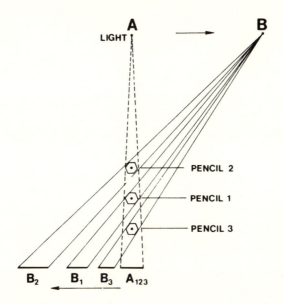

Figure 20.10. The effect of motion of a source of light on three objects lined up as shown. In position *A*, the images of all three pencils are superimposed, falling on the same spot as shadow A_{123}. In *B*, with pencil 1 at the pivot (fulcrum) level, if the light source moves toward the right and the film moves to the left at the same speed as shadow B_1, this shadow will remain sharp. The shadow of pencil 2, B_2, moves faster than the film and is blurred out. The shadow of pencil 3, B_2, moving slower than the film, is also blurred out. Note that the shadow of pencil 2 is larger than that of pencil *1*, whereas the shadow of pencil 3 is smaller than that of pencil *1*.

Substituting an x-ray beam and two structures in the body for the pencil-light model, we find an analogous situation. As shown in Figure 20.11, with the tube at *A*, the images of structures *1* and

2 are superimposed on the film. If we now shift the tube to the right (from A to B) **during the exposure** while moving the film in the opposite direction at the **same speed as the image of the desired structure, 1,** this image will maintain a constant position on the film and have good recorded detail. On the other hand, the image of structure 2 will move faster than the film and be blurred.

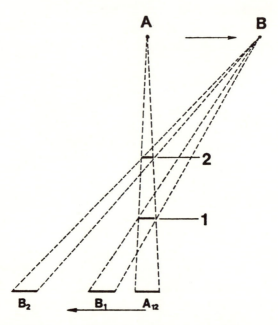

Figure 20.11. Principle of tomography. At focal spot position A, the radiographic images of structures *1* and *2* are superimposed on the film at A_{12}. When the tube is shifted to B and the film moves in the opposite direction so that the axis of rotation (pivot) of the x-ray beam is at level *1*, the image of *1* at B_1 will have a constant position on the film and will be *sharp*. In other words, A_1 and B_1 will strike the same spot on the film. But image B_2 will shift relative to B_1 and will blur out. (In practice, the tube is moved from position B through A to a corresponding point on the opposite side of A during the exposure.)

By applying the tomographic principle, we can radiograph a "slice" of desired thickness, called a **tomographic section,** at various depths in the body. Recorded detail in the section is good, whereas the images of structures lying in front of and behind it are blurred beyond recognition. As we shall see, this applies especially in the radiography of certain anatomic areas such as the first cervical vertebra, larynx, sternum, kidneys, and temporomandibular joints. It is also indispensable in uncovering pulmonary cavities in dense lung and in delineating constricted bronchi and mediastinal invasion in lung cancer.

The Three "Musts" in Tomography. To achieve true tomography, the following three conditions must prevail:

1. *Reciprocal motion of the tube and film,* that is, always in opposite directions.

2. *Parallelism of the film plane and the objective plane* (central plane of tomographic section).

3. *Proportional displacement of the tube and film.* The ratio of the SID to the source-objective plane distance must be constant throughout the excursion of the tube to keep magnification constant during tube travel.

To satisfy these conditions, the tube and the Bucky tray must be connected by a rigid rod that allows them to rotate about a pivot or *fulcrum.* There must also be movable *joints* or *linkages* between the rod and the tube carriage and between the rod and the Bucky tray. A simple drawing of a *rectilinear* ("linear") tomograph appears in Figure 20.12. Only images of points in the plane of the fulcrum (objective plane) that are parallel to the film remain in constant position on the film and show good recorded detail. Images of points lying in planes above or below the objective plane are blurred in proportion to their distance from the objective plane. Such blurred images are less visible than the sharper image of the objective plane. Figure 20.13 shows how the radiographic image of a point, *P,* outside the objective plane shifts on the film relative to the position of the image of a point, *E,* inside the objective plane.

In Figure 20.14 you see a schematic diagram of a simple rectilinear tomograph with the standard terminology and symbols used by the International Commission on Radiologic Units and Measurements (ICRU).

Thickness of Section. A tomogram is really a radiograph of a slice or *slab* of tissue rather than of a plane of zero thickness. Since blurring gradually increases at greater distances from the objective plane (ie, level of fulcrum) there is no distinct boundary between the tomographic section (slab) and its surroundings. Thus, we define *the tomographic section as the slab between two selected planes of maximum permissible blur.* Obviously, the tomographic

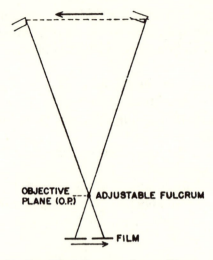

Figure 20.12. Simplified diagram of a rectilinear tomograph.

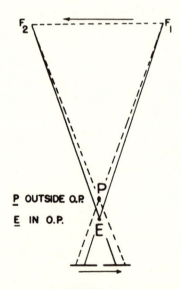

Figure 20.13. Relative shift of images of points lying outside (P) and within (E) the objective plane during tomographic excursion. Note that the image of point E retains the same location on the film, while the image of point P moves across the film and blurs. $O.P.$ = objective plane.

images of all points in the tomographic section will be sharper than the images of points beyond the boundary planes. If the maximum permissible blur has been properly chosen, say 0.5 mm, the images of all points within the slab will have acceptable recorded detail. Note that blur refers to the actual width of the blurred boundary of an image.

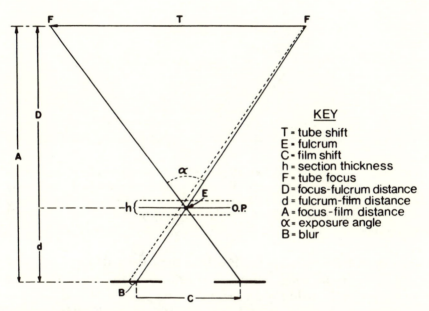

KEY

T = tube shift
E = fulcrum
C = film shift
h = section thickness
F = tube focus
D = focus-fulcrum distance
d = fulcrum-film distance
A = focus-film distance
α = exposure angle
B = blur

Figure 20.14. Standard terminology of the *ICRU* applicable to tomography.

Two factors affect the thickness of section *h* for a given maximum permissible blur *B:*

1. *Exposure Angle* —the angle through which the central ray moves during the actual exposure; that is, the angle, at the fulcrum, made by the central ray between the beginning and the end of the exposure. This definition applies regardless of whether the motion is rectilinear (straight line) or pluridirectional. It is to be emphasized that *the larger the exposure angle the thinner the tomographic section,* and conversely (see Figure 20.15). Note also that the angle through which the *tube* moves exceeds the exposure angle because tube motion starts before the exposure begins, and stops after the exposure ends.

2. *Source-Image Receptor Distance* —with a constant *exposure angle,* a change in SID does *not* affect section thickness (see Figure 20.16). But when tube excursion is stated as amplitude

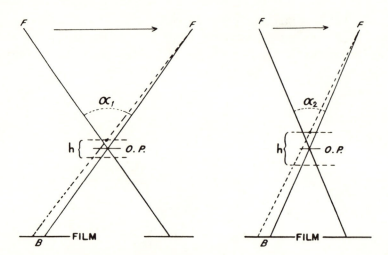

Figure 20.15. Thickness of tomographic section increases with a decrease in exposure angle. The blur *B* of the limit of the thicker section in the right-hand diagram is the same as that of the thinner section in the left-hand diagram; thus, the recorded detail is equal at the upper and lower boundaries of the thick and thin section.

(*distance* tube moves), section thickness does depend on SID; amplitude is no longer used. SID and OID influence recorded detail just as in ordinary radiography (see pages 319–322).

An equation for section thickness derived by Kieffer may be simplified as follows by the use of simple trigonometry:

$$h = \frac{B}{tan_{1/2}a} \tag{3}$$

where h = section thickness
        B = maximum permissible blur
        a = exposure angle in degrees

If we select a reasonable maximum blur such as 0.5 mm, then Table 20.1 shows the approximate thickness for various exposure angles. These data generate the curve shown in Figure 20.17. Note that section thickness decreases as exposure angle increases. At the same time, a thinner section will have less subject contrast than a thicker one.

Tomographic Image. A tomogram consists of two parts. First, it presents good recorded detail of the tomographic section, which we shall call the ***tomographic image.*** Second, it contains the blurred

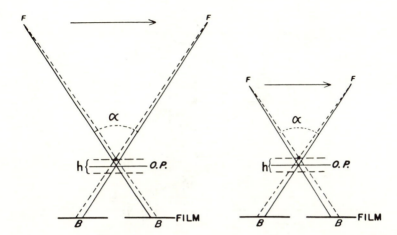

Figure 20.16. Tomographic section thickness is not affected by a change in the SID, provided the exposure angle remains constant. At the upper and lower boundaries of the tomographic sections, recorded detail is the same in both cases.

TABLE 20.1
RELATION OF SECTION THICKNESS TO EXPOSURE ANGLE

Angle	Section Thickness
	mm
50°	1.1
40°	1.4
30°	2.0
20°	3.0
10°	6.0
5°	11.0

images of points outside the tomographic section, called **redundant shadows,** which greatly influence the quality of the tomographic image. Since the redundant shadows are generated by a moving x-ray tube, tomographic image quality ultimately depends on the **path of the tube during the exposure.** The more uniform the blurring of the redundant shadows, the better the recorded detail will be. An example of nonuniform blurring is the streaking of the image in rectilinear tomography.

Thus, tomographic image quality depends ultimately on (1) the recorded detail of the tomographic section and (2) the uniformity of blurring of the redundant shadows. More complex types of tube

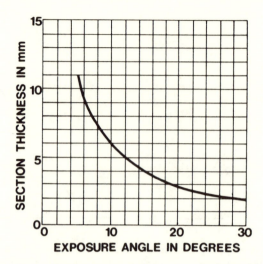

Figure 20.17. Relationship of tomographic section thickness to exposure angle. Note that the *section thickness increases as the exposure angle decreases.*

motion were developed to minimize streaking and thereby improve image quality. These included circular, elliptical, random, and hypocycloidal motion. However, their increasing complexity and cost discouraged wide acceptance, especially when the much superior CT scanners appeared on the market.

Rectilinear tomography is the simplest and most commonly used system. Here, *maximum blurring of redundant shadows and most effective tomography occur when the dominant axis of the tomographed structure lies perpendicular to the tube shift* (see Figure 20.18). Conversely, *minimum blurring and least effective tomography occur when the dominant axis is parallel to the tube shift.* For example, in anteroposterior tomography of the vertebral column, the lateral vertebral margins, being parallel to the tube shift, will appear sharp in *all* sections without blurring of structures outside each section; therefore, these margins are not truly tomographed. But recorded detail of the vertebral plates, oriented as they are perpendicular to the tube shift, will be reasonably well imaged.

Applications of Tomography. As we observed at the beginning of this section, tomography is indicated whenever the desired radiographic image would be obscured by the images of structures lying in front or behind, or whenever the desired structure lies

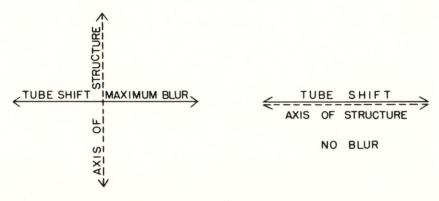

Figure 20.18. Relation of degree of blurring of redundant shadows to the direction of tube shift. Note that maximum blurring occurs when the tube is shifted perpendicular to the long axis of the object, and that no blurring occurs when the tube is shifted parallel to the axis.

within a denser structure. Included are lesions of the chest, sternum, dens, temporomandibular joints, upper thoracic spine (lateral), thoracic spine pedicles, cholecystography and cholangiography, nephrotomography, orbits, middle and inner ears, and others. Figure 20.19 shows a typical setup in rectilinear tomography. A metal rod (behind the table) connects the tube and Bucky tray through a movable linkage at each end. The rod is provided with a pivot or *fulcrum* which is adjustable at various heights above the table and determines the *objective plane.* Because the rigid rod pivots at the fulcrum, motion of the tube in one direction causes the Bucky tray to move in the opposite direction with proportional displacement of the tube and film during the entire range of motion. Tube travel is activated by a motor through a switch located in the control booth.

To obtain a tomographic series, we place the patient in the required position and localize the desired structure by appropriate radiographs. Next, we adjust the fulcrum for various heights above the table, corresponding to the series of objective planes needed to include the volume of interest. At each level of the fulcrum we expose a tomogram—commonly called a *cut.* Usually, the separation of successive fulcrum heights (and objective planes) is 0.5 or 1.0 cm, depending on the size and nature of the structure being investigated. Ordinarily, the exposure angle is 30°, but there are certain exceptions which will be discussed later. In this manner,

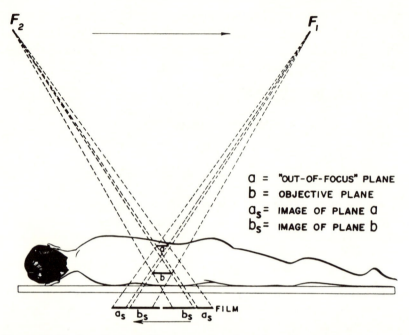

Figure 20.19. Details of rectilinear tomography. The tube is shifted during the exposure in the direction of the upper arrow, while the film moves in the opposite direction. The fulcrum level is selected at the level of the desired objective plane, *b*. The image of *b*, b_2, moves with the film, always striking the same spot. Its image is therefore sharp. But the image of *a*, a_s, constantly moving over the film during the exposure, is blurred out.

we obtain a series of tomograms in which at least one should show the desired structure with optimum recorded detail. For example, a series of tomograms of the dens (odontoid process) should include one that depicts the dens with optimal recorded detail, while the remaining ones will display it with decreasing clarity until it disappears.

Multitomography (multisection radiography). First proposed by Ziedses des Plantes in 1931, multitomography produces a set of tomograms of different sections simultaneously, with one sweep of the x-ray tube. This requires a special multisection or "book" cassette equipped with a series of pairs of graded intensifying screens, a film being placed between each pair of screens. Thus, a multisection cassette accommodates multiple films. Each pair of screens is separated from its neighboring pair by 0.5 or 1.0 cm balsa wood or polyester foam spacer, depending on the desired separation. Because of the decrease in intensity of the x-ray beam as it passes through the cassette (mainly by absorption in the films and screens), the speed of the screens is graded so that the slowest are on top and the fastest at the bottom.

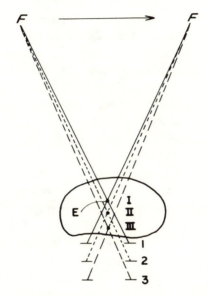

Figure 20.20. Multitomography. Simultaneous "cuts" of objective planes I, II, and III.

Figure 20.20 shows the principle of multitomography. Because x rays diverge, spacers between the screen pairs are slightly thicker than the separation of the objective planes. Spacer thickness (more strictly, separation of the films) can be calculated by the formula:

$$S_F = \frac{AS_L}{D} \tag{4}$$

where S_F = separation of films
A = focus-film distance
S_L = separation of objective planes
D = focus-fulcrum distance

Commercially available multisection cassettes provide three or five pairs of screens with 0.5 or 1.0 cm spacers.

Multitomography has the advantage of saving time, since several tomograms can be obtained with a single excursion of the x-ray tube. At the same time, it has the distinct disadvantage of producing tomograms with poorer contrast and recorded detail, especially toward the bottom of the cassette, as compared with a set of individual tomograms.

Zonography. First proposed by Ziedes des Plantes in 1931, zonography is simply tomography with a *small exposure angle* — 5° to 10° — producing a thick section. For example, with a tomographic angle of 10° section thickness is 6 mm, and with 5°, 11 mm. Such a thick section, called a *zonogram,* superficially resembles an ordinary radiograph. It has better contrast and recorded detail than a conventional tomogram made with a tomographic angle of, say, 30°. Zonography is helpful when (1) the obscuring structures are far from the objective plane and (2) thin sections are not needed. Examples of structures usually suitable for zonography include the sternum (thoracic spine far removed), temporomandibular joint (opposite joint far removed), and nephrotomography (thin sections usually not needed). Zonography is ordinarily ill-advised in studies of the lungs because small cavities may be missed. (See Ettinger for practical use of zonography.)

Patient Exposure in Tomography. In general, the surface dose for a given tomographic exposure is of the same order of magnitude as that for ordinary radiography, but is distributed over a larger area of the body. Patient exposure with multitomography is not much more than that with a sequential series of individual tomograms with the same number of films. Precautions during tomography are the same as for conventional radiography — adequate filtration (HVL 3 mm Al), high kV, collimation, shielding of gonads, and optimum processing of films. Under these conditions, and because its use is relatively limited, tomography adds an insignificant amount of radiation to the general population.

Status of Tomography. Conventional tomography can provide valuable information under limited conditions, especially in small hospitals where computed tomography is not readily available. However, in critical situations, it may not be sufficient and one must obtain computed tomographic (CT) studies as soon as possible. In departments where CT service is available, ordinary tomography has been largely replaced by CT. Still, radiographic-fluoroscopic units are usually equipped with a built-in rectilinear tomograph, mainly for excretory urography and, occasionally, in chest radiography.

COMPUTED TOMOGRAPHY

In ordinary radiography soft tissues are classified as having "water" density. Except for fat, soft tissues do not differ sufficiently in density to provide adequate contrast. In fact, anatomic structures must have a density difference of at least 10 percent to be shown in ordinary radiography, and so we cannot differentiate between such structures as liver and muscle, nor can we see soft-tissue detail within the kidneys or brain. Conventional tomography also has limited usefulness in this regard. We can, of course, use contrast media to expand the diagnostic capabilities of routine radiography.

In 1961 a neurologist, William H. Oldendorf, suggested a method of detecting small density differences among soft tissues (ie, low contrast) by elaboration of tomography, but digital computers were not available to put his ideas into practice. Then, in 1971, Godfrey Hounsfield, a research physicist at EMI, Ltd, in Middlesex, England, installed in a London hospital a *computerized tomographic unit* which he had designed. This proved to be the first major advance in radiology since the discovery of x rays.

Computed tomography (CT) is a complex application of the law of tangents which states that in conventional tomography image quality improves with the number of image-forming rays passing tangential to an anatomic object over its entire border. However, this applies to structures whose density differs sufficiently from their surroundings to provide adequate subject contrast. But with CT each slice (ie, tomographic section) is scanned by a collimated x-ray beam in a circular path around the body over a range of 180° or 360°, depending on the design of the equipment.

An x-ray tube is mounted on a support known as a *gantry* which rotates around the patient (see Figure 20.21). The exit beam for each position of the tube is received by one or more very small radiation detectors mounted on the gantry, directly opposite the tube. There are two kinds of detectors: one is solid, such as a sodium iodide or bismuth germanate crystal coupled to a photomultiplier tube (see pages 473–474); the other is a gas-filled (eg, xenon) ionization chamber. The total attenuation ("absorption") by structures that differ almost imperceptibly in density (ie, extremely low contrast) is measured for each position of the tube and detector. In other words, an extremely large amount of data—approaching

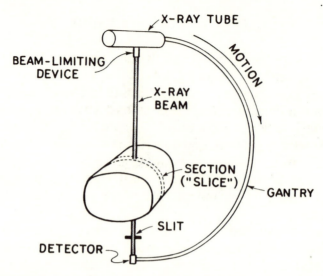

Figure 20.21. Basic arrangement of a computed tomography (CT) scanner. As the x-ray tube moves in an arc around the patient during the exposure, its narrowly collimated beam produces a tomographic section or slice. The exit or remnant radiation strikes a detector from many thousands of directions, and the resulting information is processed by a computer which reconstructs an image of the section. The tube and detector are mounted opposite each other on a rigid gantry so that the detector always remains directly opposite the tube focus. (Adapted from *The Fundamentals of Radiography*, 12th ed, Eastman Kodak Company.)

100,000 counts — is accumulated for each tomographic section and then rapidly computerized in a number of stages to yield a tomogram. Such data processing requires a digital computer. The resulting tomogram reveals exquisite recorded detail of internal structures of each section, whereas with ordinary tomography they would appear as water density without visible internal structure.

We may summarize this introductory discussion as follows:

1. Numerous x-ray projections are made of each of a series of successive contiguous tomographic sections at intervals over an arc of 180° or 360°, depending on equipment design.

2. Images are received by appropriate detectors in the form of "counts."

3. The tremendous amount of accumulated data is processed by the computer to yield a tomogram of internal structures with superb recorded detail, impossible with conventional tomography.

We shall now describe some of the important features of CT scanners.

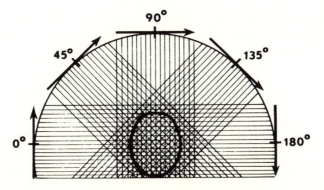

Figure 20.22. Simplified diagram of the scan pattern of the first-generation EMI CT scanner, which used rotation-translation motion. At each designated angle, at one-degree intervals, a pencil beam of x rays enters the patient and the exit or remnant radiation is picked up by a pair of collimated detectors arranged side-by-side directly opposite the x-ray tube, on the other side of the patient. Scanning x-ray exposure occurs only during the linear motion of the tube as indicated by the arrows. After each scan, the gantry rotates one degree without x-ray exposure, and exposure is again made during linear motion in the new position. With 160 measurements of transmitted radiation, multiplied by 180 degrees of rotation, the total number of measurements is 28,800. These are assigned CT numbers and processed by the computer to a final reconstructed image. In the second-generation scanner the pencil beam was replaced by a fan beam and multiple detectors to speed up scanning time.

Types of Scanning Motion. CT scanners have undergone rapid evolution in design over four "generations" or stages, in a period of about ten years. However, we shall limit our discussion to two major types of motion.

1. *Rotation and Translation.* The EMI scanner originally designed by Hounsfield combined two kinds of motion: rotation around the body, alternating with translation (ie, rectilinear motion) across it, as shown in Figure 20.22; in other words, the x-ray tube moves over an arc of a specified number of degrees (part of a circle) and then moves in a straight line. This sequence is repeated over the entire arc of 180° (one-half circle), with the x-ray tube activated *only during translation.* The closely collimated x-ray tube and detectors are maintained directly opposite each other at all times by the rigid gantry, which is free to rotate. A pair of solid detectors is mounted side by side so that with each full rotation-translation of the tube and detectors, two tomographic sections are obtained simultaneously. The original EMI scanner used a narrow "pencil"

x-ray beam and required about 5 minutes for each section pair, or 25 minutes for a whole brain scan. This scanner used an oil-immersed 120-kV x-ray tube with a stationary anode and a large focal spot.

2. *Rotation with a Pulsed Beam.* CT scanners have undergone rapid improvement in design to shorten scanning time and enhance recorded detail. In a rotating scanner (see Figure 20.23) the x-ray beam is fan shaped, the long axis of the x-ray tube being perpendicular to the width of the beam; this makes the x-ray distribution in the beam more uniform by avoiding the heel effect. At all times during a scan, the beam covers the full width of the section being scanned. Opposite the x-ray tube is mounted a row of detectors, either in a segment of a circle or in a complete circle. In the former arrangement, the array of about 30 detectors moves with the tube, but in opposite directions, because they are mounted on the same rotating gantry.

In the *full-circle* array, there is a ring of about 600 stationary detectors with the tube mounted just within the circle (see Figure 20.24). The tube moves over a full circle, 360°. Both systems, known as third- and fourth-generation scanners, respectively, have greatly increased the speed of the examination, thereby reducing motion blurring. The brain can be scanned at the rate of about 2 seconds per section. Although the fan beam increases the amount of scattered radiation in comparison with the pencil beam, each detector has its own collimator (as does the x-ray tube itself) to minimize the amount of scattered radiation reaching the detector. These scanners use rotating anode tubes with 0.6 or 1.0 mm focal spots. They are programmed to deliver x rays intermittently in 300 bursts or pulses for each tomographic section. With each pulse lasting about 3 msec (0.003 sec), the tube is only activated over a total period of 1 sec per section. This prolongs tube life, allows the use of a small focal spot for improved recorded detail, and shortens scan time by permitting a larger total exposure per scan.

The detectors in the later generation scanners are of two types. One is a bismuth germanate crystal coupled to a photomultiplier tube, which counts the x-ray photons reaching it in much the same way as a radionuclide scintillation detector. The bismuth germanate crystal has greater efficiency in absorbing photons and also has less afterglow than a sodium iodide crystal. The other type of

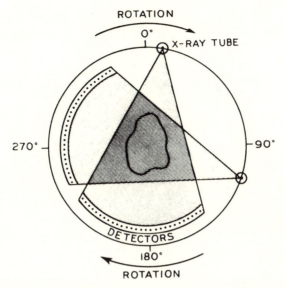

Figure 20.23. Third-generation CT rotational scanner with a fan beam and an array of thirty detectors. Mounted on a gantry, the tube and detectors rotate together around the patient. Note that the detectors always remain directly opposite the x-ray tube focus. Readings of the transmitted radiation (exit or remnant) are made continuously and processed by the computer during rotation of the gantry through a complete circle (360°). The x-ray beam is restricted by a beam-limiting device to a thin fan shape, and the detectors are also well collimated. This type of scanner markedly shortened scanning time.

detector is an ionization chamber filled with the heavy inert gas, xenon, under pressure; the gas molecules are ionized in proportion to the intensity of the detected x rays (proportional region, Figure 22.8, p. 473). Gas-filled detectors have virtually no afterglow and can be miniaturized, but they are slightly less efficient and stable than a crystal detector. The exit radiation from each tomographic section is absorbed by the detector and the resulting pulses in the photomultiplier tube counted automatically. Each count is assigned a preprogrammed number — *CT number* — by the computer. A CT number is not proportional to the number of x-ray photons absorbed in the detector but rather to the total attenuation of x rays that have traversed the tissues in that segment of the tomographic section. Thus, the greater the absorption in the tissues, the higher will be the CT number.

Although, as we have stated, CT reveals internal structures

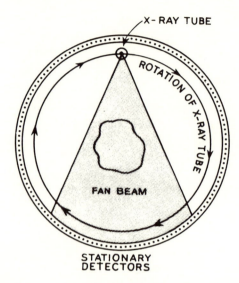

Figure 20.24. Fourth-generation CT scanner. The x-ray tube rotates full circle (360°) while about 600 detectors remain in a stationary array. Note that the beam is fan shaped.

having low inherent contrast, lesions which differ a great deal from their surroundings are more readily detected. Further improvement is achieved by intravenous injection of iodinated contrast media in the vascular system or by ingestion of opaque media, a process known as *contrast enhancement.* Contrast may be enhanced even further by delayed scanning, as with hematomas.

Image Reconstruction. This is a complex mathematical process carried out by the computer in which the image is first reconstructed as a digital image (ie, expressed in attenuation or CT numbers) and then changed to a conventional gray-scale image. The image of the section is broken up into a number of tiny picture elements called *pixels,* which are projections of solid cores of tissue known as *voxels* (volume elements) as shown in Figure 20.25; these are actually dot-sized. As we have just explained, the computer counts the light flashes in solid detectors, or ions in ionization detectors, as the radiation that has traversed each voxel reaches the corresponding detector. The computer instantaneously assigns a *CT number* to each pixel, based on the counts received by the corresponding detector. CT numbers are not absolute values but are relative to the attenuation by water, which is assigned CT number 0. CT numbers usually range from +1000 for bone to −1000 for air or gas. Thus, the computer produces a digital (numerical) image, that is, one in which the attenuation by tissue in the tomographic section is represented by numbers. Figure 20.26 shows how a single projection (tube and detector in a particular position) of a rectangular object might appear in terms of CT numbers. Note that the image pixels consist of higher CT numbers than the neighboring ones because we have chosen an object that is denser than its surroundings. But if the object were less dense than its background, its pixels would be represented by lower CT numbers.

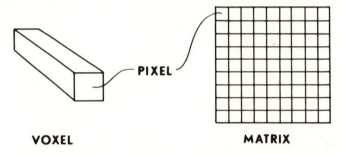

VOXEL　　　　　　　　　　　　　**MATRIX**

Figure 20.25. Explanation of imaging terms. A *voxel* is a volume element—a geometrically defined portion of the tissue section through which the beam has passed. The projection of a voxel, corresponding to its smallest side, is called a *pixel* or picture element in the *matrix*, which represents the tomographic image of the tomographic section. Pixels are actually tiny (ie, dot-sized).

We have just seen how a single projection appears in CT numbers. Reconstruction of the whole image of the section from *multiple projections* can be accomplished in several ways. These have become extremely complex and we shall not try to describe them. The simplest method, *back-projection*, is no longer used in modern CT scanners, but it does provide a relatively simple concept of the principle. Image reconstruction requires, first of all, the handling of an immense amount of data by the computer.

In Figure 20.26, the CT number for each square pixel is proportional to the sum of the object densities (ie, total absorption) in the voxel through which that portion of the radiation has passed. In back-projection, a "magic square" consisting of pixels in, for example, nine rows across and nine columns down (81 pixels altogether)

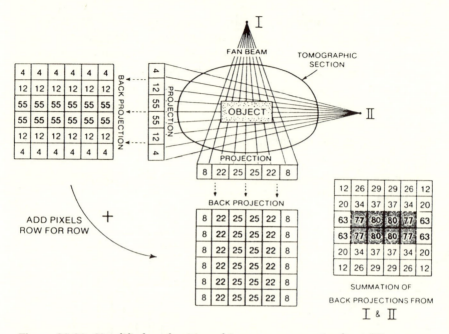

Figure 20.26. Simplified explanation of image reconstruction by a CT scanner. The corresponding pixel CT numbers (ie, numbers assigned by the computer to the radiation attenuation in each voxel) from each back projection are added as shown. The larger the number of projections and back projections resulting from rotation of the gantry around the volume of interest, the better will be the recorded detail in the reconstructed image. This is consistent with the principle in ordinary radiography that recorded detail improves with increased information, provided that noise does not increase as well. When the computer changes the CT numbers to a *gray scale* (analog image), we then have a recognizable tomogram.

forms the basis of the image and is called a matrix (see Figure 20.27). Actually, a matrix consist of a great number of pixels, such as 80 × 80 = 6400, for better recorded detail. In Figure 20.27 we see that the row of CT numbers from one projection is repeated in each succeeding row. Now let us assume that the gantry has been rotated 90° and another exposure made. The second matrix resulting from this back projection is shown on the left. Now the numbers in the corresponding squares (ie, corresponding pixels) are added to give the summated matrix shown on the far right of the figure. This procedure is repeated for each position of the tube as the gantry rotates. The more such projections that are added to the matrix, the better the recorded detail. Figure 20.26 shows the final matrix obtained by the summation of four projections; it reveals the digital image of the rectangular solid. It should be mentioned that image quality is further improved by programming the computer to make a number of important corrections.

Finally, the computer converts the digital image to a gray-scale image to resemble a conventional tomogram, except that it has incomparable recorded detail. Figure 20.26 shows the gray-scale image of the tomographic section obtained by the four projections. In practice, the computer controls allow the operator to increase or decrease image contrast at will.

Medical Applications of CT Scanning. Although CT scanning was first applied to the examination of the brain, it has also revolutionized the diagnosis of lesions in virtually all parts of the body; to name a few: head and neck, chest, abdomen, and vertebral column. In the brain, normal structures are readily depicted— gray and white matter, ventricles, and subarachnoid and subdural spaces. Lesions in the brain are also well shown—tumors, infarcts, abscesses, hematomas, edema, hydrocephalus, cerebral atrophy, and calcifications. Contrast enhancement brings out structural details that would otherwise remain invisible. While there is still an important place for radionuclide imaging and conventional angiography, CT has become the predominant noninvasive diagnostic procedure. In fact it can demonstrate lesions that would otherwise be undiagnosable, short of craniotomy.

Lesions in the abdomen and chest that are particularly amenable to detection and evaluation by CT scanning include tumors of various kinds. In the abdomen, liver, biliary tree, and kidneys can be studied by CT, but ultrasound may be superior in some instances. An important application of CT in the abdomen at present is in the detection of pancreatic tumors and enlarged retroperitoneal lymph nodes. In fact, CT has assumed a prominent role in staging Hodgkin's and non-Hodgkin's lymphoma, as well as other malignant tumors, by virtue of its ability to discover enlarged lymph nodes containing metastatic tumor.

XERORADIOGRAPHY (XEROGRAPHY)

In 1937, C. F. Carlson invented *xerography* (pronounced "zero-"), a radically different kind of photocopying process. Not long afterward the xerographic method was applied as a substitute for film in radiography where it became known as *xeroradiography.*

The xerographic *principle* depends on the fact that semiconductors such as selenium, while ordinarily behaving as insulators, become conductors when struck by light or x rays. A typical xerographic plate consists of *selenium* coated on a sheet of metal, usually aluminum. In use, the aluminum is electrically grounded.

To make a *xeroradiograph* or, in short, a *xerogram,* a positive electrostatic charge is first applied evenly to the selenium surface. Next, the charged plate is placed under the anatomic part and the proper x-ray exposure made, just as in film radiography, although the exposure factors are different. In any area of the plate struck by x-ray photons, the selenium loses positive charges through the grounded aluminum backing, the amount of charge neutralization depending on the x-ray exposure. Thus, the selenium loses much of its charge in areas beneath soft tissues which transmit large numbers of photons, whereas it loses relatively little charge in areas underlying bone, which transmits few photons.

With this system we obtain an *electrostatic latent image* consisting of areas varying in the density of positive charges remaining on the plate after exposure. These densities are then converted to a visible or manifest image by spraying the plate with a toner—a fine cloud of negatively charged blue plastic powder that sticks to the various areas on the plate in amounts depending on the size of the residual positive charge. A permanent image is obtained by pressing a sheet of clear plastic or plastic-coated paper against the plate, thereby transferring the blue powder to the plastic, which is then peeled off the plate.

After it has been cleaned and subjected to 45 seconds of heating or *relaxation,* the plate can be re-used. Heating eliminates residual images that may appear as "ghosts" on the next xerogram.

At present, xerography has its main application in mammography (see pages 344–347) but is useful also in the examination of small bones, as well as in tomography of the larynx and upper part of the trachea. It has the following important advantages over film radiography where it is applicable:

1. *Recorded Detail Enhancement.* Xerograms have a unique property called *edge effect,* an accentuation of the borders between images of different densities. This produces an apparent increase in local contrast and recorded detail. The edge effect is caused by the "theft" of powder by the denser area from the less dense area along the border (electrical "fringe" phenomenon). As a result,

details stand out strikingly, resembling bas relief. The edge effect becomes more pronounced as overall contrast is increased; therefore, low or medium kV technics are preferred. It is especially important in accentuating the image of tiny calcium deposits or *microcalcifications,* present in about 30 percent of breast cancers.

2. *Low Over-all Contrast.* Despite the apparently high contrast caused by the edge effect, the over-all contrast of xerograms is *low* in comparison with ordinary radiographs. The low contrast and the edge effect together produce xeromammograms in which details stand out boldly and all structures, from fat to bony ribs, are clearly seen with a single exposure. Although xeromammograms, after adequate experience of the observer, are easier to interpret than film mammograms, early comparative studies show them to yield about the same degree of diagnostic accuracy. However, experts still disagree about the relative merits of the two types of mammography.

3. *Wide Latitude.* Xeroradiography is characterized by wide exposure latitude, so that variation of about 10 kV at 80 kV in examinations other than mammography has only a slight effect on xerogram quality. At the same time, a wider range of densities can be depicted on a single plate, facilitating interpretation.

4. *Magnification Effect.* Not entirely explained is the phenomenon of magnification of microcalcifications and other fine structures, making them much easier to detect.

5. *Rapid Process with no Liquid Solutions.* Processing is complete in ninety seconds in the xerographic processor in a light room. No plumbing is required, since this is entirely a dry process.

6. *Less Patient Exposure in Mammography.* With suitable technic, xeromammography effects substantial reduction in patient exposure. Excellent xeromammograms are obtainable at 44 to 48 kV and 300 mAs with 2 mm filtration, giving a midplane breast dose of about 0.45 cGy (rad) for an average two-view examination. With a special low-dose film-screen system, the corresponding dose is 0.15 cGy (rad) (Jans et al, 1979). With the older method using industrial-type fine-grain film, the breast dose was 8 to 15 cGy (rad)!

7. *Positive or Negative Image Mode.* By proper selection of charge, we can obtain either positive or negative xerograms. Furthermore, these can be transferred to film or paper, depending

on the preference for viewing by transmitted or reflected light, respectively.

Xerography has several limitations. One is the artifacts caused by dust or imperfections in the plate, which may resemble microcalcifications in mammograms. Another is that xerography cannot be used at present in the examination of large parts because of the large exposures that are needed as compared with film-screen systems. Another problem is the specialized service required for proper operation of the xerographic unit, although minor repairs can be made by the radiographer after a short course in maintenance of the equipment. Refer to pages 532–533 for a discussion of breast doses during mammography, and information on dose reduction.

DIGITAL X-RAY IMAGING

Introduction

Marked improvement in computer capability and appropriate ancillary equipment have led to more sophisticated radiologic imaging, as already described for computed tomography. Now we shall turn to a method of imaging opacified arteries with simultaneous creation of subtraction images.

Subtraction technique has been used for many years in conventional arteriography to enhance the images of opacified arteries when they are concealed by superimposed bone; for example, carotid and vertebral arteries overlying cervical vertebrae. Manual subtraction must be done, after processing the arteriographic series by a *delayed* subtraction technique, according to the following steps:

1. *Copy* the best early **nonopacified** radiograph onto a special, high-contrast, single-emulsion subtraction film by exposure to light by the use of a modified glass-front cassette or subtraction printer. This copy film, after processing, serves as a **mask;** it is a **positive,** the original being a **negative.** The tones on the mask are reversed from the original, black to clear, and clear to black.

2. *Superimpose* in this order in the printer: the **mask** (positive), the selected **arteriogram** (negative), and an **unexposed** sheet of subtraction-type film. Close the printer (see Figure 20.27).

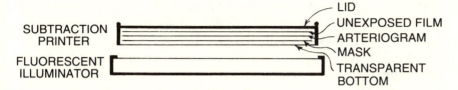

Figure 20.27. Cross-sectional diagram of a manual subtraction printer.

3. *Expose* to a source of light such as a fluorescent illuminator for an appropriate time interval, and process the subtraction film.

The subtraction radiograph will show the artery in black, while the bone has been selectively obscured by the mask; the positive mask and the negative arteriogram, when exactly superimposed, compensate for their opposite tonal scales, producing a subdued background for the opacified artery (positive). If desired, one can print the subtracted arteriogram on another sheet of film to reverse the image to a negative (artery white). The subtracted arteriogram has reasonably good image quality, especially since short-scale contrast is desirable.

Manual subtraction with conventional rapid-sequence arteriography has been virtually replaced by digital subtraction arteriography, to be described in the next section.

Digital Fluorography

The terms used in digital imaging have not been standardized. Some use *digital fluoroscopy* when the fluoroscope is used for any part of the procedure, and digital radiography for an evolving method of routine digital radiography. Others use *digital radiography* when a fluoroscope with image intensification is used in digital arteriography. The author suggests the term **digital fluorography** for arteriography with subtraction according to the method described below.

Digital fluorography makes it possible to perform subtraction automatically **during** arteriography. It is a sequential process, starting with the image in light, on the output phosphor of the image intensifier. This is an **analog image**, so-called because there is a continuous brightness range (an analog watch has hands which move continuously over the dial). In brief, the essential sequence is: output phosphor's analog image is picked up by a video camera whose analog image is digitized, processed, and converted to a gray scale analog image. A digital image is one composed of *discrete image bits* (eg, a digital watch shows a sequence of numbers as discrete symbols—numerals—with abrupt transition from one number to the next).

Now let us examine in some detail how an image in light can be digitized to form a *digital image* (see Figure 20.28). The analog image at the output phosphor of a high-quality image intensifier enters a video camera, where various degrees of image brightness undergo a change to video electronic signals whose intensities correspond to the brightness range of the original image. These analog video signals are then fed via cable to an *analog-digital converter,* which digitizes the image and stores it in the computer's main memory. There the digital image is processed to form a *matrix.* Next, a *digital-analog converter* changes the digital image to an analog image, following appropriate noise abatement and contrast optimization. Finally, the image is displayed on a TV monitor for direct viewing, and to a videodisc for later viewing and study.

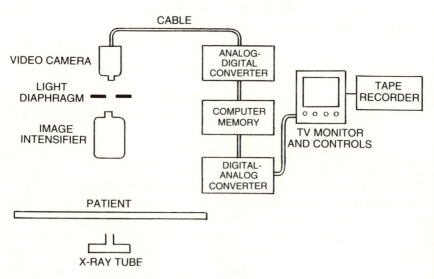

Figure 20.28. Block diagram of a digital photofluorographic unit.

What is a *matrix?* This was mentioned in the section on CT. The term matrix has several connotations, but for our purpose, it is a square or rectangular array of numeric data, each in its own box or *pixel.* Recall that a pixel is simply a picture element from which the picture is constructed. The matrix contains the pixels arranged in horizontal rows and vertical columns (see Figure 20.29). Each pixel receives an assigned number, representing the sum of the exponents of 2, which depends on the size of the video signal (output phosphor brightness) assigned to that pixel. So a pixel with number 6 has the sum of the *logs* or exponents of 2^0, 2^1, 2^2, and 2^3 (ie, $1 + 2 + 3 = 6$), since computers use the *binary system,* with 2 as the log base (review logarithms, pages 23–24).

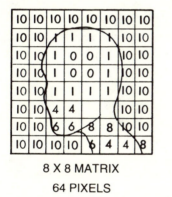

8 X 8 MATRIX

64 PIXELS

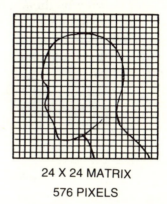

24 X 24 MATRIX

576 PIXELS

Figure 20.29. Examples of an 8 × 8 and a 24 × 24 matrix. Note that matrix size remains constant, but as pixel number is increased, pixel size decreases. In practice, 256 × 256 or 512 × 512 matrices are used; their pixels are dot-sized, resembling the dots in a half-tone print.

Pixel numbers are *relative,* not absolute; their values relate to a pixel containing zero, which represents the absence of a signal and is therefore "black." The larger the pixel number, the brighter the image (white is maximum); and the smaller the pixel number, the darker the pixel (black is minimum). The intermediate numbers indicate various shade of gray, so the range of signal brightness translates, in the end, into a gray-scale image—the number of shades of gray between white and black. It must be emphasized that the computer also assigns the *position* of each video signal to the proper pixel, corresponding to the source of the signal in the original analog image.

Matrix size equals the total number of pixels in a row times the number in a column, and *not* the dimensions of the matrix. This means that a "larger" matrix has more pixels than a "smaller" matrix, while its area remains the same. Commonly used in digital fluorography is the 512×512 matrix, although a 1024×1024 matrix has smaller and more numerous pixels (see below).

Subtraction in Digital Fluorography. Subtraction is performed as follows during image acquisition in the course of arteriography. An early image, without arterial opacification, is stored in computer memory as a *mask.* As the digital images with arterial opacification are obtained, the mask is automatically subtracted from each image and displayed on the TV monitor, along with the mask, for viewing by the radiologist. The images are also stored on a videodisc for later viewing and study. A variety of subtractions can be made by manipulation of the controls (see Table 20.2 for a comparison of manual and digital subtraction).

TABLE 20.2
COMPARISON OF MANUAL AND
DIGITAL SUBTRACTION ANGIOGRAPHY

	Manual	*Digital*
Advantages	Not so subject to patient motion	Improved contrast resolution
	Lower cost	Rapid process
		Programmed automatic immediate subtraction
		More flexible control of image quality
		Less radiopaque medium needed; eg, selective examination of carotid circulation by injection of aortic arch
Disadvantages	Delayed process	Very dependent on patient motion, especially between mask and arteriograms
	Limited manipulation of image quality (trial and error)	Artefacts possible by excessive contrast

Image Quality. As in ordinary radiography, image quality with digital fluorography is of paramount importance, since this dictates the amount of available information. Here are the factors in image quality with the digital system.

1. *Pixel Size.* As already noted, a matrix is specified according to the number of pixels it contains. The pixels are *not* really stored

in boxes, but rather in locations in computer memory corresponding to points in the analog image. If a matrix has 8 columns and 8 rows, it is designated an 8 × 8 matrix and contains 64 pixels. A 512 × 512 matrix contains 262,144 pixels. Since these occupy the same overall space as the 64 pixels in the 8 × 8 matrix, they would have to be about 41,000 times smaller! In other words, the larger the matrix, the larger the number of pixels and the smaller their size.

The *width* of a pixel can be easily determined by dividing the width of the image on the intensifier's *input* phosphor by the number of pixels in the horizontal row of the matrix:

$$\text{pixel width} = \frac{\textbf{image width on input phosphor (mm)}}{\textbf{number of pixels in row}} = 1 \text{ mm}$$

Thus, with a 6-in. (152 mm) input phosphor and 512-pixels,

$$152 \text{ mm}/ 512 = 0.3 \text{ mm pixel width}$$

which is the theoretical limit of resolution of the system. With a 1024 × 1024 matrix, pixel width is 0.15 mm. To put this into perspective, ordinary radiography with a film-screen combination can image an object as small as 0.1 mm, nearly the same as the digital image. Pixels are actually about the *same size as the dots in a half-tone print.*

Look at pixel size this way with regard to image resolution. The larger the pixel size, the larger will be the fragments into which the image will be "broken," so the image will appear as though made up of square blocks with few shades of gray between white and black. In fact, the image becomes unrecognizable. But with a small pixel size, the image is broken into smaller fragments which become less distinct, while the image becomes clear, with good recorded detail and resolution. In fact, it resembles a conventional radiograph. The computer must be designed to be able to handle the increased information presented to it by a large matrix, such as 1024 × 1024.

2. *Contrast.* Just as in ordinary radiography, contrast plays a major role in image quality. You have learned earlier that contrast scale should be neither too short or too long, although at times short scale contrast is desirable, as in arteriography. Generally, though, excessive contrast may lose information on account of the absence of enough intermediate shades of gray. On the other

hand, extremely long-scale contrast may impair visibility of recorded detail. In digital fluorography, the range of pixel numbers, or *numeric range,* expresses the spread of signal intensities and determines the contrast scale. The numeric range increases with matrix size; in a 512×512 matrix the numeric range is 2^9 ($= 512$), this being designated a 9-bit range. Similarly, a 1024×1024 matrix has a numeric range of 2^{10}, or a 10-bit range. A 2^1 numeric range, with a 1-bit range, would have only two tones, white and black—an extreme case. By computer manipulation, one can modify the numeric range to shorten or lengthen contrast scale as needed. Moreover, there is virtually *no loss of information* in the process.

3. *Noise.* As is true of any system that plays on our senses, digital fluorography is afflicted by *noise.* Here, the main source of noise is spurious electric signals in the *video camera,* which must therefore be of high quality. These non-information-bearing signals obscure informational signals and are difficult to eliminate. The relation between the intensities of signal and noise is called the *signal-to-noise (S/N) ratio.* The higher the S/N ratio, the more the signal drowns out the noise; and the lower the S/N ratio, the more the noise obscures the signal. Also, the higher the S/N ratio, the higher the fidelity of the system, which means that the received information corresponds closely to the transmitted information. For example, static in radio reception is noise, which is virtually eliminated by a high fidelity (hi-fi) system.

Noise has a severely detrimental effect on image quality in digital fluorography. Anything that increases the signal without also increasing noise to the same degree, improves the ratio. The following factors improve the S/N ratio:

1. *Increase Signal Intensity.* High mA yields a brighter image, which results in enhanced signal intensity. Recall that mA is the most important carrier of information in radiography. Moreover, high mA reduces quantum mottle (see pages 338–339).

2. *Reduce Scattered Radiation.* The factors in limiting scattered radiation were considered in detail under *radiographic quality;* they include beam limitation, grids, and lower range of kVp. Scattered radiation on the input phosphor eventually shows up as noise in the digital images. Thus, a decrease in scatter increases the S/N ratio.

3. *High Fidelity Video Camera.* As noted above, the video camera is the main contributor of noise to digital imaging. It is there-

fore imperative that the video camera have a high S/N ratio of no less than about 1000:1, an ordinary one having a ratio of about 200:1.

A method of reducing detectable noise involves a process called *smoothing*. Computer manipulation can be used to average the values of groups of adjoining pixels in a regular sequence throughout the matrix. Noise becomes less perceptible, but there is some decrease in image resolution.

Digital subtraction image quality is especially subject to the detrimental effect of noise. The mask and opacified images must have similar high fidelity. Since noise is a random process, the two digital images have a high probability of possessing different noise distributions and S/N ratios in various pixels. This again shows the need for high mA and a hi-fi video camera, as well as optional smoothing. Table 20.2 compares manual and digital subtraction.

X-Ray Generator. Digital fluorography starts with an under-the-table x-ray tube (see Figure 20.28). However, it must have a high enough rating to operate at several hundred mA, like an angiographic tube, capable of firing short bursts of x rays in rapid sequence. This requires on-and-off switching at the rate of 1 or a few per sec. The high mA produces a bright image without appreciable quantum mottle (see pages 338–339), to permit digitization yielding high-quality images during the repeated short exposures. A three-phase power supply is required to provide the high-intensity, short, repeated exposures with rapid on-and-off switching.

Future Prospects. A number of methods of digitizing radiography are in the process of development; they may properly be called *digital radiography.* One of these uses a closely restricted fan-shaped x-ray beam similar to that in moving-slit radiography (see pages 394–395) and in CT scanning (see pages 430–432). The beam is narrowed to a thickness of about 1 cm and an angle of about 45 degrees by two coaxial (in tandem) beam restricters, one being located between the x-ray tube and patient and the other on the exit side of the patient. As a result, scattered radiation is markedly reduced, thereby enhancing contrast. During radiography the fan beam moves broad-side (at a right angle to its flat side) across the anatomic part. The image receptor consists of an array of closely-spaced, tiny electronic detectors whose output signals are digitized and then processed by a computer into an analog image. Such images are stored in the computer memory in digital form, to be retrieved as needed.

Other methods of digital radiography use a conventional x-ray beam and a special ***photostimulable phosphor*** as the image receptor which, when subjected to intense laser energy, emits a blue light; this acts on a photomultiplier tube whose output light is digitized and stored in a computer memory bank, later retrievable as an analog image.

It is believed that film radiography as we know it will eventually become obsolete and give way to digital radiography, with the images stored in computer memory banks in digital form. These can be retrieved for immediate or later use in analog form. The digital images lend themselves to easy manipulation for contrast enhancement, and can also be transmitted over telephone lines by means of a modem.

QUESTIONS

1. Discuss fluoroscopic image intensification. Compare it with ordinary fluoroscopy. What is the advantage in using a cesium iodide phosphor?
2. What is meant by stereoscopic vision?
3. How do we determine the correct tube shift in stereoradiography?
4. List the steps in making and viewing stereoradiographs.
5. Describe the principle of tomography.
6. Define tomographic section; fulcrum; objective plane; tomographic image; redundant shadows; exposure angle.
7. Discuss the various types of tomographic motion. Compare their advantages and disadvantages.
8. Explain the main modifications of tomography: zonography; multitomography.
9. What is the comparative exposure with multitomography and with a series consisting of an equal number of individual tomograms?
10. How should tomographic equipment be activated, for maximum safety of personnel?
11. How is tomographic section thickness defined? Explain the factors that influence section thickness.

12. How does computed tomography resemble and differ from ordinary tomography?
13. Describe briefly the main features of three types of CT scanners.
14. Define voxel; pixel; back projection.
15. Explain the xerographic process and list the advantages and disadvantages of xeroradiography as compared with ordinary film radiography.
16. Briefly compare manual and digital subtraction.
17. State the steps in digital fluorography (digital fluoroscopy).
18. Define matrix; matrix size; pixel.
19. How is the number of pixels related to matrix size?
20. What is the main source of noise in digital fluorography? State the three ways that the signal/noise ratio can be improved.
21. What kind of generator is needed for digital fluorography?
22. Briefly describe digital radiography, and state its prospects for the future.

Chapter 21

RADIOACTIVITY AND RADIUM

O NE OF THE OUTSTANDING ACHIEVEMENTS of the late nineteenth century was the discovery that certain elements had the unique property of giving off penetrating and ionizing radiation. In 1896, Henri Becquerel found that the heavy metal *uranium* emitted rays that passed through paper and darkened a photographic plate in a darkened room. Soon after, Marie Curie and her husband Pierre began looking for other elements that might behave in a similar way. In 1898, after two years of extraordinary hardship and personal sacrifice, they announced their discovery of *polonium* and *radium*, new elements which emitted radiation. Not only did this discovery open new therapeutic opportunities in medicine, but it also expanded immeasurably our knowledge about the structure of matter. The term *radioactivity* was applied by Marie Curie to the special behavior of these elements.

RADIOACTIVITY

Unstable Atoms

The accepted designation of nuclear structure is *nuclide, defined as a nuclear species with a particular number of protons and neutrons, and having a particular energy state.* Radioactive nuclides, known also as *radionuclides,* are exceptional in that they have *unstable nuclei,* which means that their nuclear structure is inherently changeable. In the process of undergoing such spontaneous change, called *radioactive decay* or *disintegration,* they give off *ionizing radiation.*

Why are certain nuclides naturally radioactive? The precise answer is unknown, but there are a number of theories. Nuclear particles exist in energy levels, analogous to those of orbital electrons that occupy various energy levels outside the nucleus. Transition of particles within the nucleus to lower energy levels in radionuclides are also accompanied by the release of radiation. In naturally occurring nuclides with a high atomic number (above 81), nuclear energy is large enough to allow random escape of nuclear particles, usually accompanied by electromagnetic radiation (gamma rays). Such nuclides are unstable and as they give off radiation they change into other nuclides. It is uncertain which particular nucleus of a sample of nuclide will decay at any particular instant, but a *constant fraction of those present will decay in a given time interval* (eg, per sec). This fraction, called the *decay constant* (other terms: *deexcitation, disintegration, or transformation constant*), *is the same for all samples of a given radionuclide but is different for different radionuclides.* The decay constant of a radionuclide cannot be changed without extreme difficulty.

Radioactivity, then, may be defined as the ability of certain nuclides—radionuclides—to (1) emit ionizing, penetrating radiation while (2) undergoing spontaneous, uncontrolled decay.

There are two kinds of radioactivity—natural and artificial—but, as we shall see, they both obey the same laws. *Natural radioactivity* is a property of naturally occurring radionuclides such as radium. *Artificial radioactivity* pertains to artificial radionuclides such as radiocobalt; these are nuclides that have been made radioactive by irradiation of the stable form with subatomic particles (neutrons, deuterons, etc) in special high energy devices such as the nuclear reactor or cyclotron.

The *artificial radionuclides,* to be discussed more fully in the next chapter, make up an impressive array of material that has important application in biologic research, clinical medicine, and industry.

Radioactive Series

The naturally occurring radionuclides can be grouped in three series or families, based on their parent elements: (1) *thorium,* (2) *uranium,* and (3) *actinium.* Each of these gives rise, by spontaneous decay, to a series of nuclides characteristic of that series. The chain of breakdown of the successive nuclides always occurs in exactly the same way and at a fixed rate. Some daughter nuclides, such as thorium A (^{216}Po), decay at extremely rapid rates and are therefore difficult to isolate; others, such as radium, in the uranium series, decay very slowly and can more readily be obtained in pure form. Radium has had an important place in the treatment of certain types of cancer. Radium and its offspring comprise the last descendants of the uranium series.

RADIUM

Introduction

Although radium (Ra) has been virtually abandoned in radiotherapy and replaced by artificial radionuclides such as cesium 137 and iridium 192, it serves as a useful model for the behavior of radionuclides in general. Therefore, it will be discussed briefly as an introduction to the subject of radioactivity and radioactive substances.

Properties

Radium, a heavy metal having a silvery-white appearance when pure, behaves chemically like barium and calcium. It belongs to the uranium radioactive series and has atomic number 88 and mass number 226. Because its radioactivity persists unchanged when combined with other elements, it is available only in the form of one of its salts—usually *radium chloride* ($RaCl_2$)—which are far more easily prepared than the pure metal.

Types of Radiation

There are three different kinds of radiation emitted by radionuclides: (1) alpha particles, (2) beta particles, and (3) gamma rays. A given radionuclide may emit all of these radiations or only one or two of them. Various members of the radium series, as will be shown below, give off one or more of these radiations. Alpha and beta particles ionize matter directly, whereas gamma rays do so indirectly, like x rays.

1. **Alpha particles** are simply *helium nuclei* (ie, helium ions) consisting of two protons and two neutrons. Since this gives them an atomic number of 2 and a mass number of 4, they can be represented by the symbol $^4_2$He. Alpha particles, having a large

mass and two positive charges, strongly ionize the atoms of matter through which they pass. Upon accepting two electrons, an alpha particle becomes a neutral helium atom:

$$\text{alpha particle} + 2 \text{ electrons} \rightarrow \text{helium atom}$$
$$\text{He}^{++} + 2\text{e}^- \rightarrow \text{He}$$

Alpha particles leave the nucleus at high speeds, 1.5×10^7 to 3.0×10^7 meters per sec (9,000 to 18,000 miles per sec) but do not penetrate matter for any appreciable depth because of their strong ionizing ability; in fact, they are absorbed by an ordinary sheet of paper.

2. **Beta Particles,** often called beta rays, consist of high speed *electrons* represented by the symbol e^- or β^-. Ejected from the nucleus at speeds approaching that of light rays, they are able to penetrate matter to a greater depth than alpha particles, their maximum range in tissue being 1.5 cm. In fact, 0.5 mm of platinum or 1 mm of lead is required to stop these beta particles. They are used in therapy, mainly for superficial skin lesions in small areas. Beta particles are less strongly ionizing than alpha particles.

3. **Gamma rays** are high energy electromagnetic waves (photons). For example, the gamma rays of radium have an average energy of 0.83 MeV, equivalent to 1.5 million volt (MV) x rays. They travel with the speed of light and are much more penetrating than alpha or beta radiation. Gamma rays are physically identical to x rays, the difference being that gamma rays originate within atomic *nuclei,* whereas x rays arise from electronic processes outside nuclei. Since gamma rays are electromagnetic waves (and photons), they are not deflected by a magnetic field. They ionize matter in the same way as x rays by first releasing fast electrons, which, in turn, cause much stronger ionization along their paths. The application of radium in therapy most often depends on the gamma radiation, the alpha and beta particles being removed by appropriate absorbing filters.

The Radium Series

When an atom of radium decays, it does not disappear but changes to another nuclide, that is, it undergoes *radioactive transformation.* Let us "look into" an atom of radium and see what happens during its radioactive transformation. The atom ejects an alpha particle, which carries a charge of +2 because of its two protons. Since the atomic number of radium is 88, that of the atom remaining after the emission of an alpha particle is 88 − 2 = 86. Furthermore, the mass number of radium is 226 and that of the ejected alpha particle is 4, so the mass number of the daughter atom is 226 − 4 = 222. Thus, the daughter atom differs from its parent radium; it is the element *radon,* which has atomic number 86 and mass number 222 and is the first decay product of radium:

1 radium atom →1 radon atom +1 alpha particle

$$^{226}_{88}\text{Ra} \qquad\qquad ^{222}_{86}\text{Rn} \qquad\qquad ^{4}_{2}\text{He}$$

Note that the sum of the mass numbers (superscripts) of the products radon and alpha particle equals the mass number of radium. The same holds for the atomic numbers (subscripts), so the equation is balanced.

The radon atom is radioactive, decaying to a different radionuclide, which in turn decays to still another radionuclide, and so on. This chain of decay goes on until a stable atom is left (in this case a metal, **lead**). Figure 21.1, showing the radium series, is included mainly to indicate pertinent data and not to tax your memory. Note that the radiation emitted by the radium element itself is limited to alpha particles; radium element also emits a gamma ray, too weak to be of use in therapy. Beta and penetrating gamma rays first appear with radium B (lead isotope) and C (bismuth isotope). Since a sealed source of radium reaches a state of *equilibrium* with its descendants after approximately thirty days (ie, the same number of atoms of a particular daughter is appearing as is disappearing in a given time interval) and then actually consists of a mixture of these descendants, we speak loosely of radium as emitting all three types of radiation. However, it is the *radium series* as a whole that emits all three types of radiation. Whether we start with radium or radon gas there is soon a sufficient accumulation of radium B and C to give appreciable intensities of beta and gamma radiation.

Radioactive Decay

As we have already indicated, radionuclides such as radium undergo nuclear decay (synonyms *transformation, disintegration*), emitting one or more types of radiation while changing to another nuclide. The radioactive *decay scheme* is a diagrammatic method

	URANIUM $\xrightarrow{alpha}$	RADIUM $\xrightarrow{alpha}$	RADON $\xrightarrow{alpha}$	RADIUM A $\xrightarrow{alpha}$ POLONIUM	RADIUM B $\xrightarrow[gamma]{beta}$ LEAD	RADIUM C $\xrightarrow[gamma]{\substack{alpha \\ beta}}$ BISMUTH	LEAD
Atomic No.	92	88	86	84	82	83	82
Mass No.	238	226	222	218	214	214	206
Half Life	4.5×10⁹yr	1622 yr	3.8 da	3 min	26.8 min	19.7 min	stable

Figure 21.1. Successive decay products of radium, shown from left to right. These make up the radium series which is the last part of the uranium series. Data on radium D, E, and F have been omitted.

of representing the process (see Figure 21.2). The parent radionuclide radium (Ra) should be drawn at the top. An arrow points downward and to the *left*, because the daughter nuclide radon (Rn) has a smaller atomic number than its parent. Each nuclide symbol has a superscript and a subscript indicating its mass number and atomic number, respectively. Finally, the type of radiation should be indicated by its symbol near the arrow, in this case by an α for the alpha particle.

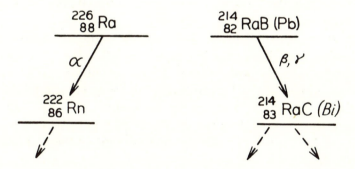

Figure 21.2. Decay schemes of radium 226 and radium B (lead 214). Only the first decay product is shown in each instance. By convention, when the product nuclide has a lower atomic number than the parent, the arrow points downward and to the left, as is the case when radium decays to radon. When the product nuclide has a larger atomic number than the parent, the arrow points downward and to the right, as occurs with the decay of radium B (lead 214) to radium C (bismuth 214). Note the two arrows below radium C—this is an example of branching decay by alternate routes.

In the case of *beta decay*, the daughter nuclide has a larger atomic number than the parent, so the arrow points downward and to the *right*. With *gamma-ray emission* the atomic number of the daughter nuclide remains unchanged, so the arrow points *vertically downward.*

Decay Constant

Radioactive transformation occurs as a random process, that is, we cannot predict which particular atom will undergo nuclear decay at any instant. Starting with a given quantity of radium (or other radionuclide), the activity (number of disintegrations per second) decreases with time, but the *rate at which the activity decreases remains constant* (does not change). There are two ways of specifying the decay rate. One is the *decay constant* which is derived by a special branch of mathematics called calculus; to find the amount of activity remaining after a particular time interval requires the use of a special (exponential) equation.

The other way of indicating decay rate is *fractional decay,* that is, the fractional decrease in activity per given time interval. In other words, fractional decay means that if one gram of a radionuclide decreases 10 percent in one day, 0.9 g remains. At the end of the second day the 0.9 g will also decrease by 10 percent, leaving 90 percent of 0.9 g or 0.81 g, etc.

The decay constant is not the same as fractional decay unless fractional decay happens to be very small, about 0.2 or less (Pfalzner). Both are constant and unique for any particular radionuclide (see page 449). For example, the decay constant λ (Greek lambda) is 4.27×10^{-4} per yr for Ra, and 0.181 per day for Rn. But note that fractional decay for Rn is 16.6 percent per day; its decay constant 0.181 per day can be used only in the equation for exponential decay.

Half-Life

The caption *half-life* appears in Figure 21.1. This is a term which *denotes the length of time required for one-half the initial amount of a radionuclide to decay.* For example, suppose we have 1 gram of radium at the start. Figure 21.1 shows its half-life to be 1622 years (half-lives of artificial radionuclides used in medicine are extremely shorter). Thus, at the end of 1622 years the 1 g of Ra will have slowly decayed to 0.5 g. In another 1622 years it will have decayed to one-half; that is, it will be one-fourth the original amount, or 0.25 g. Notice that the half-life is constant for any given radionuclide, and that the half-lives of different radionuclides show wide differences. Thus, the half-life of uranium is 4.5 billion years, while that of radium A is three minutes.

In any radioactive series, after equilibrium has been reached among all the daughter nuclides, the quantity of any one member of the series is larger if its half-life is longer. Thus, if we start with a radium sample, radium having the longest half-life in its series, will be present in largest amount; whereas radium A (polonium 218) with a half-life of three minutes will be present in smallest amount.

As we have already indicated, there is no way of predicting which radium atom will decay at a given instant. However, we know statistically that in 1622 years one-half the atoms in a given amount of radium will have decayed. Thus, half-life is really a *statistical* concept indicating the time it takes for radioactive decay of one-half the atoms on the basis of pure chance.

It is of interest that the decay constant λ of Ra (or of any radionuclide) can be derived from its half-life T by the following equation:

$$\lambda = 0.693/T$$

Substituting 1622 yr, the half-life of Ra,

$$T = 0.693/1622 = 4.27 \times 10^{-4}/\text{yr}$$

This is the value of Ra half-life quoted above.

Average Life

Although we cannot determine which atom of a radionuclide will decay at any particular instant, we do have a simple way of computing the probable survival time or life expectancy of an individual atom, similar to the determination of life expectancy by life insurance companies. This, the average life T_a, is simply the reciprocal of the decay constant

$$T_a = 1/\lambda$$

For example, the decay constant of Rn is 0.181/day, so the average life is $1/0.181 = 5.5$ days.

Radon

When an atom of radium decays and emits an alpha particle, an atom of radon remains (see page 452). Radon is a radionuclide that exists as a colorless heavy gas. Figure 21.3 shows the decay curve of radon, as well as the derivation of its half-life. As already noted, fractional decay is constant with time.

Because of its convenient half-life of 3.8 days and the availability of gamma rays from its daughter radionuclides, radon was at one time used in implant therapy in the form of "seeds." These were prepared by collecting the radon gas in thin gold tubes and then pinching off and sealing short segments about 3 to 5 mm long. However, because of problems with radiation safety during manufacture and possible leakage of the seeds, radon is no longer available in the United States, having been replaced by artificial radionuclides such as gold 198, iridium 192, and tantalum 182.

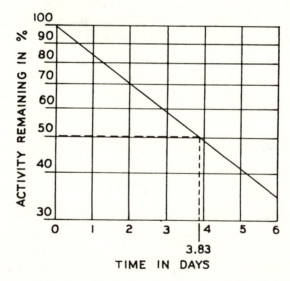

Figure 21.3. Decay curve of radon, semilogarithmic plot. The half-life is 3.83 days.

RADIOACTIVE EQUILIBRIUM*

A radionuclide decay series is said to be in *radioactive equilibrium* when, in a sealed source. the rate of appearance of each daughter equals its rate of decay; that is, the amount being formed per second equals the amount decaying per second. However, a distinction is made between *secular* and *transient equilibrium.*

In *secular equilibrium* the parent nuclide has a much longer half-life than any of its daughters; radium is an example. Because of its long half-life, radium furnishes a virtually inexhaustible supply of daughter nuclides. These accumulate until secular equilibrium has been reached (after about thirty days), provided the radium was first sealed in a leakproof container such as a needle or tube. Now the daughters are present in constant amounts. Since the use of radium in gamma-ray therapy depends on RaB (radiolead) and RaC (radiobismuth), and their amounts are constant at equilibrium, the gamma-ray output of the sealed source is constant, thereby simplifying dosimetry. Furthermore, the half-life of the sealed source is the same as that of the parent radium — 1622 years.

Radon is an example of a radionuclide which reaches *transient equilibrium* with its descendants because it has a relatively *short half-life* of 3.83 days. Upon reaching transient equilibrium at about five hours after being sealed, radon and its decay series continue to undergo rapid decay, with the same half-life as the parent radon. Therefore, the gamma-ray output decreases appreciably in a period of days. On the other hand, the gamma-ray output of a radium source remains constant because of its extremely long half-life relative to treatment time.

QUESTIONS

1. Define radioactivity. By whom was it discovered? Named?
2. Give an example of a decay scheme. What is meant by the statement "decay rate is constant?"
3. What makes some nuclides radioactive? Where do the ionizing radiations come from during radioactive decay?
4. Discuss radioactive series.

---

*Portions of this section are from Selman: *The Basic Physics of Radiation Therapy*, 3rd Ed., 1991. Courtesy of Charles C Thomas, Publisher, Springfield, Illinois.

5. If a radioactive atom emits a beta particle, what change occurs in the atomic number? Atomic mass number?
6. How does the emission of a gamma ray affect the atomic number? Atomic mass number?
7. Briefly discuss radioactive equilibrium.
8. How does artificial radioactivity differ from natural?
9. To what radioactive series does radium belong?
10. Name and describe the three types of radiation emitted by various radionuclides. In the radium series, which members emit gamma rays that are useful in therapy?
11. Define half-life; average life.
12. Define curie; millicurie; becquerel. State the symbols for each.
13. Discuss the use of artificial radionuclides as radium substitutes.
14. Which ones are available and in what forms?
15. What are the advantages of radium substitutes over radium itself?

Chapter 22

ARTIFICIAL RADIONUCLIDES

I N CHAPTER 4 we learned that matter is made up of extremely small particles called *atoms.* The concept of atomic structure has become extraordinarily complex, but for our purpose we may now review it in much simplified form—a central core or *nucleus* containing almost the entire mass of the atom and, surrounding the nucleus, the *orbits* or paths in which the negative *electrons* are in continual motion. The nucleus has two main constituents, *protons* and *neutrons,* collectively referred to as *nucleons.* A proton is a positively charged particle whose mass is nearly 2000 times that of an electron. A neutron has nearly the same mass as a proton but carries no charge.

All matter is made up of one or more simple entities called *elements.* The atoms of any particular element have a characteristic and distinct number of positive charges or *protons* in the nucleus and an equal number of negative charges or *electrons* in the orbits around the nucleus. The number of nuclear protons is called the *atomic number.* Notice that the atomic number is specific for a given element; all atoms of any one element have the same atomic number. Atoms of different elements have different atomic numbers.

Since the nucleons are responsible for the mass of an atom, the *mass number* of a particular atom has been defined as the total number of nuclear protons and neutrons. For example, hydrogen with one nuclear proton has a mass number of 1, and helium with two protons and two neutrons has a mass number of 4.

Isotopes

As we have stated before, all the atoms in a given sample of a particular element have the same atomic number, that is, the same number of nuclear protons (their chemical behavior is also identical). However, all the atoms in this sample do not necessarily have the

459

same mass number—total number of protons **and** neutrons. Since the number of protons is the same, the difference in mass number must be due to a difference in the number of **neutrons.** Such atoms whose nuclei contain the same number of protons, but different numbers of neutrons, are called **isotopes.**

The following equations show how a radionuclide differs from its isotope only in the number of nuclear neutrons:

$$\text{nuclide} = x \text{ protons} + y \text{ neutrons}$$
$$- (\text{isotope} = x \text{ protons} + z \text{ neutrons})$$

$$\text{nuclide} - \text{isotope} = y - z \text{ neutrons}$$

Note that there is not an unlimited variety of isotopes of any particular element; as they occur in nature, there may be only two, or three, or several. However, the numerical ratio of the different isotopes is nearly constant for each element.

Since all the isotopes of an element have the same atomic number, they have the same chemical properties. The gas hydrogen, for example, has three isotopes (see Figure 4.5): ordinary hydrogen or **protium** with mass number 1, **deuterium** (heavy hydrogen) with mass number 2, and **tritium** with mass number 3. As would be anticipated, these three forms of hydrogen have the same atomic number—1—and identical chemical properties.

The term **nuclide** refers to a species of atom having in its nucleus a particular number of protons and neutrons. Accordingly, we can say there are three distinct hydrogen nuclides.

Artificial Radionuclides

In recent years many artificial radionuclides have been produced, a number of which happen to be radioactive. These decay to entirely different elements while emitting ionizing radiation, as described above. Radioactive isotopes have the same **chemical** properties as their stable counterparts, differing only in their mass number (number of nuclear neutrons) and radioactive nature. We prefer to speak of radioactive nuclides or **radionuclides** rather than radioisotopes, since they are nuclides that happen to be radioactive.

Artificial radionuclides are produced by irradiation of stable nuclides by subatomic particles such as neutrons or deuterons in an atomic reactor or cyclotron, respectively. A neutron is an uncharged particle having a mass number of 1. A deuteron is a heavy hydrogen nucleus having a single positive charge and a mass number of 2. The irradiated nucleus captures a neutron or deuteron, becoming unstable, that is, radioactive.

When the daughter element differs from the parent, the process is called **transmutation.** An example of the transmutation of an ordinary, stable element to a different radioactive element will make this process more easily understandable. Ordinary sulfur is changed, during neutron irradiation, to radioactive phosphorus:

$$\text{sulfur} + \text{neutron} \rightarrow \text{phosphorus} + \text{hydrogen}$$
$$\underset{(stable)}{{}^{32}_{16}\text{S}} + {}^{1}_{0}\text{n} \rightarrow \underset{(radioactive)}{{}^{32}_{15}\text{P}} + {}^{1}_{1}\text{H}$$

In each term of the equation, the lower number or **subscript** represents the atomic number, and the upper number or **superscript** represents the mass number. The sum of the superscripts on one side of the equation equals that on the other side. The same holds true for the subscripts. Knowing that the atomic number of phosphorus is always 15, we can express radioactive phosphorus as ${}^{32}\text{P}$. Since it has a specific number of protons (15) and neutrons (17) in its nucleus and therefore a particular species of atom, it is an example of a **radionuclide** or **radioactive nuclide.** ${}^{32}\text{P}$ is unstable because its nucleus contains a surplus neutron (ordinary nonradioactive phosphorus is ${}^{31}\text{P}$) and it therefore decays to ordinary sulfur, emitting a beta particle (negative electron) in the process:

$$\underset{}{{}^{32}_{15}\text{P}} \rightarrow {}^{32}_{16}\text{S} + \underset{beta\ particle}{{}^{0}_{-1}\text{e}} + \underset{antineutrino}{\tilde{v}}$$

At the present time, there is no way of altering the type of radiation emitted by a given radionuclide, natural or artificial.

Another method of inducing artificial radioactivity is to irradiate stable atoms with deuterons (heavy hydrogen nuclei) in a cyclotron. However, this is not a major commercial source of medical radionuclides.

Nuclear Reactor

As mentioned above, radionuclides can be produced by exposing stable nuclides to a flow of neutrons—*neutron flux.* In fact, this is how artificial radioactivity is often induced both for medical and industrial use. One type of nuclear reactor (see Figure 22.1) depends on the fission (splitting) of nuclei of uranium 235 (^{235}U), an isotope of ordinary uranium (^{237}U), by extremely slow, environmental neutrons (see Figure 22.2). These neutrons have a very small average kinetic energy, about the same as that of the molecules of air at the prevailing temperature and are therefore called **thermal neutrons.** When a thermal neutron is captured by a ^{235}U nucleus, the nucleus splits into fragments of various sizes while emitting one to three neutrons plus 200 million electron volts (MeV) of energy:

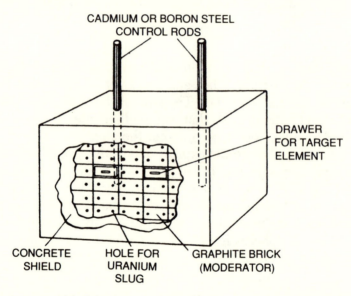

Figure 22.1. Simplified diagram of nuclear reactor using uranium 235 as the fuel.

$$^{235}U + {}_0^1n \rightarrow \frac{\text{sum of masses of}}{\text{nuclear fragments}} + 1 \text{ to } 3\,{}_0^1n + 200\,\text{MeV}$$

The splitting of the ^{235}U atom in this manner is termed *fission.* Fast neutrons emitted during fission must be slowed to thermal energy to facilitate their capture by other ^{235}U atoms which, in turn, undergo fission. Graphite blocks are used as moderators to slow the neutrons. The liberation of one to three neutrons each time a neutron is captured and the rapid multiplication of this process constitute a *chain reaction,* accompanied by the release of a tremendous amount of energy in an exceedingly small interval of time.

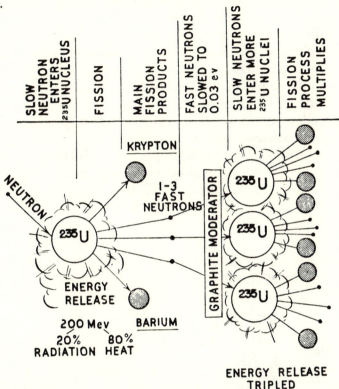

Figure 22.2. Fission of uranium 235 by slow (thermal) neutrons, resulting in a nuclear chain reaction. This differs from radioactive decay in the production of large nuclear fragments, as well as radiation, during fission.

If a piece of ^{235}U is large enough to permit more neutrons to be emitted than can escape from the surface, the mass will explode. Thus, for the reactor to operate properly, the mass of ^{235}U must be *critical*, that is, just large enough for the fission process to continue at a steady rate, but still not so large as to go out of control or *go supercritical*. Fission rate can be varied by the use of boron or cadmium *control rods;* these have a strong affinity for neutrons, and adjustment of the depth to which the rods are inserted into the reactor regulates the size of the *neutron flux* (flow). Normally, the reactor's operation is adjusted by the cadmium control rods to maintain the fission process at the desired level without letting the reactor go out of control. Note that the "fuel" of the reactor is not pure ^{235}U, but rather natural uranium, of which only about $1/140$ is ^{235}U, enriched by the addition of ^{235}U.

Since the neutron carries no charge, it is not repelled by the positively charged nucleus; in fact, as a neutron approaches a nucleus there is actually an attractive force between them. Neutrons entering stable nuclei render them unstable. Thus, when placed in drawers and inserted into the reactor (see Figure 22.1), stable nuclides are exposed to neutron flux and changed to radionuclides through *neutron capture.*

The use of the nuclear reactor for the production of radionuclides will now be exemplified by typical equations.

1. *Transmutation Reaction* — a different element is produced from the one entering the reaction, the latter being called the *target element* (see Figure 22.3). This reaction is designated as an (n,p) reaction because a neutron enters the nucleus and a proton (hydrogen nucleus) leaves it. The ^{32}P can be separated chemically from sulfur in *carrier-free* (pure) form.

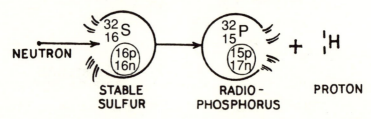

Figure 22.3. Transmutation of sulfur to radiophosphorus by neutron capture.

$$^{32}_{16}S \quad + \quad ^{1}_{0}n \quad \rightarrow \quad ^{32}_{15}P \quad + \quad ^{1}_{1}H$$

stable sulfur   neutron   radioactive phosphorus   proton

2. *Radiative Capture* — the product nuclide is a radioactive isotope of the target nuclide and can be separated from it only with great difficulty. Contaminated by its stable isotope, the product radionuclide resulting from neutron capture is not carrier-free.

$$^{59}_{27}Co + \quad ^{1}_{0}n \quad \rightarrow \quad ^{60}_{27}Co \quad + \quad \gamma$$

stable   neutron   radioactive   gamma ray
cobalt   cobalt

This is an (n, γ) reaction because a neutron enters the nucleus and a gamma ray is emitted. On decaying, ^{60}Co changes to ^{60}Ni, while gamma rays and beta particles are given off.

3. **Fission Products** — various radionuclides representing the fragments of the ^{235}U nuclei that have undergone fission. They are actually "waste" products of the nuclear reactor, some of which can be separated chemically and used in medicine. In fact, they provide the main source of iodine 131 (^{131}I), cesium 137 (^{137}Cs), and strontium 90 (^{90}Sr), all three having medical application.

Properties of Artificial Radionuclides

Some of the more important properties of artificial radionuclides will now be discussed. They decay in much the same manner as radium (see Chapter 21), and they differ widely in their decay rates.

Types Of Radiation. The radiations emitted by radionuclides have been described on pages 450–451. But those artificial radionuclides that are of medical interest emit only beta particles or gamma rays with beta particles.

Beta particles are fast-moving electrons ejected from the nuclei of certain radionuclides, actually arising from neutron decay:

$$_{0}^{1}n \rightarrow {}_{1}^{1}H + {}_{-1}^{0}e + \tilde{v}$$

$$\text{neutron} \quad \text{proton} \quad \text{electron} \quad \text{antineutrino}$$

For example, ^{32}P is a pure beta emitter, decaying as follows:

$$_{15}^{32}P \rightarrow {}_{16}^{32}S + {}_{-1}^{0}e + \tilde{v}$$

$$\text{beta particle} \quad \text{antineutrino}$$

A few radionuclides emit **positrons** — positively charged electrons. Beta particles cause significant ionization in tissues where they are for the most part absorbed. They cannot be detected directly by special external counting devices, although the brems radiation resulting from interaction with atomic nuclei can be so detected.

Gamma rays are electromagnetic in nature and are identical to x rays. Carrying no charge, they do not ionize directly but release primary electrons in the tissues by interaction with atoms. Gamma-ray emission is frequently accompanied by beta particles; thus, ^{60}Co and ^{131}I give off both gamma and beta radiation. External radionuclide detection by Geiger or scintillation counters depends on gamma radiation which usually has sufficient penetrating ability to pass completely out of the body.

Units of Activity. The rate at which a radioactive sample decays is its *activity*, measured in units called *curies* (Ci). *One curie is an activity of 3.7 (10)^{10} disintegrations per second (dps).* Smaller units representing decimal fractions of the curie are usually more convenient. One millicurie (mCi) is an activity of 3.7 (10)^7 dps, and one microcurie (μCi) is 3.7 (10)^4 dps.

The S.I. unit of radioactivity is the *becquerel (Bq)*, defined as 1 dps; thus, 1 Ci = 3.7 (10)^{10} Bq. In this country, the S.I. system has not yet met with wide acceptance except for the gray, which is the unit of absorbed dose.

Specific Activity. A given sample of a radionuclide may not be carrier-free; that is, it may also contain the stable isotope. The *specific activity* of such a sample is defined as its activity (due to the radionuclide present) divided by its total weight. For example, if 10 g of stable cobalt irradiated in a nuclear reactor for a specific time acquires an activity of 55 Ci, then the specific activity of the resulting ^{60}Co sample is 55/10 = 5.5 Ci/g. In actual practice, the specific activity of ^{60}Co teletherapy sources may reach about 200 Ci/g.

Radioactive Decay

All radionuclides undergo a decrease in activity with time. The rate of decrease is exponential and is specific for any given radionuclide. So, regardless of the number of such atoms initially present in a sample, a certain fixed fraction will decay in a particular time interval. For example, if the initial number of atoms is 1 million, and the decay rate for the radionuclide is 0.5 per sec, then 500,000 will have decayed in 1 sec. In the next second, 0.5 of the remaining 500,000 atoms, or 250,000, will have decayed, etc. At present, there is no practicable method of changing the decay rate of radionuclides (see page 454 for a discussion of fractional decay *vs* decay constant).

Half-life. In nuclear medicine there are three kinds of half-lives: physical, biologic, and effective. These depend on the method of measuring a radioactive sample over a period of time. Such curves for ^{131}I are shown in Figure 22.4.

1. *Physical Half-life* (*T*) is defined as the time required for a given radioactive sample to decay to one-half its initial activity. For example, with an initial activity of 10 mCi (3.7 × 10^8 Bq) the

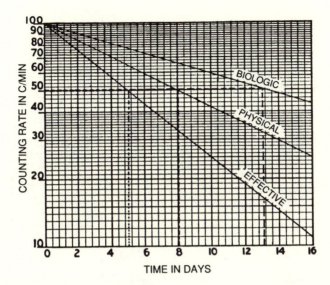

Figure 22.4. The three types of decay curves and corresponding half-lives for [131]I. In this example, the effective half-life is about five days and the biologic half-life about thirteen days. The physical half-life is constant — 8.05 days.

physical half-life is the *time* it takes for the activity to decrease to 5 mCi (1.85×10^8 Bq) as measured outside the body (*in vitro*). Examples of physical half-lives include: [131]I 8.1 days; [32]P 14.3 days; [60]Co 5.2 years.

Let us see how physical half-life is actually measured. First we measure the activity of a sample of the selected radionuclide by counting the number of decay events (disintegrations) per min by means of a suitable counter (see pages 471–475) under standard conditions, and subtract the background counting rate. This is the initial net counting rate. We repeat the procedure several times at suitable intervals over a period longer than the half-life. Finally, we plot the data, with counting rate as a function of time elapsed since the initial count; the resulting curve is shown in Figure 22.5. To find the half-life, we locate the point on the vertical axis corresponding to one-half the initial counting rate, then trace a horizontal line to the curve and, from the point of intersection, a vertical line to the horizontal axis where the final point of intersection locates the half-life. The physical half-life is constant for a particular radionuclide and is related to the decay constant; in fact, half-life T can be derived from the decay constant λ by the following equation:

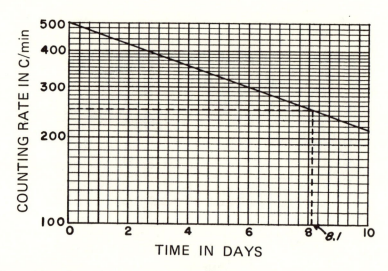

Figure 22.5. Physical decay of ^{131}I (semilogarithmic plot to obtain a straight-line curve.) With an initial net counting rate of 500 counts per min the half-life is the time required for the counting rate to decrease to 250 c/m. For ^{131}I this is 8.1 days.

$$T = 0.693/\lambda \qquad (1)$$

Or, knowing the half-life, we can derive the decay constant by rearranging equation (1) as follows:

$$\lambda = 0.693/T \qquad (2)$$

There is also a useful formula for the fraction of a radioactive sample remaining after n half-lives: $1/2^n$. Thus, if a sample of ^{131}I has an activity of 100 mCi, what fraction remains after twenty-four days? Since the half-life of ^{131}I is 8.1 days, the stated time approximates $24/8 = 3$ half-lives. The fraction remaining at the end of this time is

$$1/2^n = 1/2^3 = 1/8$$

The activity remaining is then 100 mCi $\times$ $1/8$ = 12.5 mCi.

2. *Effective Half-life* (T_e) is that due to both physical and biologic decay and is always less than the physical half-life. It is obtained from the curve plot of a series of absolute counting rates over a particular part of the body at suitable intervals after administration of a radionuclide. Effective half-life relates to physical half-life T and biologic half-life T_b as follows:

$$1/T_e = 1/T + 1/T_b \qquad (2)$$

This can be rearranged as follows:

$$1/T_b = 1/T_e - 1/T \qquad (3)$$

and in words

$$\text{effective half-life} = \frac{\text{physical half-life} \times \text{biologic half-life}}{\text{physical half-life} + \text{biologic half-life}}$$

3. *Biologic half-life* (T_b) is the time required for a given deposit of radionuclide in the body or a particular organ to decrease to one-half its initial maximum activity, due only to biologic utilization and excretion, and disregarding physical decay. It can be computed from equation (2), which may be rearranged as follows:

$$1/T_b = 1/T_e - 1/T \qquad (4)$$

For example, we know that ^{131}I has a physical half-life of 8.1 days. Suppose we find by absolute counting that the effective half-life of sodium iodide 131 in the thyroid gland is six days. Substituting these values in equation (3),

$$1/T_b = 1/6 - 1/8.1 = 0.167 - 0.123 = 0.044$$

$$T_b = 1/0.044 = 23 \text{ days, } \textit{the biologic half-life}$$

APPLICATION OF RADIONUCLIDES IN MEDICINE

A number of artificial radionuclides, when prepared in suitable compounds called *radiopharmaceuticals,* react in the body just like their stable isotopes. For example, the body cannot tell the difference between radioactive iodine and stable iodine. But once a radionuclide has entered the body, it acts as a radioactive tag that allows it to be detected, and its course through the body traced, by special instruments; or, alternatively, it can be detected in the blood, secretions, or excretions within or outside the body. Furthermore, it can be measured with reasonable precision in most instances. In this chapter, only basic principles will be covered. We shall also touch upon the use of radionuclides in therapy, which depends on the fact that sometimes, as with iodine in the thyroid gland, a radionuclide will concentrate selectively in certain organs. Under such circumstances, a sufficiently large dose can be given to irradiate effectively the target organ.

The following outline summarizes the medical uses of radionuclides according to various indications:

1. ***Diagnosis*** — general procedure
   a. Compound tagged with radionuclide (radioactive isotope of a normal element substituted for it in the compound).
   b. Injected or ingested in tracer amount — small enough to be diagnostic, but not large enough to deliver excessive radiation exposure.
   c. Body cannot distinguish between radioactively tagged compound and its stable counterpart.
   d. Fate of tagged element determined by counting with special instruments (GM counter; scintillation counter; gamma camera)
      (1) Externally as ^{123}I, ^{131}I, or ^{99m}Tc, thyroid gland; ^{99m}Tc in various compounds, brain, lung, skeletal system, liver, kidneys.
      (2) Excretions as in counting urine in Schilling test for pernicious anemia.
      (3) Secretions as in measuring ^{131}I output in saliva.

2. ***Therapy*** — methods
   a. *Chemical*
      (1) Selective absorption as ^{131}I in thyroid gland: ^{131}I given orally in therapeutic amount; selectively absorbed in target organ — thyroid — which is irradiated by deposited radionuclide, in this instance, 90 percent of therapeutic dose from beta particles, 10 percent from gamma rays.
   b. *Physical*
      (1) External as ^{60}Co and ^{137}Cs in teletherapy; ^{90}Sr in contact therapy.
      (2) Interstitial as ^{137}Cs needles; ^{192}Ir seeds; ^{198}Au grains; ^{125}I seeds; ^{182}Ta wires.
      (3) Intracavitary as ^{137}Cs tubes; ^{32}P-chromic phosphate in pleural or peritoneal malignant effusions.

In recent years diagnostic nuclear medicine has evolved into two separate branches, **laboratory tests** and **organ** imaging. The remainder of this chapter will deal mainly with imaging, but first it will introduce basic instrumentation and a few examples of radionuclide counting procedures.

RADIONUCLIDE INSTRUMENTATION

The basic procedure in any quantitative diagnostic radionuclide study is to count, with an appropriate detecting or counting instrument, the number of photons or ionizing particles it receives and detects from a radioactive source in a specified interval of time and with a defined geometric arrangement. The source may be either an organ in the body or a specimen outside the body, depending on the examination being conducted.

A general formula for radionuclide counting may be expressed as follows, ***provided the sample and the standard are counted under identical conditions:***

$$\frac{\text{mCi in sample}}{\text{mCi in standard}} = \frac{\text{net counting rate in sample}}{\text{net counting rate in standard}}$$

The net sample counting rate, R_S, is the difference between the total counting rate, R_T, and the background counting rate, R_B.

$$R_S = R_T - R_B \tag{5}$$

We obtain the background counting rate after having removed the sample and all other radioactive sources from the counting range of the detector. The counting rate is the total count divided by the elapsed time. In any radionuclide procedure, the background counting rate must be subtracted from the total to obtain the net counting rate, unless the background is negligibly small.

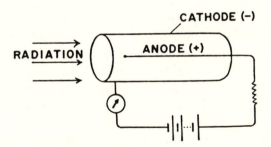

Figure 22.6. Simplified diagram of an end-window GM counter.

We must have a method of counting the individual radiations emitted by a radioactive source. The same devices, with modifications, can be used to count both photons and charged particles. These devices consist of (1) the radiation detector plus its related electric circuitry and power supply and (2) the instrument for registering the number of counts.

Radiation Detectors. The two main types of detectors, which differ completely in principle, are the ***Geiger-Muller*** or ***GM tube*** and the ***scintillation detector.*** A detector with its related electrical circuitry is often referred to as a ***counter;*** thus, GM counter; scintillation counter.

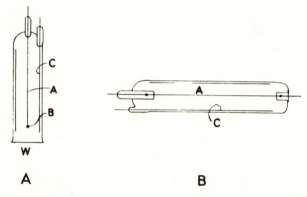

Figure 22.7. A. End-window type of GM tube. B. Side-window type. A is anode, B is bead, W is window.

1. ***GM Tube.*** This is an electronic pulse-type device that can detect individual ionizing particles or photons. It consists of a central wire ***anode*** and a cylindrical ***cathode*** enclosed in glass (see Figure 22.6). Within the tube is a suitable gas, often argon and alcohol under reduced pressure of about one-eighth atmosphere. The GM tube operates within a range of about 1,000 to 1,500 volts. There are two general types of GM tubes—end-window and side-window (see Figure 22.7). Radiation enters the end-window tube at one end and is therefore said to be ***directional.*** On the other hand, radiation enters a side-window GM tube at any point around its circumference and is therefore ***nondirectional.***

The characteristic curve of a GM tube is shown in Figure 22.8. Although this curve is typical of all GM tubes, it may vary in numerical detail from one tube to another.

Photons or beta particles entering the tube cause ionization of the contained gas. With an applied potential of 800 to 1100 volts, the ions undergo sufficient acceleration to cause further ***ionization***

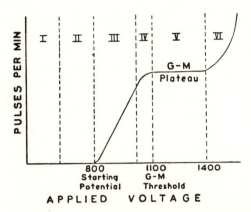

Figure 22.8. Characteristic curve of one type of GM tube. The voltage may be different for other GM tube. The operating voltage should correspond to about the middle of the pleateau.

by collision with other atoms, resulting in an electrical discharge along the central wire anode. This discharge is manifested by a "pop" in the audio circuit. In this voltage range (800 to 1100 for this particular GM tube), the number of pops or counts per second is independent of the applied voltage but increases with the number of photons or beta particles entering the tube—this is the *GM region* of the characteristic curve (see Figure 22.8). Here, each entering photon or beta particle is detected as an individual count.

Such a counter can be used to detect the presence of photons or beta particles and, under suitable conditions, determine their quantity. A GM counter displays high efficiency in detecting beta particles but low efficiency (about 1 to 2 percent) in detecting gamma radiation. Its efficiency in detecting gamma rays can be improved by coating the cathode with bismuth.

While the GM counter is still used in certain types of survey meters (see pages 539–540), it has been replaced by the much more efficient and dependable scintillation counter in nuclear medicine.

2. *Scintillation Detector.* When x or gamma rays enter certain crystals, they interact photoelectrically with crystal atoms; the characteristic radiation is in the form of visible and ultraviolet light, the latter appearing as *scintillations* (light-flashes). Application of this principle gave us the scintillation counter and, as will be shown later, the *gamma camera* (also see pages 493–494).

In a scintillation detector (see Figure 22.9) all sides are encased

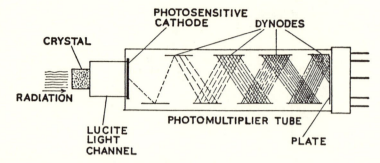

Figure 22.9. Diagram of a scintillation counter. Incoming radiation produces flashes of light (scintillations) in the detector crystal, one flash per photon. The light flashes are conducted by the Lucite channel to the light-sensitive cathode of the photomultiplier where photoelectrons are released. These are then accelerated by a series of dynodes (anodes) and are detected as pulses by the counting circuit after they strike the plate.

except one; this clear side is placed in close contact with the *photocathode* of a *photomultiplier,* which also contains a series of about ten or eleven accelerating anodes called *dynodes,* as shown in Figure 22.9. The light-flashes from the crystal interact with the photocathode where they liberate electrons; these undergo serial acceleration and multiplication by the dynodes, to about one million times their original number. The electrons are collected to produce a voltage pulse in the circuit. One pulse occurs for each entering gamma photon, the size of the pulse being proportional to the energy of the photon entering the crystal. Since the scintillation counter has a dead time of only about 5 microsec, it is capable of tremendous counting rates. Also, its high degree of sensitivity permits the counting of sources having extremely small activities.

Most scintillation counters have a thallium-activated sodium iodide detector crystal. However, anthracene crystals are available, especially for beta particles. The larger the volume of the crystal, the greater is the sensitivity or counting rate for a given activity. Crystals vary in diameter from a few mm for the measurement of exposure rates near sealed radioactive sources, to 22 in. (45 cm) or more in gamma cameras.

Scintillation counters are incorporated into a digital counting system that displays the number of counts in a pre-set time. In addition, selected "windows" confine the detected counts to a specified energy range, to count the photons from a particular radionuclide in a mixture containing those with various energies other than the desired one.

The **well counter** is a scintillation counter specially modified for measuring samples of low activity. A hole bored in the detector crystal of a well counter can accommodate a small bottle containing the radioactive sample (see Figure 22.10). Because the crystal surrounds the sample on the sides and bottom, the well counter is extremely sensitive; in fact, it can count a source with an activity as low as 10^{-5} μCi (microcurie). At the same time, the high sensitivity requires heavy lead shielding (usually about 5 cm Pb) to minimize background effects.

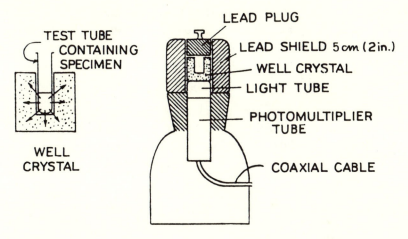

Figure 22.10. Basic well counter, using a well crystal in a scintillation counter. Thick lead shielding is required to reduce the background counting rate.

Sources of Error in Counting

Various errors are inherent in the counting process, the most important ones being (1) statistical errors and (2) recovery time of the counter.

Statistical Errors. Since radioactive decay—a purely random process—obeys the laws of chance, statistical errors affect the accuracy of a given counting rate which, after all, reflects the rate of decay of a radioactive source. As a general rule, it is well to remember that the *reliability* (or reproducibility) *of a given count depends on the total number of counts recorded.* If the same sample is counted repeatedly under identical conditions, the counts will differ from each other because of chance variations (statistical fluctuation). How, then, can we tell which count is correct and how much it differs from the ideal or "true" count? Actually the answer lies in specifying a certain statistical reliability of a given count. Only the important elementary concepts will be included here.

1. **Standard Deviation,** σ (Greek letter "sigma"), is defined as follows:

$$\sigma = \pm \sqrt{N} \text{ (approx.)} \tag{6}$$

where N is the total number of counts recorded in a given run. The standard deviation means that if we take the range $N \pm \sqrt{N}$, or the range of values from $N - \sigma$ to $N + \sigma$, there is about a two to one chance that the true count would lie somewhere within this range. For example, if the total count is 6400,

$$\sigma = \pm \sqrt{6400} = \pm 80$$

There is about a two to one chance that the true count falls between $6400 - 80$ and $6400 + 80$, or between 6320 and 6480.

2. **Percent Standard Deviation** $(\%\sigma)$. It is often more convenient to express σ as a percentage of the total count:

$$\%\sigma = \frac{\pm \sqrt{N}}{N} \times 100$$

$$\%\sigma = \pm \frac{100}{\sqrt{N}} \tag{7}$$

or,

$$\%\sigma = \pm \frac{100}{\sigma}$$

In the preceding numerical example,

$$\%\sigma = \pm \frac{100}{\sqrt{6400}} = \pm \frac{100}{80} = \pm 1.3\%$$

The final count may then be stated as $6400 \pm 1.3\%$; there is, again, a two to one chance that the true count lies between these limits. Figure 22.11 shows a curve plot of the percent standard deviation as a function of the total number of counts.

We can easily demonstrate that the percent standard deviation becomes smaller (ie, reliability becomes greater) as the total num-

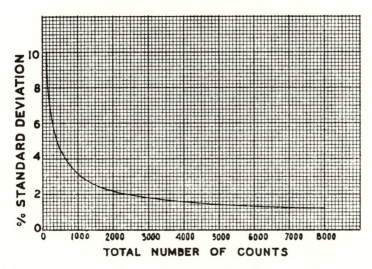

Figure 22.11. Variation in percent standard deviation relative to the total number of counts.

ber of recorded counts is increased. Suppose we count a given sample long enough to accumulate 2500 counts. Then,

$$\%\sigma = \pm \frac{100}{\sqrt{2500}} = \pm \frac{100}{50} = \pm 2\%$$

If the same sample is counted long enough to accumulate only 100 counts, then,

$$\%\sigma = \pm \frac{100}{\sqrt{100}} = \pm \frac{100}{10} = \pm 10\%$$

Thus, the percent standard deviation is simply 100 divided by the square root of the number of counts. The percent standard deviation for a given **total count** also applies to the **counting rate** based on that total count.

In actual practice, with preset timing, we count a sample for a predetermined interval of time; and we also count the background, but not necessarily for the same interval. We then compute the counting rate by simply dividing each count by its corresponding time. The true sample counting rate R (without background), is equal to the gross sample counting rate (including background) R_T, minus the background counting rate R_B,

Recovery Time of Counter. As already mentioned in the discussion of GM and scintillation counters, a high counting rate may incur an error owing to the inability of the counter to detect radiation during a tiny interval of time following a pulse. This interval is called the ***dead time*** or ***recovery time*** of the counter — the higher the counting rate, the greater the likelihood that two ionizing events may occur so closely in time that only one is detected. A simple formula gives the approximate percentage correction for the dead time at extremely high counting rates:

$$\%C = 100\,\frac{R_t}{1 - R_t} \tag{11}$$

where $\%C$ is the percentage correction that must be added to the observed counting rate R, to obtain the corrected counting rate, and τ (Greek letter "tau") is the dead time of the counter in question. For example, with a GM counter having a τ of $2(10)^{-4}$ sec, an observed counting rate of 200 cps should be corrected as follows:

$$\%C = 100\,\frac{200 \times 2 \times 10^{-4}}{1 - 200 \times 2 \times 10^{-4}}$$

Performing the indicated multiplication first,

$$\%C = \frac{4}{1 - 0.04} = \frac{4}{0.96} = 4\%$$

The corrected counting rate is then

$$200 + 200(0.04) = 208 \text{ cps}$$

With a scintillation counter having a dead time of $(10)^{-6}$ sec, the correction factor is only about 1 percent even with an extremely high counting rate of 500 cps (the corresponding correction with a GM counter would be approximately 11 percent).

Efficiency and Sensitivity of Counters

The ability of a particular counter to detect radiation is generally known as its ***efficiency.*** But as we have seen, radioactive decay is a random process accompanied by the emission of radiation. In order to detect all of the radiation, the source would have to be placed completely within a special type of counter.

Sensitivity of Counters. In practice, a counter is placed at some distance from the source, which may be an organ or sample having sufficient activity. Because the radiation is emitted in all directions, the counter cannot receive it all, but can "see" only some of the radiation coming toward it, as in Figure 22.12. Furthermore, of the radiation actually entering the counter, only a fraction produces ionizing events and is detected. For example, an ordinary GM counter will detect about 1 to 2 percent of the entering gamma rays, although this can be improved by coating the cathode with bismuth. These two negative aspects of efficiency—the inability of a counter to "see" all the radiation and its failure to detect all the radiation that enters it—are embodied in *counter sensitivity f* defined by the following equation:

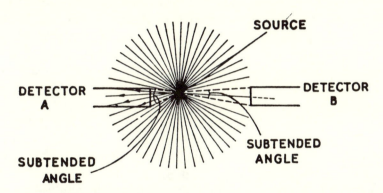

Figure 22.12. Angle of acceptance. The closer the counter is to the source, the greater is the number of photons entering it. Thus, detector *A* would record a higher counting rate than detector *B* with the same radioactive sample.

$$f = \frac{\text{counting rate}}{\text{activity}} \qquad (12)$$

Bear in mind that sensitivity is related to a given geometric arrangement of the source and the counter. Note that sensitivity may be expressed as counts per sec per μCi, or counts per min per mCi, or any other convenient combination of units. Sensitivity can be used to convert counting rate to activity of a sample, by rearrangement of equation (8),

$$\text{activity} = \frac{\text{counting rate}}{f} \qquad (13)$$

provided the same geometric setup, counter, and radionuclide are used as in the initial determination of sensitivity. It also permits the computation of the sample activity needed to obtain a desired counting rate.

The preceding discussion applies primarily to the determination of the sensitivity of a particular counting system. In actual practice, there are two additional factors that influence the counting rate when counting is done over the body. One is attenuation of the radiation as it passes through tissues or other matter, on its way to the detecting instrument. The other is the scattering of radiation from nearby objects, with entrance of this scattered radiation into the detector. We can minimize these factors by using radionuclides whose radiation is not appreciably attenuated, and by keeping the counting system isolated from scattering material. The influence of scattering can be further controlled by narrowing the energy window of the detecting system to match the energy of the primary radiation.

Minimum Detectable Activity. We sometimes need to know the smallest quantity of a radionuclide that a counter can detect to a reasonable degree of reliability, with a given geometric arrangement and type of counter. According to the ICRU, the minimum detectable activity is that activity which, in a given counting time, records a number of counts on the instrument equal to three times the standard deviation of the background count in the same time interval. The equation specifying minimum detectable activity, A_{min}, is

$$A_{min} = \frac{3}{f} \sqrt{R_B/t_B} \tag{14}$$

where $\sqrt{R_B/t_B}$ is the standard deviation of the background counting rate, and f is the sensitivity of the counter. Since $R_B = N_B/t_B$, we can rearrange as $t_B = N_B/R_B$. Substituting this value of t_B in equation (10),

$$A_{min} = \frac{3 R_B}{f \sqrt{N_B}} \tag{15}$$

N_B is the number of background counts recorded in a preselected interval of time, usually about ten minutes. For example, if, with a particular geometric arrangement, f is 2000 cps per μCi, R_B 4 cps, and N_B 2500 counts in 10 min, then the minimum detectable activity is

$$A_{\min} = \frac{3 \times 4}{2000 \times \sqrt{2500}} = \frac{3}{500 \times 50} = 0.0001 \, \mu\text{Ci}$$

Note that the sensitivity and minimum detectable activity both depend on the geometric arrangement of the source and counter, the type of counter, and the radionuclide being counted. Therefore, calibration of the equipment should be carried out separately for each counting setup and each radionuclide in use in the department.

Geometric Factors in Counting

In counting a radionuclide sample, we must take into account (1) *the distance between the source and the counter,* (2) *the presence of scattering material* in the vicinity of the source, (3) *the collimation of the counter,* and (4) *the size of the source.* Figure 22.12 showed how a source gives off radiation in all directions. The inverse square law of radiation applies (approximately because not a point source), so that the nearer the counter is to the source, the more radiation it "sees." Matter near the source gives rise to scattered radiation, which may be detected by the counter and give a falsely high counting rate. Collimation affects the fraction of emitted radiation reaching the counter. Figure 22.13 shows the response of the same counter with two different degrees of collimation. As the collimator is lengthened, the counting rate decreases, but the area of maximum sensitivity depends only on the diameter of the collimator hole. On the basis of this principle, a counter can be so collimated that it can detect, in the thyroid gland, small nodules whose activity is different from that of the surrounding gland. But collimation must not be so severe that the counter fails to see the entire organ, as in measuring radioiodine uptake by the thyroid gland.

A *focused collimator* (see Figure 22.14) provides maximum sensitivity for a particular degree of collimation, making it much more efficient than a single hole collimator.

Because of the fact that sensitivity depends so intimately on the geometric arrangement of the source and counter, equipment should be standardized and used under an identical geometric setup.

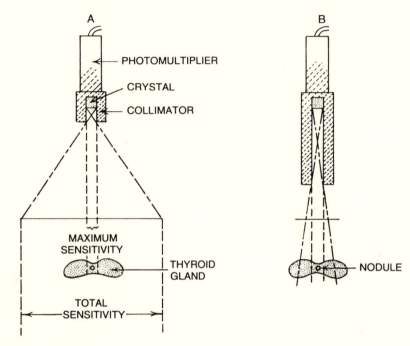

Figure 22.13. Relation between collimator length and field of view. In *A*, the field of view is too wide, allowing excess scatter (noise) from tissues surrounding the thyroid gland.

In *B*, with a longer collimator, the field of view is too restricted to include the entire gland, but scatter is reduced and maximum sensitivity is the same as in *A*. Thus, image resolution of the nodule improves because of reduced scatter.

Methods of Counting

Two methods of counting, whether this is to be done on specimens outside the body, or over certain areas of the body, include (1) absolute counting and (2) comparative counting, both of which we shall now describe. But note that health authorities require calibration of all test doses before they are given to the patient.

1. **Absolute Counting.** In principle, this is similar to the method of determining the sensitivity, *f*, of a counter. First we measure sensitivity with a standard known radioactive source (similar to the one to be counted in the patient or specimen) in the same geometric setup to be used with the clinical material. Suppose the

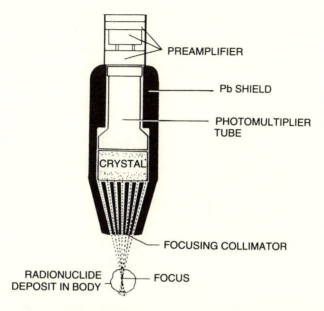

PREAMPLIFIER

Pb SHIELD

PHOTOMULTIPLIER
TUBE

CRYSTAL

FOCUSING COLLIMATOR

RADIONUCLIDE
DEPOSIT IN BODY — FOCUS

Figure 22.14. A hand-held scintillation counter, with a converging collimator.

standard source under these geometric conditions has an activity of 1 μCi and a net counting rate of 3000 cps. Then f = 3000 cps per μCi. If, now, we count the specimen under identical conditions (same radionuclide and geometric setup), and find a net counting rate of 1000 cps, then according to equation (13),

$$\text{activity of sample} = \frac{\text{counting rate}}{f} \mu\text{Ci}$$

Therefore,

$$\text{activity of sample} = \frac{1000}{3000} = 0.33 \, \mu\text{Ci}$$

2. **Comparative Counting.** In clinical radionuclide studies utilizing the tracer method, we usually count the sample and standard separately under identical conditions. These net counting rates are then compared directly, the specimen counting rate being expressed as a percentage of the standard. Comparative counting may be exemplified by tracer studies of the radioiodine uptake by the thyroid gland. A predetermined dose of radioiodine is given to the patient and, after the appropriate time has elapsed, the counting rate over the thyroid gland is measured at a fixed distance (for

example, 20 cm) with the "window" set for the proper energy range. The same dose as that given the patient is now measured separately in a neck "phantom" (see Figure 22.15) at the same distance. After correction of both counting rates for background,

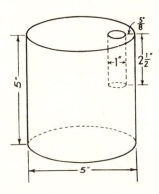

Figure 22.15. Diagram of a standard neck phantom, made of Lucite. The standard radioiodine capsule is placed in the 1-inch well.

$$\% \text{ thyroid uptake} = \frac{\text{net thyroid cpm}}{\text{net standard cpm}} \times 100 \qquad (16)$$

This procedure is carried out at two, six, and twenty-four hours, but additional studies are usually indicated, including chemical laboratory tests and thyroid scintiscanning (see Figure 22.15).

Important Medical Radionuclides

We shall now summarize the properties of some of the more important radionuclides used in medicine.

Radioactive Iodine. Iodine 131 (^{131}I), the most commonly used radioisotope of iodine, has a physical half-life of 8.1 days and emits 364-keV gamma rays and 183-keV (av.) beta particles. Obtained mainly by separation from the fission products of the nuclear reactor, ^{131}I as sodium iodide is used especially in the diagnosis and treatment of diseases of the thyroid gland. The therapeutic effect of ^{131}I in hyperthyroidism and certain thyroid cancers depends mainly on the *beta particles*, whereas tracer and scanning studies depend on the *gamma rays*, which are much more penetrating and can be detected outside the body with a GM or scintillation counter.

Another radioisotope of iodine — [123]I — is preferred for uptake and imaging studies of the thyroid gland. Produced in the cyclotron by deuteron irradiation of tellurium, it has a half-life of 13.2 hr. Iodine 123 decays by electron capture and then emits 159-keV gamma rays, but no particles. Its main advantage, aside from the favorable energy of its gamma rays, lies in its short half-life which permits the administration of larger activities than with ^{131}I, but with a substantially smaller dose to the thyroid gland. For example, a 400-μCi imaging dose of ^{123}I gives an absorbed dose of about 20 rads to the normal thyroid gland, as contrasted with about 130 rads from a 100-μCi dose of ^{131}I, a really significant difference.

Radioactive Phosphorus. Phosphorus 32 has a half-life of 14.3 days and emits 700-keV (av.) beta particles. It is prepared mainly by neutron irradiation of stable sulfur in a nuclear reactor (see page 464). After administration in suitable form orally or intravenously, ^{32}P tends to concentrate in the bone marrow, spleen, liver, and lymph nodes. It can therefore be used in the treatment of certain blood disorders, mainly polycythemia rubra vera, a condition in which the bone marrow produces an excess of red blood cells. Phosphorus 32 suppresses the overproduction of these cells. Intravenous ^{32}P as phosphate delivers a whole body dose of about 1.5 rads per mCi during the first three days, and thereafter, about 2.5 rads per mCi to soft tissues and 25 rads to bone (Shapiro).

Technetium 99m. The radionuclide ^{99m}Tc has a short, though convenient, half-life of six hours. Its gamma radiation, having a relatively low energy — 140 keV — is easily collimated and at the same time has adequate penetrating ability to pass through the body and reach the detector. Technetium 99m is at present used mainly as the pertechnetate salt in aqueous solution. Because of the short half-life and the absence of beta radiation, large doses can be given with relatively little exposure to the patient. As a result we can obtain high count rates with excellent image resolution, using [99m]*Tc-tagged compounds* that have a more or less specific affinity for particular organs. Examples of ^{99m}Tc-tagged compunds and target organs include:

Diphosphonates (phosphorus compounds) — bones
Sulfur colloid — liver and spleen
Albumin aggregate — lungs

*MAG*3^R* (mertiatide) and *DTPA* — kidneys
*Ceretec*** — brain perfusion
Pertechnetate — thyroid gland

We should mention the **technetium "cow,"** a molybdenum (^{99}Mo) generator which decays to ^{99m}Tc. Each day 0.9 percent sodium chloride solution is passed through the generator, dissolving out the ^{99m}Tc; this is called "milking" the "cow." The resulting ^{99m}Tc is filtered during the milking process and, after calibration, is ready for use.

Radioactive Cobalt. Cobalt 60 (^{60}Co), a radioisotope of cobalt 59, is produced in a nuclear reactor. It has a half-life of 5.2 years and emits gamma rays of two energies—1.17 and 1.33 MeV (average 1.25 MeV)—and a low energy beta particle (av. 96 keV). Cobalt 60 is available in a variety of forms for brachytherapy (short distance therapy)—needles, beads, and wires—as a substitute for radium. More recently, cesium 137 has been found to surpass ^{60}Co for this purpose. Large ^{60}Co teletherapy units are designed for use in external irradiation.

Vitamin B$_{12}$ or cyanocobalamin contains cobalt as a part of its molecular structure; when this compound is tagged with the radionuclide ^{57}Co, it is symbolized by ^{57}CoB$_{12}$ and is the agent that is widely used in the Schilling test for pernicious anemia (see pages 488–489).

Radioactive Chromium. The radionuclide chromium 51 (^{51}Cr) is used in the form of chromate to determine red cell volume in polycythemia and in patients undergoing extensive surgery and to determine red cell survival in hemolytic anemias. Chromium 51 has a physical half-life of 27.8 days, and emits a 320-keV gamma ray and 242-keV (av.) beta rays. In the chromate form, this nuclide tags red cell, whereas in the chromic form it labels blood plasma.

Examples of the Use of Radionuclides in Medical Diagnosis

It is beyond the scope of this book to give a detailed account of the numerous diagnostic procedures in the Nuclear Medicine

---

*Mallinckrodt Medical

**Amersham Corporation

Department. Instead, we shall include examples of the more important procedures based on various principles; these will include (1) **uptake by an organ** and (2) **excretion.** Organ imaging will be discussed in the next section.

1. **Uptake or Tracer Studies.** As mentioned earlier, the thyroid gland has a strong affinity for iodine. If radioactive iodine is taken it will normally be trapped in the thyroid gland over a period of twenty-four hours, its concentration depending on the functional capacity of the gland. Obviously, before such a test can be run, the patient must avoid the intake of ordinary stable iodine (^{127}I), as well as drugs that are known to suppress the activity of the thyroid gland. Such restriction may require weeks or months, depending on the type of drug; radiopaque media containing iodine may interfere with further iodine uptake for several months.

Depending on the type of equipment, a tracer activity of 2 μCi to 5 μCi of ^{131}I is taken orally in the morning by the fasting patient. A phantom simulating the neck should be available, made of Lucite or hard paraffin according to ICRU specifications (see Figure 22.15). A plastic test tube containing a standard of the same activity as that given the patient is placed in the hole in the phantom. At prescribed intervals, preferably two, six, and twenty-four hours, counting rates are taken at the same fixed distance—20 cm—from the thyroid and from the phantom (comparative counting). The window is set for a range of 354 to 375 keV. Each time, a body background count should also be made over the lower end of the thigh, which resembles the tissues of the neck without the thyroid; the thigh counting rate should be subtracted from the thyroid counting rate each time. The percent uptakes at each counting session (two, six, twenty-four hr) are computed from the following equation.

$$\% \text{ thyroid uptake} = \frac{100 \ (\text{neck } R_\text{T} - \text{thigh } R_\text{T})}{\text{phantom } R_\text{T} - \text{room } R_\text{B}} \qquad (17)$$

where R_T is the total counting rate of each item, and R_B is the background counting rate.

Figure 22.16 shows typical uptake curves obtained in various functional states of the thyroid gland. There is significant overlap of normal and abnormal values at both ends, this being more serious in the hypothyroid (low) range (see Table 22.1). Still, the

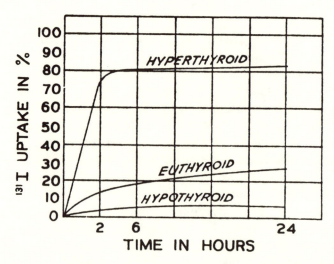

Figure 22.16. Representative iodine 131 uptake curves in the three functional states of the thyroid gland. Actually, there is no sharp dividing line between the normal (euthyroid) and abnormal (hypo- and hyperthyroid) states. The upper curve represents severe hyperthyroidism.

thyroid uptake of radioiodine is an extremely useful measure of thyroid gland function, especially when combined with supplementary chemical and radiochemical studies, and radionuclide imaging.

When available, ^{123}I is preferred for uptake and imaging studies of the thyroid gland because of the much smaller absorbed dose as compared with ^{131}I (see page 485).

2. **Excretion of a Radionuclide by the Body.** An example is the Schilling test for *pernicious anemia,* performed with cyanocobalamin (vitamin B_{12}) labeled with the radionuclide cobalt 57 (Co^{57}). Normal persons readily absorb this vitamin from the gastrointestinal tract, storing it in the liver within four to seven days. However, patients with pernicious anemia cannot absorb it because they lack *intrinsic factor,* which is normally secreted by the stomach. The fasting patient is first asked to void, the urine specimen being saved as a pretest control. The patient then swallows a capsule containing 0.5 μCi of $^{57}CoB_{12}$ with one-half glass of water (patient must not have received ordinary B_{12} orally or parenterally during the preceding two days). One hour later, a large dose of ordinary, stable B_{12} — 1000 μg — is given intramuscularly to flush any absorbed

Table 22.1
RANGE OF 24-HR UPTAKES OF [131]I
BY THE THYROID GLAND.
IN VARIOUS FUNCTIONAL STATES*

Euthyroid (normal)	5%–30%
Hyperthyroid	more than 30%
Hypothyroid	less than 5%

*These values are typical; actually, there may be much more overlap of the normal and abnormal than that shown here. other tests are usually necessary to establish the diagnosis.

radioactive B_{12} through the kidneys and into the urine, which is collected for 24 hours in a one-gallon plastic bottle with a screw cap, and should include the last bladder contents. After measurement of the total urine volume, a 4 ml sample of the pre-test urine (from previous day) and of the collected urine, and of a [57]Co standard should each be counted for 2 min in a well counter. A 2-min background count should also be made. Calculations are as follows after subtracting the pre-test urine count from the test sample count to obtain the net urine count:

$$\% \ ^{57}Co \ \text{in urine} = \frac{\text{net urine count} \times \text{total urine volume}}{4 \times \text{net standard count}}$$

Values greater than 6 percent are normal, and 0–3 percent are indicative of pernicious anemia. Intermediate values are questionable and require further study.

IMAGING WITH RADIONUCLIDES — THE GAMMA CAMERA

A diagnostic imaging system called a ***gamma camera*** or ***scintiscanner*** can display the distribution of gamma-ray emitting radiopharmaceuticals (RPCs) that have been deposited in the selected target organ. By target organ is meant the one that selectively takes up a particular RPC. The gamma camera is a greatly expanded and modified version of the basic hand-held scintillation detector described on pages 481–483; it is capable of using large numbers of radioactive decay events (counts) to create an image of the target organ. In a sense, imaging with a camera is analogous to radiography, if you can imagine that the RPC in the target organ is a miniature x-ray machine that produces an image from within outward with the gamma camera serving as the image receptor.

In a typical procedure, an appropriate RPC is first administered by injection, ingestion, or inhalation, depending on the designated examination. After an appropriate delay to allow distribution of the RPC, the patient is placed on the table directly under the camera at a specified distance. Figure 22.17 presents a block diagram of a gamma camera and its auxiliary devices, which will be described in the following sections.

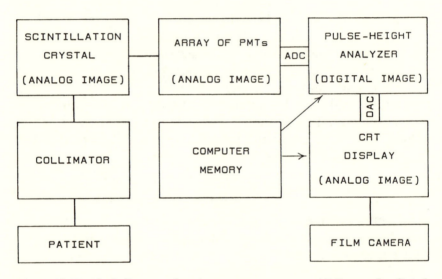

Figure 22.17. Block diagram of an Anger gamma camera. ADC = analog-digital converter. DAC = digital-analog converter. PMT = Photomultiplier tube.

Collimators

Just as in radiography, collimators absorb scattered radiation arising in the body by Compton interactions with atoms in the tissues. As will be explained shortly, the collimator acts like a grid. Radiation arising in the target organ must pass up through holes in the collimator toward the scintillation crystal (see Figure 22.14, page 483); scattered photons approach the collimator at various angles and are absorbed in the septa which separate the holes. As you already know, scattered radiation is a form of *noise* that obscures the final image.

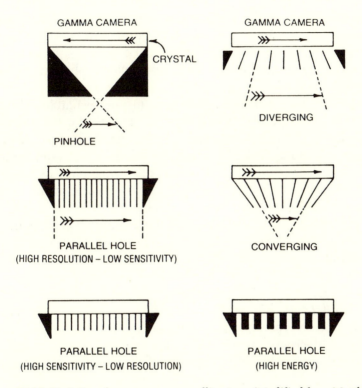

Figure 22.18. Varieties of gamma-camera collimators. (Modified from Mettler FA, Guiberteau MJ. *Essentials of Nuclear Medicine Imaging,* 3rd ed. New York, Grune & Stratton, 1990. By permission.)

Several types of collimators are available (see Figure 22.18). One must be aware of their *sensitivity,* the fraction of gamma photons passing through the collimator holes, relative to the number incident on its face; and their *spatial resolution,* the ability of the collimator to enhance contrast by absorbing as many of the Compton-scattered photons as possible. The collimator is situated between the patient and the camera's scintillation crystal.

1. *Pinhole Collimator.* This, the simplest type of collimator, consists of a small hole in the center of a lead (Pb) plate of

required thickness. The hole serves as a conduit for gamma photons emitted by the radionuclide in the target organ. It provides high resolution in imaging small organs like the thyroid gland, but sensitivity is very low—relatively few photons pass through the hole, so the examination has to be prolonged to allow the camera to accumulate a sufficient number of counts. The use of the pinhole camera has gone into decline since the advent of the more modern gamma cameras with which parallel-hole collimators are used.

2. *Multihole Collimators.* Several forms are available. They all consist of lead plates in which multiple holes have been drilled. Unavoidable in such collimators is the penetration of photons through the septal walls between the holes; this should be kept well below 25 percent because septal-penetrating photons resemble scattered radiation in radiography by being a noise factor. Newer types of collimators have hexagonal holes to improve sensitivity, but this entails some loss of spatial resolution because having thinner septa than round-hole collimators permits increased septal penetration.

There are three kinds of multihole collimators: (a) parallel-hole, (b) converging, and (c) diverging.

a. *Parallel-Hole Collimator.* The holes (and septa) are parallel to each other (perpendicular to the crystal face), resembling tubes. Sensitivity decreases markedly with increasing septal length. In fact, sensitivity is inversely proportional to the square of the septal length:

$$S \propto \frac{1}{L^2}$$

where S is sensitivity and L septal length.

The parallel hole collimator is the most popular today, especially in its low-energy high-resolution form. Low energy refers to its use with radionuclides whose gamma energy is less than 150 keV; this fits its application to technetium 99m (^{99m}Tc), by far the most widely used radionuclide.

The next two types of collimators are no longer important in imaging, but will be mentioned for the sake of historical interest.

b. *Converging Collimator.* Here the directions of the holes come together at a focus located at a distance beyond the collimator (ie, toward the patient). Figure 22.14 shows a hand-held version of a converging collimator to explain its principle. In use, the distance of the collimator's focus should be such that it lies within the target organ at the site of interest (eg, thyroid nodule). Note that maximum resolution prevails at the focal point and that there is a zone of acceptable image resolution or *focal depth* within a short distance above and below the focus. The converging collimator produces a slightly enlarged image.

c. *Diverging Collimator.* The directions of the holes move farther apart at increasing distances (ie, away from the camera). This type of collimator has been used in imaging large organs such as the liver or lungs. But sensitivity and resolution both decrease as distance increases. With the larger rectangular crystals available today, the diverging collimator is becoming obsolete.

Crystal-Photomultiplier Complex

As already mentioned, the collimator is in close contact with a special *scintillation crystal* of thallium-activated sodium iodide; this has the property of emitting flashes of light when gamma photons interact with its atoms by *photoelectric absorption.* Such interactions resemble those with x-ray photons, but in this case, the interactions occur in the outer shells of the crystal's atoms so the characteristic photon energy is in the visible rather than the x-ray range.

The circular crystals range from 10 to 22 in. (about 25 to 45 cm) in diameter, and 1/4 to 1/in. (about 7 to 13 mm) thick. The newer cameras come with rectangular crystals measuring 20 × 25 in. (about 51 × 64 cm). Thicker crystals possess increased sensitivity (more light emitted because more atoms present), but decreased resolution (more light scattered within crystal). If you could observe a crystal in the dark, you would see a light-image corresponding to the gamma-photon image emitted by the radionuclide in the target organ. This is equivalent to the light-image created by x rays in an intensifying or fluoroscopic screen.

Next, an array of many (about 40 to 90 or more) small, closely spaced PMTs, upon receiving the light flashes from the closely applied crystal, converts them to a pattern of voltage signals (see pages 439–440) as an analog image. An analog-digital converter changes this to a digital image (see pages 440–441 for digitization). Next, the pulse-height analyzer, set to the selected energy "window," accepts a narrow range about the peak energy of the radionuclide in use, and rejects random pulses as well as those generated by scattered radiation that may have penetrated the collimator. Finally, the digital image is converted to an analog image for display on a CRT (cathode ray tube), from which a film can be obtained. At the same time, the digital image is stored on the computer's hard disc for retrieval later.

Single-Photon Emission Computed Tomography (SPECT)

A major advance took place in the early 1980s when single-photon emission computed tomography (SPECT) was introduced as a modification of the conventional gamma camera. A number of advances in engineering design have made SPECT a powerful tool in radionuclide imaging. One can set the window to accept photons of only a *single energy* according to the radionuclide being used. As a result, resolution is improved because extraneous signals such as scattered radiation and random noise in the system are rejected. By tying the camera into a self-contained CT system, images are created in single "slices" as with conventional CT in various planes, and reconstructed into three-dimensional images! An important factor in enhancing resolution is use of a digitizer for each individual PMT.

Another innovation has been the introduction of the *dual camera,* one above and the other under the patient. These produce simultaneous images that are combined into one by computer processing, which results in both improved resolution and reduced procedure time. For example, the usual whole-body skeletal imaging in which the patient's table moves caudad during the imaging process requires about one hour during which the patient must lie still; this time is reduced to about 15 minutes with the dual head unit. Also, a small organ like the thyroid gland can be imaged life-sized. Other types of examinations, previously impossible with a gamma camera, can now be accomplished with the dual-camera SPECT system.

Available Radionuclide Imaging

The process of radionuclide imaging is often called *scintigraphy,* and the images *scintigrams* or *scintiphotos,* although these are sometimes referred to, loosely, as *scintiscans.* It must be emphasized that scintigraphy may not always provide a positive diagnosis, and must often be used in conjunction with other diagnostic modalities. In any event, it provides a wealth of diagnostic information when used appropriately.

A summary will now include some of the more frequently used imaging studies. (See also pages 484–486 for properties of various radionuclides.)

Bone — mainly for detection of metastatic cancer, although osteomyelitis and aseptic necrosis may also be imaged. Radiopharmaceuticals include ^{99m}Tc diphosphonate, ^{99m}Tc medronate, and ^{99m}Tc oxidronate; these compounds concentrate in areas of bone reacting to a destructive process. Lesions appear black, as so-called "hot" areas.

Liver-Spleen — to reveal primary or metastatic lesions, hemangiomas, and cysts in the liver, and defects such as infarcts in the spleen. The radiopharmaceutical (RPC) is usually ^{99m}Tc sulfur colloid, which deposits in the reticuloendothelial system of the liver, spleen, and bone marrow. Lesions in the liver and spleen appear light, as so-called "cold" areas.

Hepatobiliary system — for diagnosis of acute or chronic cholecystitis and biliary tree obstruction. Different segments of the biliary tree and duodenum may or may not be imaged, depending on the site of obstruction. The RPC is ^{99m}TC IDA (*i*mino*d*iacetic *a*cid compounds).

Lung perfusion — the leading RPC ^{99m}Tc-MAA (macroaggregated albumin) consists of minute albumin particles, larger than about 5 μm which, after intravenous injection, block pulmonary capillaries. Although this occurs with only about one out of 1000 particles, there results a diffuse distribution except where an occluded artery (by infarct or embolus) prevents the radioactive particles from reaching the pulmonary zone it normally supplies. This area appears "cold" (white), but there is sufficient lack of certainty that the radiologist reports it in terms of low, intermediate, or high probability (see next paragraph).

Lung ventilation — to help differentiate between infarct or embolus, and a zone of diminished perfusion from bullae, localized emphysema, or airway obstruction. A "mismatch," that is, a high-probability perfusion study (suggesting infarction or embolus) in the face of a normal ventilation scintiphoto supports the impression of pulmonary infarct or embolus. The ventilation RPC is xenon 133, an inert, artificially radioactive gas.

Myocardium —evaluation of coronary artery perfusion of the heart muscle. The RPC is thallium 201 as thallous chloride. To reveal myocardial *infarction,* RPCs include ^{99m}Tc pyrophosphate and Cardiolite® (^{99m}Tc sestamibi).

Kidneys —detection of renal lesions and evaluation of renal perfusion. Several RPCs include ^{131}I hippuran, ^{99m}Tc DTPA, and ^{99m}Tc MAG3®.

Thyroid gland —detection and evaluation of nodules. I 123 and ^{99m}Tc pertechnetate are preferred. For uptake studies, I 123 (rarely, I 131 because of higher dose to thyroid and normal organs).

Cerebral Perfusion —determination of adequacy and pattern of blood flow to the brain, a truly dynamic study (blood in motion). Ceretec™ (^{99m}Tc exametazime) is widely used; it is a lyophilic complex, meaning that it has an affinity for brain lipids (fatty substances).

Cancer detection —especially in lungs and mediastinum, to help differentiate from infection, but abscess may also be positive. RPC is ^{62}gallium citrate.

Leukocyte and platelet labeling —for demonstration of pyogenic abscess, indium 111.

You can see that ^{99m}Tc is the most versatile and widely used radionuclide because it serves as a convenient radioactive tag for a large number of RPCs. Table 22.2 summarizes the physical date for presently available medical radionuclides.

Calibration of Radiopharmaceuticals and Gamma Camera

Radiopharmaceuticals. Before administration of an RPC, the following precautions must be taken: (1) patient must be positively identified, (2) RPC must be positively identified, (3) activity of RPC must be measured in a special calibrator, and dose verified. The accuracy of the calibrator itself must be checked daily to within ±5 percent, and the RPC to within ±10 percent.

TABLE 22.2
SOME RADIONUCLIDES OF MEDICAL INTEREST

Nuclide	Physical Symbol	Half Life	Main Radiation		Yse in Medicine
			$E\gamma$	$E\beta$	
			MeV	MeV	
Cesium 137	^{137}Cs	30 years	0.662	0.188	Teletherapy. Needles.
Chromium 51	^{51}Cr	27.8 days	0.320	0.242	Red cell volume and survival.
Cobalt 57	^{57}Co	270 days	0.122	—	Tagged vitamin B_{12} ($^{57}CoB_{12}$) in diagnosis of pernicious anemia.
Cobalt 60	^{607}Co	5.2 years	1.17 1.33	0.096	Teletherapy.
Gold 198	^{198}Au	2.7 days	0.412	0.316	Implant therapy as seeds or grains.
Iodine 123	^{123}I	13.2 hr	0.159	—	Diagnosis of thyroid disorders
Iodine 131	^{131}I	8.1 days	0.364	0.183	Diagnosis and treatment of thyroid disorders. Localization of thyroid cancer metastases. Renal function test as iodohippourate.
Phosphorus 32	^{32}P	14.3 days	—	0.700	Treatment of polycythemia rubra vera. Treatment of certain metastatic bone tumors (esp. breast and prostate). Malignant pleural and pertioneal effusions.
Strontium 90	^{90}Sr	28 years	—	0.196 0.937°	Contact therapy of superficial lesions of eye.
Technetium 99m	^{99m7}Tc	6 hours	0.140	0.084	Imaging of brain, thyroid, liver, lungs, bones, and kidneys

$E\gamma$ is energy of principal gamma rays. $E\beta$ is average energy of beta rays.
°Refers to yttrium 90, always associated with strontium 90.

Gamma Camera. Imaging quality must be checked periodically for:

1. *Image Resolution.* Requires a Hine-Duley or similar phantom with parallel lead (Pb) bars covering the entire field of view of the camera. The phantom is placed between the collimator and a flood-field phantom (see 2).

2. *Field Uniformity.* This can be determined by use of a *flood-field* phantom. It consists of a plastic (lucite or plexiglass) plate with a flat space in its center parallel to the surface; this space is filled with a radioactive liquid such as ^{99m}Tc. Alternatively, one can use a sheet of plastic that uniformly incorporates a radioactive material. Either of these phantoms should be imaged daily with-

out a collimator, and field uniformity evaluated simply by visual inspection.

3. *Image Distortion.* This refers to the curvature of straight lines as they appear in the image. It can be evaluated by means of a phantom like the Hine-Duley, whose parallel Pb bars should normally appear straight on the scintiphoto. An image of the phantom is obtained by placing it between the flood-field phantom and the camera.

Quality assurance with a SPECT system requires greater care because small changes in quality with a conventional gamma camera are exaggerated with SPECT. Special phantoms are used and a greater number of counts (scintillations) accumulated, with the collimator in position.

QUESTIONS AND PROBLEMS

1. Define isotope; radioisotope; nuclide; radionuclide; nucleon.
2. What is radioactivity? Artificial radioactivity?
3. Describe a nuclear reactor, including fission, chain reaction, and control.
4. Discuss the two main methods of producing radioisotopes in the nuclear reactor, with equations showing how ^{32}P and ^{60}Co are obtained. Name the principal source of ^{131}I.
5. Discuss the three types of half-life.
6. State the simple formula that gives the fraction of a radionuclide remaining after four half-lives have elapsed; also, after five half-lives.
7. Explain the construction, operation, and use of GM and scintillation counters. Define dead time.
8. Name the sources of error in counting.
9. Suppose you find the thyroid counting rate to be 200 cpm and the background counting rate 50 cpm. If the counting rate of the standard is 2000 cpm, calculate the percent uptake in the thyroid.
10. A sample has a counting rate of 400 cps. What is the counting rate, corrected for dead time, with a GM counter having a recovery time of $2(10^{-4})$ sec? What would this be with a scintillation counter having a recovery time of 10^{-6} sec, and the same sample counting rate?

11. Enumerate and explain the geometric factors in counting.
12. State three basic procedures in radionuclide diagnosis, and cite examples of each.
13. Summarize the application of three radiopharmaceuticals diagnin di
14. Why is ^{99m}Tc so widely used in scintigraphy?
15. Describe what happens when gamma photons enter a scintillation crystal.
16. With the aid of a simple diagram, explain how a photomultiplier works.
17. Prepare a block diagram of a gamma camera, between the patient and the CRT tube.
18. Compare the steps in image production in a gamma camera system, with those in radiographic imaging.

Chapter 23

PROTECTION IN RADIOLOGY—HEALTH PHYSICS

THE HARMFUL EFFECTS of exposure to ionizing radiation were largely unsuspected at the time of the discovery of x rays and radium. However, the pioneer workers in this field soon recognized the injurious potentiality of such radiation. It is therefore surprising that so many of the early radiologists and technicians carelessly exposed themselves to x rays and radium, incurring serious local and general radiation injuries, which all too often resulted in death.

In the intervening years we have learned a great deal about radiation hazards and their prevention. As a result, protective measures have become ever more stringent, especially with the extensive use of radionuclides in medicine and industry, and nuclear reactors in the generation of electric power.

Every radiologic technologist should have access to pertinent *National Council on Radiation Protection and Measurements (NCRP) Reports; Bureau of Radiologic Health (BRH) publications;* and *Code of Federal Regulations 21.* These are listed in the Bibliography, together with information about their availability.

What is the "safe" limit of exposure to radiation? It is doubtful whether any amount of radiation is really safe. However, an attempt will be made to answer this question on the basis of acceptable exposure limits, as well as to point out what steps can be taken to eliminate the risk of overexposure. The subject of protection or **health physics** will be taken up principally as it affects the **radiologic technologist** and the **patient.** Attention will also be directed to the use of artificial radionuclides.

BACKGROUND RADIATION

A sensitive radiation detector, such as a GM counter, will indicate the presence of environmental ionizing radiation even in the absence of all known sources, such as radium, x rays, and radioac-

500

tive nuclides. As it turns out, man has virtually no control over environmental radiation, which we call *natural background radiation,* and to which all matter, living and nonliving, is continuously exposed.

Natural background includes not only external sources of radiation, but also radioactive materials located within the body and even the detector itself. Let us survey the sources of natural background radiation, which may be classified as *external* and *internal.*

1. *External sources* comprise three categories. The first is the *cosmic rays,* which are of two types. The *primary cosmic rays,* arising in the sun and other stars, consist mainly of high energy protons (more than 2.5 billion electron volts), but also include alpha particles, atomic nuclei, and high energy electrons and photons. *Secondary cosmic rays* are produced by interaction of primary cosmic rays with nuclei in the earth's atmosphere and consist mainly of mesons (a type of nuclear particle), electrons, a small number of neutrons (absorbed mainly at high altitudes), and gamma rays. Most cosmic rays detected on earth are secondary and are extremely penetrating. However, human exposure indoors is about 80 percent of that outdoors due to residential shielding.

A second external source of background radiation includes the *naturally radioactive minerals* within the earth. These occur in minute amounts almost everywhere in the earth's crust, including building materials, and in larger amounts where the minerals of uranium, thorium, and actinium have been deposited. It must be emphasized that terrestrial (earthly) background radiation varies from place to place; for example, in Kerala, India, it is about twenty times greater than that in most parts of the United States.

A third external source includes the radionuclides produced by interaction of cosmic rays with nuclides in the earth's atmosphere. The main products are carbon 14 and hydrogen 3 (tritium) resulting from capture of slow cosmic-ray neutrons by stable atmospheric nitrogen; sodium 22, and beryllium 7.

2. *Internal sources* comprise the naturally radioactive nuclides incorporated in the tissues of the body, and in the materials of which the detector itself is constructed. The main naturally radioactive nuclides are potassium 40, carbon 14, and lead 210-polonium 210.

To the natural background radiation must be added the **artificial radiation** arising from the following sources: medical and dental x rays; occupational, such as x-ray, radium, and radionuclide technology; environs of atomic energy plants; and atomic bomb fallout. In contrast to natural background, man can exercise some measure of control over exposure to artificial sources.

Table 23.1 summarizes the **average** exposures of the general population of the United States on an annual basis as to the site of exposure, dose per exposed person, and the prorated exposure to the population as a whole (based on *BEIR Report, 1980*). Obviously, the concern with these small exposures is with genetic effects appearing in future generations from damage to the reproductive cells, and with injury to the bone marrow. Bear in mind that these data are approximate and do not necessarily apply to a particular individual.

DOSE EQUIVALENT LIMIT

Our knowledge about the deleterious effects of ionizing radiation on living cells demand that an upper limit of absorbed dose be established for personnel and for the population at large. Even when radiation exposure is required for medical diagnosis, it must be kept as low as possible.

The concept of dose limitation implies **risk limitation.** The dose equivalent limit will be detailed later, but at this point we shall consider it on the basis of stochastic and nonstochastic effects:

A **stochastic effect** pertains only to the **statistical chance** or probability that a particular effect will occur, usually from a small absorbed dose. This can be plotted graphically as incidence of effect as a function of dose. Note that the severity of a radiation injury is not included. A stochastic effect is assumed *not* to have a threshold dose (ie, minimal dose below which there is no effect). Examples of stochastic effects include cancer induction (solid cancer and leukemia) and genetic effects.

A **nonstochastic effect** is one which increases in **severity** with increasing absorbed dose by way of injury to increasing numbers of normal cells. It may be regarded as a **certainty effect** in that there are threshold doses for particular clinically significant radiation injuries. Such thresholds are high enough that they may be avoided. Examples include opacities (cataracts of

TABLE 23.1
ANNUAL DOSE RATES FROM IMPORTANT SIGNIFICANT SOURCES OF RADIATION EXPOSURE IN THE U.S. (*BEIR 1980*).*

Multiply each value by 0.01 to convert to mSv.

Source	Exposed Group	Body Portion Exposed	Average Dose (exposed persons)	Prorated over Population
			mrem/yr	mrem/yr
NATURAL BACKGROUND				
cosmic radiation	total population	whole body	28	28
terrestrial radiation	total population	whole body	26	26
internal sources	total population	gonads	28	28
		bone marrow	24	24
		Total (rounded)	80	80**
MEDICAL X RAYS				
Patients				
medical diagnosis	adults	bone marrow	100	77
dental diagnosis	adults	bone marrow	3	1.4
		Total (rounded)	103	78
Personnel (occupational)				
medical	adults	whole body	320	0.3
dental	adults	whole body	80	0.05
RADIOPHARMACEUTICALS				
*Patients (diagnosis)***	adults	bone marrow	300	13.6
Personnel	adults	whole body	300	0.1
ATMOSPHERIC WEAPONS TESTING	total population	whole body	4-5	4-5

*NCRP REPORT No. 105 (1989) gives natural background values that differ slightly from BEIR.
**Based on shielding by dwellings and other buildings; otherwise, average 100 mrem/yr. Range on Atlantic and Gulf coasts 15 to 35 mrem/yr, and Colorado Plateau 75 to 140 mrem/yr.
***The *Environmental Protection Agency (Report ORP/CDS 72-1)* estimates the exposure of patients during nuclear medicine procedures to be about 20 percent of the total radiation exposure in medical diagnosis.

the ocular lens), decrease in blood count, and decrease in sperm production.

DERIVATION OF UNIT FOR RISK ASSESSMENT

Before taking up the special dose unit for radiation protection, we must define two underlying concepts: (1) linear energy transfer (LET), and (2) relative biologic effectiveness (RBE).

1. **Linear energy transfer** is the amount of energy deposited per unit length of path by an ionizing particle, expressed in keV/μm. In general, the greater the LET, the more severe the biologic effect. For example, neutrons have a stronger effect than x rays, for an equal absorbed dose (cGy or rad).

2. **Relative biologic effectiveness** is used primarily in radiobiology to express the ratio of the **absorbed dose** of a standard radiation—200-kV x rays—to that of a given kind of radiation, to produce the same intensity of biologic effect, expressed by

$$\text{RBE} = \frac{\textbf{dose of standard radiation (200-kV x rays)}}{\textbf{dose of radiation in question}} \quad \begin{array}{l} \textit{for a} \\ \textit{given effect} \end{array}$$

We may therefore conclude that equal absorbed doses of radiations having different LETs would pose different degrees of risk. Hence, in assessing risk, it would be misleading to consider only the absorbed dose of single or mixed radiations when one or more has an RBE exceeding unity.

Dose Equivalent

To express on a common scale for protection purposes, the risk of radiation injury from all kinds of ionizing radiation, the quantity **dose equivalent H** was introduced by the ICRU many years ago and adopted by the NCRP. The dose equivalent is obtained by simply multiplying the absorbed dose of a given type of radiation by its relevant **quality factor** $\overline{Q}$. Derived experimentally from LET in water, $\overline{Q}$ for a given kind of radiation is then related to an arbitrary value of unity for low-LET radiation (x, beta, gamma) (see Table 23.2).

TABLE 23.2
RECOMMENDED VALUES OF $\overline{Q}$ FOR RADIATION
OF VARIOUS KINDS*

Radiation	Approximate value of Q
X-rays, gamma rays, beta particles, and electrons	1
Thermal neutrons	5
Neutrons (other than thermal), protons, alpha particles, and multiple-charged particles	20

*NCRP Report No. 91

The following equation states how dose equivalent H is derived:

$$H = \textbf{absorbed dose} \times \overline{Q} \qquad (1)$$

$$rem \qquad rad \qquad \times \overline{Q}$$

$$cSv \qquad cGy \qquad \times \overline{Q}$$

The appropriate value of $\overline{Q}$ converts cGy to cSv, and rad to rem; the units are related as follows:

$$1 \text{ sievert (Sv)} = 100 \text{ rem}$$
$$\text{and } 1 \text{ gray (Gy)} = 100 \text{ rad}$$

Therefore,

$$1 \text{ centisievert (cSv)} = 1 \text{ rem}$$
$$\text{and } 1 \text{ centigray (cGy)} = 1 \text{ rad}$$

As already noted, x, gamma, and beta (electron) radiation have been arbitrarily assigned a $\overline{Q}$ value of 1, so

$$H = \text{absorbed dose} \times 1$$
$$= 1 \text{ cGy (rad)} \times 1 = 1 \text{ cSv (rem)}$$

Therefore, 1 cSv (or rem) = 1 cGy (rad) for these specific kinds of radiation. Actually, in protection surveys for these low-LET radiations, we may use the roentgen (R) in place of cSv or rem because they are numerically about the same; moreover, survey instruments are still calibrated in R. But organ doses should still be expressed in H units—cSv or rem.

For fast neutrons, $\overline{Q} = 20$, that is, 1 cGy of fast neutrons is 20 times as effective as 1 cGy of x rays for induction of ocular cataract; conversely, the cataractogenic dose of fast neutrons is $1/20$ that of x rays. This can be seen from equation (1),

$$H = \text{absorbed dose} \times 20$$
$$= \qquad 1 \text{ cGy} \times 20 = 20 \text{ cSv } \textit{for fast-neutron}$$
$$\textit{cataract induction}$$

You can see clearly how the dose equivalent expresses the relative risk from radiations having different $\overline{Q}$ values.

It is obvious that the dose equivalents of various kinds of radiation (ie, different values of $\overline{Q}$) can be added, whereas absorbed doses cannot. For example, if a person were exposed to a mixture of radiations, receiving 1 cGy of x rays and a cGy of fast neutrons, the total dose would be 2 cGy. But this would seriously underestimate the radiation hazard. Instead, conversion to H would give

x rays	1 cGy =	1 cSv
fast neutrons	1 cGy =	20 cSv
	Total	21 cSv (rems)

Thus, the value of H in cSv gives a more realistic measure of radiobiologic hazard when $\overline{Q}$ differs from unity. In medical radiology and radiotherapy, however, by far the most commonly used radiation comprises x and γ rays whose $\overline{Q} = 1$.

Effective Dose Equivalent. With uniform irradiation of the whole body, stochastic effects (cancer induction and genetic injuries) may be expressed by the dose equivalent to the whole body, symbolized by H_{wb}. However, truly total body irradiation, such as one might expect in a radiology department, does not entail uniform tissue dosage because of variations in absorption of the low-energy x rays, as well as uncertainties introduced by shielding with lead aprons and other protective devices.

Because of this inherent nonuniformity of a single exposure to ionizing radiation, and the different stochastic limits for various organs, the ICRP introduced, and the *NCRP Report No. 91* pursued, a new concept—*effective dose equivalent* H_E—to provide a more realistic assessment of stochastic risk by assigning *weighting factors* to individual organs. This attempts to correct for the appreciable differences in dose equivalent H to individual organs, especially when a mixture of radiations is involved.

At present, there is some question about the relevance of H_E in diagnostic radiology where the main source of exposure to personnel is in fluoroscopy and special procedures, and in diagnostic nuclear medicine—these all deal with low-LET radiation. Recall that we are concerned with stochastic limits. The attempt to relate exposure monitoring with film badge and TLD, to H_E, is complicated by uncertainties in shielding and in absorbed dose to various organs, so the readings obtained with these monitors worn on the collar outside the lead apron remain the most reliable index of exposure. It turns out that *NCRP Report No. 107* (p. 9) assumes the monitor readings adequately represent H_E when reported for low-LET radiation (x and gamma rays) in units of either dose equivalent H (mSv or rem), absorbed dose (cGy or rad), or exposure (R).

NUMERICAL DOSE EQUIVALENT LIMITS

Occupational Limits

As explained in the preceding section, effective dose equivalent may be replaced by the simple *dose equivalent H* for setting exposure limits for radiation workers in *diagnostic imaging.* According to *NCRP Report No. 91,* a uniform dose equivalent of 1 rem (10 mSv) to the whole body is estimated to entail a lifetime fatality risk of 10^{-4} (ie, 1 in 10,000) as a *stochastic* effect for adults. But data from the Environmental Protection Agency (quoted in *NCRP Report No. 91*) indicate an average of only about 0.23 rem (2.3 mSv) per year for monitored radiation workers with measurable exposure; that is, 1/20 the earlier NCRP prospective annual whole body exposure limit of 5 rem (50 mSv). Thus, a reduction in occupational exposure limits is achievable on the basis of current practice, so the NCRP makes the following recommendations (see Table 23.3 for summary):

TABLE 23.3
DOSE LIMIT RECOMMENDATIONS

OCCUPATIONAL EXPOSURE (*ANNUAL*)		
Effective Dose Equivalent Limit°°		
(*stochastic effects*)	5 rem	(50 mSv)
Dose Equivalent Limits for Tissues and Organs		
(*nonstochastic effects*)		
Lens of eye	15 rem	(150 mSv)
All other organs (eg, red marrow, breast, lung,		
gonads, skin, and limbs)	50 rem	(500 mSv)
Guidance: cumulative exposure	1 rem × age in years	(10 mSv × age)
PUBLIC EXPOSURE (*ANNUAL*)		
Effective Dose Equivalent Limit		
continuous or infrequent exposure°°	0.1 rem	(1 mSv)
infrequent exposure	0.5 rem	(5 mSv)
Dose Equivalent Limits: lens of eye, skin, and limbs	5 rem	(50 mSv)
EMBRYO-FETUS (CONCEPTUS) EXPOSURES°		
Total dose equivalent limit	0.5 rem	(5 mSv)
Dose equivalent limit in any month	0.05 rem (50 mrem)	(0.5 mSv)

°After *NCRP Report No. 91*
°°Sum of external and internal exposures

1. For **Stochastic Effects:**

a. *Abandon the previous dose limit 5 (age − 18 rem (50 mSv).*

b. Set *annual* whole body dose equivalent limit of 5 *rem (50 mSv)*° *for routine procedures* (occupational). This is a limiting condition — one that may be approached only when the cost (financial or otherwise) becomes prohibitive.

c. Set *lifetime* dose equivalent limit to the whole body at tens of mSv, that is, in rem, not to exceed the radiation worker's age in years. For example, at age 25 years, the cumulated dose shall be no greater than 25 × 1 rem = 25 rem, or 25 × 10 mSv = 250 mSv. This does not mean that it is acceptable for younger individuals to receive larger exposures than older ones. Note that this limit is less than the old MPD at age 25. Moreover, it has been estimated to result in about the same lifetime fatality rate as in so-called "safe" industries and in auto travel to and from work (ICRP, 1977a).

d. *During pregnancy* the specific dose equivalent limit applies to the **conceptus** (embryo and fetus) and **not** to the mother. Therefore, all pregnant women are subject to the dose equivalent limit for the conceptus. Naturally, radiation workers are continually monitored while the general public is not. Monitor badges over-estimate the dose to the conceptus because they do not take into account attenuation of x rays in the lead apron and abdominal tissues. This can be estimated in abdominal radiography of fertile women as described on pages 527–528. The recommended dose equivalent limit is **0.5 rem (5 mSv) for the entire gestational (pregnant) period,** but at a rate not exceeding **0.05 rem (0.5 mSv) in any one month.**

2. For **Nonstochastic Effects.** The annual dose equivalent limits include:

a. *Ocular lens 15 rem (150 mSv),* and

b. All *other organs* such as bone red marrow, breast, lung, gonads, skin, thyroid gland 50 *rem (500 mSv)*. However, experience with personnel monitoring by film badge or TLD shows that radiation workers receive only about one-tenth (or less) of the annual dose equivalent limit for stochastic effects in diagnostic imaging. Therefore, the lifetime dose equivalent to these organs should be expected to be **no greater than [age in years × 1 rem (10 mSv)]** (see paragraph 1c above).

Radiation workers should know that exposure to x rays for their

---

°*International Commission on Radiologic Protection* (ICRP) in 1990 reduced this to 2 *rem per year,* but it has not yet been implemented in the U.S.

own medical diagnosis or treatment does not count as an occupational exposure, but during pregnancy the conceptus should be protected to the greatest possible degree.

Pregnant Radiation Workers

As already stated, the *conceptus* (embryo or fetus) is the primary object of protection in the pregnant radiation worker, generally the radiologic technologist. The conceptus is an *embryo* from the time of implantation in the uterine wall to the end of the second month, and a *fetus* thereafter until delivery.

The monitored surface dose to the skin of the abdomen will be higher than the dose equivalent to the conceptus; the technologist should be wearing an acceptable lead apron (0.5 mm Pb), which reduces the exposure rate at the skin to approximately 10 percent of that at the outside surface of the apron in the usual fluoroscopic kVp range. Further attenuation occurs in the intervening soft tissues of the abdomen and amniotic fluid.

According to the recommendations in *NCRP Report No. 105*, all fertile (premenopausal) woman technologists **shall*** be informed by a responsible person, in advance, about the risks to which a conceptus may be exposed. If the technologist is uncertain, counseling *should** be available. Personnel records, exposure records, departmental radiation surveys, sources of radiation, and availability of protective devices *should* be reviewed to aid in evaluating risk.

If conditions are such that the conceptus could receive a dose equivalent greater than 0.5 rem (5 mSv) during gestation, the technologist should discuss this with her employer. Once a technologist learns she is pregnant, she should notify him or her.

NCRP Report No. 105 does not advocate restriction of the pregnant technologist in fluoroscopy, mobile radiography, and special procedures, other than not being allowed to hold patients during these or any other x-ray examinations. However, she must consistently wear a lead protective apron and two personnel monitors (film badge or TLD), one on the collar for the face and neck, and the other on the anterior abdomen beneath the apron for the conceptus. She should wear a wrap-around lead apron in C-arm fluoroscopy and radiography, and in special procedures.

A U.S. Supreme Court ruling in March 1991 emphasizes the

---

*Shall means mandatory. *Should* means advisable.

necessity of radiology departments to have carefully written policies regarding the employment of pregnant women and restrictions in their duties. There is a conflict between the exposure of a conceptus and discrimination against the mother. Moreover, the ICRP in 1990 recommended lower the annual limit for all radiation workers from 5 rem to 2 rem. It is advisable for every radiology department and office to remain current on future developments. The ACR (American College of Radiology) discusses this problem in its *Bulletin* of September 1992.

Nonoccupational (General Public) Limit

It is extremely difficult to estimate accurately the radiation risk to the general public because of the confounding effect of the great variety of risks from nonradiation sources. Radiation sources include: (1) natural background (see pages 500–502) and (2) man-made radiation. For the general public, *NCRP Report No. 91* recommends:

1. An ***annual dose equivalent limit of no more than 0.1 rem (1 mSv) for continuous or frequent exposure,*** to limit the annual fatality risk to the level of common nonradiologic risk of about 10^{-5} (1 per 100,000) per year.

2. A ***maximum annual dose equivalent limit of 0.5 rem (5 mSv) for infrequent exposures.***

We are using dose equivalent instead of effective dose equivalent because the exposures here involve diagnostic x rays, other than those used for medical procedures performed on the individual. Moreover, they do not include background radiation. For other kinds of radiation such as high-energy x or gamma rays, or artificial radionuclides, effective dose equivalent applies to the above limits.

ALARA Concept

The *NCRP* in *Report No. 107*, and the *ICRP*° in *Publication 26*, have both addressed the ***ALARA*** (As Low As Reasonably Achievable) concept, which deals with the problem of reducing exposure limits further than presently recommended. They maintain that the "reasonably achievable" aspect of ALARA implies a

---

°*International Commission on Radiological Protection.*

balancing of benefit (risk reduction) *vs* cost (financial and social). In other words, further reduction should yield a definite net benefit after the resulting economic and social factors have been carefully evaluated.

It is difficult to determine ALARA quantitatively, but we can approach the problem in two ways: (1) planning radiology departmental protection in advance of construction, and (2) promoting awareness within the department of the need to carry out all radiologic procedures utilizing all appropriate protective measures, with *rare*, deliberate exceptions. This resembles the constant attention of operating room personnel to aseptic technique, which requires a certain "frame of mind."

The day-to-day procedures in diagnostic radiology demand risk-limitation not only in the department itself but also in mobile radiography, mammography, angiography, cardiac catheterization, and interventional procedures wherever radiography or fluoroscopy is done. Technologists and other personnel must keep "patient holding" to a bare minimum; protective lead gloves, aprons, and shields, and other appropriate devices should be readily available and used properly.

It is especially important in C-arm fluoroscopy and radiography that the technologist wear a lead protective apron and gloves, and particularly to avoid exposure to the direct beam. A physicist should be involved in establishing risk-limiting procedures under these conditions.

Thus, in x-ray imaging procedures, ALARA does not require expensive modifications of rooms, equipment, or devices (provided department has adequate built-in protection), if the monitored personnel exposures are below the recommended limits. This depends on continuous vigilance in using the necessary protective devices. Whenever there is an appreciable rise in the monitor readings, a search must be made promptly for "breaks" in technique, and corrected promptly.

The principles of ALARA for other radiologic personnel—nuclear medicine, radiation oncology, and dentistry—appear in detail in *NCRP Report No. 107.*

PERSONNEL PROTECTION FROM EXPOSURE TO X RAYS

Let us now consider the protection of personnel against exposure to x rays, under three main headings:

1. Determination of radiation safety in the Radiology Department, called *radiation monitoring.*

2. Deleterious effects to the individual resulting from excessive exposure.

3. Protective shielding of radiographic and radiotherapeutic installations.

Radiation Monitoring. In the Radiology Department we can monitor whole-body exposure without special equipment or training (a qualified health physicist can do this more precisely, but not ordinarily required). Several devices are available. The most sensitive instrument, the *pocket dosimeter,* outwardly resembles a fountain pen but contains a thimble ionization chamber at one end. The exposure may be read by means of a separate electrometer; or, in the self-reading type, by means of a built-in electrometer. A dosimeter of this type, while sensitive to exposures up to 0.2 R (200 milliroentgens), is easily damaged and is unreliable in inexperienced hands. Besides, it does not provide a permanent record.

The *film badge* offers the most convenient method of personnel monitoring under average conditions. Commercial laboratories specialize in furnishing and servicing the film badges. The x-ray film badge consists of a dental film, covered by a copper and plastic filter to show the quality of the radiation, and allow conversion to tissue dose. The film is backed by a sheet of lead foil to absorb scattered radiation from behind the badge. It is usually worn on the clothing over the front of the hip or chest, but during fluoroscopy it should be worn at the neck, clipped to top of the front of the apron, to monitor exposure of the thyroid gland and eyes. Pregnant personnel should wear a second badge over the abdomen beneath the leaded apron during fluoroscopy and mobile radiography.

The laboratory regularly supplies a fresh badge, which, after being worn for the prescribed interval (usually one month), is returned to the laboratory for processing under standard conditions. Dosimetric comparison is then made with standard films exposed to known amounts of radiation. The resulting film exposure in roentgens is reported to the Radiology Department for permanent filing. Badges should never be exchanged among personnel and should be clearly marked for identification. With exposure levels well below the recommended dose equivalent limit, the badge should be worn and returned *monthly,* especially if the work load is reasonably constant. The one-month period is preferred because of convenience and greater accuracy of calibration.

Another kind of personnel monitor, the ***thermoluminescent dosimeter*** (TLD), depends on the ability of certain crystalline materials to store energy on exposure to ionizing radiation because of the trapping of valence electrons in crystal lattice defects. When the crystals are heated under strictly controlled conditions, the electrons return to their normal state, the stored energy being released in the form of light. Measurement of the light by a photomultiplier device gives a measure of the initial radiation exposure, since the two are very nearly proportional. The dosimeters should be returned to a commercial laboratory for readout, unless this can be done accurately on the premises. Lithium fluoride (LiF) is at present the most prevalent material for TLD. The many advantages of TLD over film badge monitoring include: (1) detectors are cheap and small—for example, 1 mm × 6 mm—and can be sealed in Teflon® to avoid mechanical damage, (2) direct reading available at any time, (3) response to radiation proportional up to about 400 R, (4) response almost independent of radiation energy from 50 kV to 20 MV (film badge is not), (5) response very similar to that of human tissues, (6) wide exposure range detectable—1 mR to 1,000 R, (7) accuracy about ±5 percent (film badge accurate to about ±50 percent, (8) feasibility of incorporating detector in jewelry to make it unobtrusive. For these reasons, it seems likely that the TLD will eventually become the method of choice in personnel monitoring.

On termination of employment all technologists must receive an appropriate summary of occupational exposure, regardless of the monitoring system. If reemployed, the technologist should present this summary report to the new employer.

Harmful Effects of Radiation. Why has health physics become so important in radiology? The answer lies in the fact that radiation is always harmful to tissues, the amount and nature of any damage being dependent on radiation quality (LET), absorbed dose, size and location of the treated region, and fractionation or number of days over which the treatment is administered. It is the differential damage to a given type of tumor as compared with that to normal tissues that underlies successful irradiation therapy. The term *therapeutic ratio* has been applied to the ratio of the dose the normal tissues can tolerate, to the dose needed to destroy the tumor. Remember that the therapeutic exposure of patients is a matter quite different from the small exposure that technologists may receive over a prolonged period of time. The latter problem concerns us here. *The objective of a program in health physics is to keep the exposure of personnel as low as reasonably achievable so as to avoid the slightest radiation injury.* But suppose overexposure is incurred; what sort of harmful effects may be anticipated? These are three in number: *local, general, and genetic.*

1. *Local Effects.* Small occupational exposures should not cause local changes in the skin. Several hundred rem of low energy x rays are required to produce erythema (skin reddening), but such a dose would be very unlikely except in a therapy patient. However, even with very small exposures repeated over a period of years, late irradiation effects may occur in the form of dry skin, which is subject to cracking, ulceration, and possible cancer induction. Another harmful effect is cataract (opacity) formation in the highly radiosensitive lens of the eye by a dose equivalent of a few hundred rem, a nonstochastic effect unlikely in x-ray personnel.

2. *General Effects.* Whole body exposure to x or gamma radiation always entails the possibility of harmful effects. The blood-forming organs (bone marrow and lymphatic tissues) are especially susceptible. No certainty exists as to the presence of a threshold; that is, a minimum level of exposure below which there is no damage, and above which damage abruptly appears. Such *a threshold or lower limit has not been demonstrated.* We know that a dose equivalent of as little as 5 to 25 rem to the whole body in one exposure can produce chromosomal abnormalities; and in some individuals, 50 to 75 rem may cause detectable changes in the blood picture. Lymphocytes, leucocytes, and platelets are most susceptible, in that order. Small doses to a large part of the body increase slightly the chance of inducing leukemia.

Evidence in laboratory animals indicates that chronic exposure to radiation causes *shortening of life span,* increasing in severity with increasing exposure. Statistical studies have failed to show conclusively a similar effect in humans.

3. *Pregnancy.* Exposure during pregnancy is especially hazard-

ous to the *conceptus* (embryo and fetus). From the first to the tenth day after fertilization, a dose of about 10 to 15 rads (cGy) will probably induce abortion, but virtually no congenital anomalies will result from exposure during this period. Maximum sensitivity of the human conceptus appears during the 11th to the 41st day of gestation (pregnancy), the period of major organogenesis (organ development). According to *NCRP Report No. 91,* mental retardation in humans is proportional to dose, from the eighth to the fifteenth week of gestation, and decreases thereafter until delivery.

Insofar as patients are concerned, radiologic procedures should be kept at a minimum consistent with diagnostic necessity, particularly during the vulnerable early months of pregnancy. In fact, Hammer-Jacobson (1959) held that a dose of 10 cGy during the first six weeks of pregnancy is so likely to cause serious congenital abnormalities that abortion should be induced; this has not been universally accepted.

If the recommended exposure limits are enforced—0.5 rem (5 mSv) during entire gestation, with no more than 0.05 rem (0.5 mSv) in any one month—the risk of cancer (including leukemia) and malformations induced by radiation is very small, by comparison with the natural incidence of about 6 percent (*UNSCEAR, 1986*). It must be emphasized that the severe congenital malformations (eg, microcephaly, exencephaly, heart defects) seen in experiments with mice and in radiotherapy of humans occur with large doses in the range of 200 to 250 cGy.

In planning treatment with internal *radionuclides,* the physician must take into account the effective dose equivalent to the conceptus and consider alternative treatment options.

4. *Acute Whole Body Irradiation.* Serious radiation injury results from large dose equivalents to the whole body in a single exposure, in the range of about 100 to 1000 rem (1 to 10 Sv). Such massive exposures may occur accidentally in atomic energy installations, and deliberately (eg, 1000 rems [10 Sv]) in the treatment of leukemia, followed by compatible bone marrow transplants and various medications.

Acute whole body exposure (ie, in a single dose) to the above levels gives rise to the so-called *acute radiation syndrome,* whose severity depends on the dose. A supralethal exposure of about 600 to 1000 rem (6 to 10 Sv) to the whole body causes the *gastrointestinal syndrome,* marked by nausea, vomiting, diarrhea, and prostration within one to two hours after exposure. There are associated anorexia (loss of appetite), dehydration, and a rapid profound drop in white blood count. Hemorrhages occur in the skin and intestinal tract, terminating in death from sepsis in one to two weeks. However, intensive treatment with blood transfusions, fluids, antibiotics, and compatible bone marrow transfusions make recovery possible.

With a *midlethal exposure* of about 450 rem (4.5 Sv) the effects are less pronounced, though still severe; about one-half the exposed persons die within one month from pancytopenia (marked loss of all three kinds of blood cells and platelets). This is called the *bone marrow syndrome.*

A *sublethal exposure* of 100 to 300 rem (1 to 3 Sv) gives rise to mild nausea, vomiting, and malaise lasting about twelve to twenty-four hours. After a latent period of about three to five weeks, loss of hair occurs, although it usually regrows. There is a drop in all of the blood cell counts, and many individuals develop cataracts in two to six years. Most recover, but some enter a chronic stage, remaining feeble and anemic for months. Statistically, there is an increased incidence of leukemia in the survivors—about three to five times that in the unexposed population, as found in the survivors of the atomic bombing of Hiroshima and Nagasaki in World War II.

5. *Genetic Effects.* A serious problem associated with exposure to radiation is the *genetic hazard*, that is, the appearance of abnormalities in future generations. This results from the extreme vulnerability of the genetic material—*genes*—(particularly in the chromosomes of the sex cells) to radiation damage. There is no known safe lower limit of dosage, the damage being approximately proportional to the absorbed dose. Because of the transmission of genetic material to offspring during reproduction, it is obvious that any radiation-damaged genetic material will be similarly handed down. This becomes especially serious when large populations are exposed individually or in groups, so we must shield the gonads (ovaries and testicles) during diagnostic procedures whenever possible. Radiation damage to the genetic material is expressed as *chromosome aberrations* and *gene mutations.* The latter, being recessive, tend to be hidden by pairing with normal genes during reproduction, only to become manifest in subsequent generations. Many such mutants die very early in pregnancy and therefore

never appear. Insofar as technologists are concerned, they are not subject to appreciable genetic damage if whole-body exposure remains within the recommended limits.

Gonadal exposure risk is expressed in terms of the *genetically significant dose* (GSD), which may be defined as that dose of ionizing radiation to the gonads of *exposed* individuals which, if spread over the entire population, would induce the same total genetic risk to their offspring. This assumes that the genetic material of all persons constitutes a common mass called the *genetic pool.* Thus, no matter how small the gonadal exposure of one individual, it contributes to the genetic pool and therefore creates an equal total genetic risk there. For example, in an assumed genetic pool of 1 million (10^6) persons, a total gonadal dose of 10 rem (100 mSv) to *exposed* persons would be equivalent (GSD) to 0.001 rem (0.01 mSv) to the 1 million *unexposed* population, according to:

$$10 \text{ rem} \times 100 \text{ (exposed)} = X \text{ rem (GSD)} \times 1{,}000{,}000 \text{ (unexposed)}$$

$$10^3 \text{ (exposed)} = X \text{ rem (GSD)} \times 10^6 \text{ (unexposed)}$$

$$\text{GSD } X = \frac{10^3}{10^6} = 10^{-3} = 0.001 \text{ rem (0.01 mSv)}$$

With careful monitoring, the likelihood of incurring radiation injury is virtually nonexistent. When, under *unusual* circumstances, damage to the blood-forming organs is suspected, corrective steps should be taken immediately. First, the affected individual should be removed from further exposure to ionizing radiation. Second, a trained health physicist should survey the Radiology Department and recommended the best means of eliminating the hazardous conditions. However, it must be emphasized here that even under suitably controlled conditions, each technologist must observe certain rules of conduct. In other words **even in the best controlled department, the technologist may incur overexposure, unless full use is always made of all available protective facilities.**

Protective Measures

General Principles

Protection in the diagnostic radiology department involves three basic principles: (1) exposure time, (2) distance, and (3) lead barriers.

1. **Exposure Time.** The total dose equivalent H to a person for a particular H rate is directly proportional to the exposure time:

$$H_{total} \text{ rem} = H \text{ rem/sec} \times \text{time (sec)} \qquad (3)$$

Note that the unit sec appears in both the numerator and denominator of the *right* side of the equation and therefore cancel out, leaving the answer in rem (if mSv is used, the answer will be in mSv).

2. **Distance.** The inverse square law applies closely enough to be valid for protection purposes (see pages 333–337). Whenever possible, the technologist should stand at least 2 m (6 ft) from the patient, x-ray tube, and direct beam; *preferably,* in the control booth or behind a mobile lead (Pb) shield. During fluoroscopy, if the technologist stands 1 m (3 ft) from the x-ray tube, moving back only one good step (about 1 m) reduces the exposure rate to about one-fourth. Thus, distance is a powerful factor in radiation protection.

3. **Lead (Pb) Barrier.** Lead is an efficient absorber of x rays in the kV region for diagnostic radiology because of the low photon energy. Therefore, placing a relatively small but appropriate thickness of Pb between the x-ray source and the personnel contributes greatly to exposure reduction. The thickness of protective barrier is usually stated in *half-value layer* (*HVL*) for kilovoltage x rays (tenth-value layer or TVL for megavoltage photons). Recall that HVL applies to any material and is that *thickness* which reduces the exposure rate by one-half; for kV x rays, this is specified in mm or cm Pb. For any protective barrier, 1 TVL = 3.3 HVL.

Application of the HVL concept means that if the initial dose equivalent (H) is 10 rem (100 mSv), and the HVL is 1 mm Pb for the given kVp, then 1 mm Pb will reduce H to 5 rem (50 mSv). A second HVL of 1 mm Pb will reduce H to another one-half, that is, to 2.5 rem (25 mSv), or one-fourth the initial value. Note that adding 2 HVLs has the same effect as doubling the distance.

Protective Barriers in Radiography and Fluoroscopy

Radiation Sources. All x-ray tubes must be ray proof, that is, the manufacturer must enclose the tube in a metal housing that reduces leakage radiation to a prescribed acceptable level (leakage radiation is that radiation which penetrates the protective housing). The specifications for housing appear in *NCRP Report No. 102:* leakage radiation from a diagnostic tube, when operated at its maximum kV and mA, shall not exceed 0.1 R/hr at a distance of 1 meter.

Wall Protection. This requires the use of built-in *protective barriers* of suitable radiation-absorbing material(s) to restrict exposure to the recommended limit. Wall protection should be planned in advance by a radiation physicist qualified in this field, to avoid expensive alteration after completion of the building; poor planning may result in excessive wall protection, which becomes unnecessarily expensive. However, it costs relatively little to reduce the exposure rate to 10 mR/week, a policy consistent with the ALARA concept. A basic principle in permanent barriers is that joints and holes must be covered by the same or equivalent protective barrier as the wall. Proper wall protection varies with the *energy of the radiation* modified by certain factors described in *NCRP Report No. 49.*

Terms that often apply to wall protection are "controlled" and "uncontrolled" areas. These may be defined as follows:

Controlled area is one that is under the supervision of the Radiation Safety Officer and in which the occupational exposure limit is 1 rem (10 mSv) per year.

Uncontrolled area is one that is *not* under supervision and in which the annual exposure limit to the general public prevails, that is, 0.5 rem (5 mSv) for infrequent or occasional exposures.

There are four main types of radiation to consider in radiation protection:

Useful beam is the radiation passing through the tube aperture and the beam-limiting device, previously called the primary beam.

Leakage radiation includes all the radiation passing through the tube housing, other than the useful beam.

Scattered radiation is that which has undergone a change in direction during passage through matter.

Stray radiation is the sum of scattered and leakage radiation.

There are two kinds of wall barriers. A *primary protective barrier* is one whose atomic number and thickness suffice to reduce the exposure rate of the useful beam to the occupational exposure limit. A *secondary protective barrier* is one whose atomic number and thickness suffice to reduce the exposure rate of the stray radiation to the occupational exposure limit.

Although, as stated above, wall barriers should be planned before construction, typical approximate values are as follows: the *primary protective barrier,* that is, where the useful beam can hit the wall *directly,* in radiography up to 140 kV is about $1/16$ in. lead (Pb) extending 7 ft up from the floor when the x-ray tube is 5 to 7 ft from the wall. This barrier also takes care of leakage radiation, that is, radiation coming through the tube housing. A *secondary protective barrier* (covering areas exposed only to scattered and leakage radiation) is about $1/32$ in. Pb under the same operating conditions. This secondary barrier extends from the top of the primary barrier to the ceiling, but should overlap the primary barrier at least $1/2$ in. at the seam. Ordinary plaster often suffices as a secondary barrier without added lead, especially in large radiographic rooms. Control booths should have the same protection as walls, and radiation should scatter at least *twice* before reaching the opening of the booth.

The leaded glass observation port in the control booth should have the same lead equivalent as the adjacent wall and should be overlapped about $1/2$ in. by the lead in the wall. Leaded glass must be about four times as thick as sheet lead for equivalent protection; for example, $1/4$ in. leaded glass is equivalent to $1/16$ in. sheet lead.

Working Conditions

Certain general rules govern personnel safety. These should be impressed repeatedly on all Radiology Department personnel, since their cooperation is mandatory.

1. Never expose a human for demonstration purposes alone.
2. Never remain in a radiographic or radiotherapy room while an exposure is in progress. In general, approximately 0.1 percent of the useful beam is scattered perpendicular to the beam at a distance of 1 meter from the patient.
3. Never hold a patient for therapy, because of the high intensity of the scattered radiation. Patients should *rarely* be held for radiography, since many efficient devices are available, in addition to the technologist's ingenuity. However, if a patient must be held for radiography, this should preferably be done by a person not habitually exposed to ionizing radiation; in any case, protection by a lead apron and lead gloves is mandatory, and exposure to the direct beam must be avoided.
4. Give yourself the same protection as a loaded cassette!
5. In *fluoroscopy,* personnel protection is extremely important

and you should be sure that the following items are available and in use:

a. A ***protective drape*** having a lead (Pb) equivalent of at least 0.25 mm, or preferably 0.5 mm, situated between the fluoroscopic assembly and the fluoroscopist to block scattered radiation, mainly from the patient.

b. A ***protective apron*** to be worn by personnel in the fluoroscopy room. This apron must have a Pb equivalent of 0.5 mm. When not actually assisting the radiologist, stand either in the control booth or behind the radiologist.

c. A ***lead shield*** at least 0.5 mm thick covering the ***Bucky slot*** to give additional protection to the personnel's gonads from scattered radiation.

d. Special precautions against the increased scattered radiation engendered by large format image intensifiers (12 in. × 12 in. or larger).

6. Check ***lead-protective gloves*** (same Pb requirements as for aprons) periodically for cracks by means of a radiographic test using medium speed intensifying screens and exposure factors of 100 kV and 2.5 mAs at a 40-in. (100 cm) distance, with a 400-speed film-screen combination.

7. When on duty, always wear a film badge or TLD monitor as explained on pages 512–513.

PROTECTION SURVEYS

The evaluation of actual or potential exposure hazards in radiology requires periodic protection surveys. A complete survey must be made ***initially*** by a ***qualified expert***, then repeated after any change in conditions that might affect exposure of personnel. If more shielding is indicated, the area should be resurveyed after installation of the additional barrier. If, in the opinion of the qualified expert, there is reasonable probability that a person in a controlled or uncontrolled area may receive more than recommended equivalent dose limit, then one or more of the following pertinent corrective actions should be undertaken:

1. Determine the cumulative dose in the area in question.
2. Use personnel monitoring in the area in question.

3. Add barrier material to comply with authoritative recommendations in *NCRP Report No. 105.*

4. Impose restrictions on the use of the equipment, or on the direction of the beam.

5. Impose restrictions on the occupancy of the area if this is controlled.

A tissue-equivalent phantom such as stacked acrylic plates must be used as a source of scattered radiation to check the adequacy of the secondary barrier.

All "on-off" control mechanisms (control panel, entrance door, emergency cutoff switch) should be checked semiannually and repaired if necessary.

Official yellow warning signs reading RADIATION AREA (see Figure 23.1) in black letters should be mounted on the outside of all doors having access to radiographic/fluoroscopic rooms. This would cover prevailing exposure rates of 5 to 100 mR per hour. Other signs are available for areas in which the exposure rate could exceed 100 mR per hour, but such levels are highly unlikely in diagnostic radiology departments.

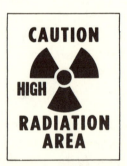

Figure 23.1. Yellow radiation warning labels. A badge must be posted on the entrance door to a room, the badge shown on the left when prevailing exposure rates are 5 to 100 mR/hr, and the one on the right when exposure rates are higher, but less than 500 mR/hr.

Whenever any restriction has been placed on the equipment, such as limitation of beam angulation to prevent penetration of a secondary barrier by the useful beam, the specified restriction must be enforced.

All reports of calibrations and surveys should be made in writing, signed by the qualified expert, and filed *permanently.* The report should indicate whether a resurvey is needed, and when.

Instructions must be posted conspicuously in the control area describing the steps to be taken in the event equipment cannot be turned off.

The name, address, and phone number of the Radiation Safety Officer should be posted in a conspicuous place in the control area for prompt service in the event of an emergency. A responsible substitute should be available, on call, if the Radiation Safety Officer cannot be reached.

PROTECTION OF THE PATIENT IN DIAGNOSTIC RADIOLOGY

All x-ray exposures, regardless of how small they may be, entail some degree of risk to the patient. In other words, there is no threshold dose for stochastic effects. However, this should not deter us from using x rays for diagnostic purposes when medically indicated; you must keep in mind the benefit/risk ratio in each case, although as a technologist you have no control over the ordering of x-ray examinations. At the same time, you should continually strive to minimize patient exposure; this is especially important in fertile women who may unknowingly be in the first trimester of pregnancy.

Because of the urgency of protecting the general public from unnecessary radiation, the U.S. Congress in 1968 passed the Radiation Control for Health and Safety Act which set up the Bureau of Radiological Health (BRH) under the Food and Drug Administration (FDA). The BRH has the responsibility, through its five divisions, of (1) establishing manufacturing standards for all x-ray-producing electronic equipment, (2) following up on compliance with these standards, (3) studying the biologic effects of radiation, (4) developing programs for training and medical applications, and (5) overseeing the field of radioactive materials and nuclear medicine.

In 1964, and again in 1970, the BRH surveyed a number of installations to find possible trends in the exposure of patients during diagnostic radiologic procedures. The results of these surveys are summarized in Table 23.4. Note the 20 percent increase in the utilization of diagnostic x rays during the six-year interval. Also, while the mean ratio of the beam area to the film area has shown a significant decrease of 30 percent, the average skin

The Fundamentals of X-ray and Radium Physics

exposure for a plain radiograph of the abdomen has increased by 25 percent! According to the later survey, the average skin exposure for a posteroanterior radiograph of the chest is about 30 mR (0.03 R), and for an anteroposterior radiograph of the abdomen about 600 mR (0.6 R).

TABLE 23.4
SELECTED RESULTS OF THE X-RAY EXPOSURE STUDIES
OF THE BUREAU OF RADIOLOGICAL HEALTH

	1964	1970	Change
# having 1 or more examinations	108×10^6	130×10^6	+20%
# radiographs per examination	2.2	2.4	+10%
mean ratio of beam area to film area	1.9	1.2	−30%
av. skin exposure per film (AP or PA abdomen)	480 mR	620 mR	+25%
av. skin exposure per film (PA chest)	28 mR	27 mR	0
genetically significant dose per year*	16 mrads	20 mrads	+25%

*According to NCRP Report No. 93 (1987) the annual GSD is 22–30 mrem (0.22 to 0.030 mSv) from diagnostic x rays, especially of hips and pelvis in men, and lumbar spine and barium enema in women. This may be overestimated because of gonadal shielding.

Because of possible genetic damage, gonadal exposure must be minimized by appropriate beam limitation (collimation), filtration, protective Pb shielding, and high speed film-screen imaging systems. Rare earth screens have contributed much to the reduction of gonadal as well as whole-body exposure of patients.

The last line in Table 23.4 shows that the genetically significant dose (GSD) has increased about 25 percent, to 20 mrads per year, between the two surveys. The GSD is the quotient of the total gonadal dose received in diagnostic radiology by all fertile individuals in the population, divided by the total population, and corrected by the average number of expected offspring.

In 1987, *NCRP Report No. 93* estimated the annual GSD from diagnostic x rays to be 22 to 30 mrem (0.22 to 0.30 mSv) for women (mainly from lumbar spine and barium enema examinations), and for men (mainly from hip and pelvis examinations). However, this is probably overestimated because of the prevalence of gonadal shielding in men.

Table 23.5 has been adapted from *NCRP Report No. 91 (1987)* to give the dose equivalent H_E for a number of typical x-ray examinations. Recall that H_E is the weighted (effective) dose equivalent to an organ, which entails the same detriment or risk as though the same dose equivalent were given to the whole body. For example, in Table 23.5, the average H for a lumbar spine examination is 130 mrem (1.3 mSv); this is equivalent to 130 mrem to the whole body. Thus, H_E estimates the level of risk to the whole body from H_E mrem to the lumbar spine. However, as noted on pages 506–507, *NCRP Report No. 107* states that monitor readings in terms of exposure, absorbed dose, or dose equivalent in diagnostic radiology may serve, instead of effective dose equivalent, as a measure of risk on account of uncertainties in determination of weighting factors and the low LET of x rays and medical radionuclides.

TABLE 23.5
AVERAGE EFFECTIVE DOSE EQUIVALENT H_E FOR THE U.S. IN 1980
FROM DIAGNOSTIC MEDICAL EXAMINATIONS*

Examination	Annual number of examinations (in thousands)	Average effective dose equivalent per examination
		*mrem***
Computed tomography (head and body)	3,300	110
Chest	64,000	6
Skull and other head and neck	8,200	20
Cervical spine	5,100	20
Biliary tract	3,400	190
Lumbar spine	12,900	130
Upper gastrointestinal tract	7,600	245
Kidney, ureters, bladder	7,900	55
Barium enema	4,900	405
Intravenous pyelogram	4,200	160
Pelvis and hip	4,700	65
Extremities	45,000	1
Other***	8,400	50

*Adapted from *NCRP Report No. 91 (1987)*.
**1 mrem = 0.01 mSv.
***Other* is estimated from mean value of examinations such as thoracic spine, full spine, mammography, etc.

The values of H_E from nuclear medicine tests, included in Table 23.6, are adapted from *NCRP Report No. 91.*

Under the Radiation Control for Health and Safety Act, the standards for medical x-ray equipment went into effect on August 1, 1974. Of major importance is the recommendation that the

TABLE 23.6
AVERAGE EFFECTIVE DOSE EQUIVALENT H_E FOR THE U.S. IN 1980
FROM DIAGNOSTIC NUCLEAR MEDICINE TESTS*

Examination	Annual number of examinations (in thousands)	Average effective dose equivalent per examination
		mrem**
Brain	810	650
Hepatobiliary	180	370
Liver	1,400	240
Bone	1,800	440
Lung	1,200	150
Thyroid	680	590
Kidney	240	310
Tumor	120	1200
Cardiovascular	950	710

*Adapted from *NCRP Report No. 93 (1987)*.
**1 mrem = 0.01 mSv

components of a diagnostic x-ray unit must, together, make up an integrated system. Furthermore, the manufacturer must test and adjust the equipment at the factory as the final step in the production line. Finally, the installers of the equipment must follow closely the manufacturers' instructions. Some of the more significant of the numerous specifications include:

1. Ability of the equipment to reproduce an exposure for any permissible combination of kV, mA, and time.

2. Proportionality between exposure and time.

3. Provision of automatic collimation for each film size in stationary equipment.

4. Automatic collimation of the fluoroscopic beam at the patient's entrance port, for spotfilm radiography.

5. Built-in filtration to remove radiation that is too soft to be radiographically useful.

Patient Protection in Radiography

We must emphasize that every effort should be directed toward minimizing exposure of the patient during radiologic examinations, especially to the gonads and the pregnant uterus, because of the danger of inducing injury to the genetic material or embryo. Protection can be achieved by lead shielding and by tight collimation whenever these do not hamper the examination. At the same time, unnecessary exposure to other parts of the body should be avoided because of possible somatic injury by radiation, especially to the blood-forming bone marrow.

Of great importance are the estimated absorbed doses to the **gonads** during radiographic examinations. These can be readily determined by using the x-ray output of the radiographic unit to derive the gonadal absorbed dose from the appropriate table in *Handbook of Selected Organ Doses for Projections Common in Diagnostic Radiology* by M. Rosenstein (HEW/FDA 76-8031). Table 23.7 has been adapted from Table 7 in that publication to show in particular the absorbed doses in the ovaries, uterus (no larger than 2-month's pregnant), and testes, resulting from an anteroposterior view of the abdomen.

Note that the basic data include the organ dose per 1000 mR (1 R) in air exposure at the skin, for an SID of 40 in. (100 cm), film size 14 in. × 17 in. (35.6 cm × 43.2 cm), and a variety of x-ray beam HVLs (mm Al). X-ray output in mR/mAs having been measured at 40 in. using the selected kVp, this value is recorded and kept in an accessible place.

To find the ovarian or uterine (normal size) dose for an antero-posterior view of the abdomen in a particular patient, suppose that with exposure factors 90 kVp, 100 mAs, 40-in. FFD, and 2.5 mm Al filter, the output is 5 mR/mAs, then for 100 mAs the exposure at the skin (entrance exposure) would be

TABLE 23.7
ABDOMINAL-ORGAN DOSE (mrad) FOR 1,000 mR
ENTRANCE SKIN EXPOSURE IN AIR
AP ABDOMEN, 40-IN FFD, 14 in. X 17 IN. FILM

	Organ Dose in mrads				
Beam Quality HVL (mm Al)	1.5	2.0	2.5	3.0	3.5
Ovaries	97	149	203	258	313
Embryo (uterus)	133	199	265	330	392
Testes	<0.01	<0.01	<0.01	<0.01	<0.01

Data abbreviated from Table 7 in *Handbook of Selected Organ Doses for Projections Common in Diagnostic Radiology*, by M. Rosenstein (HEW/FDA 76-8031), 1976.

$$5 \text{ mR/mAs} \times 100 \text{ mAs} = 500 \text{ mR}$$

In Table 23.7 we see that for a radiographic beam having a HVL of

2.5 mm Al, the uterine dose is 265 mrads for an entrance exposure of 1000 mR. Since we have just found that the required skin exposure in this case is 500 mR, the uterine dose for a KUB would be

$$500 \text{ mR}/1000 \text{ mR} \times 265 \text{ mrads} = 123 \text{ mrads}$$

On the assumption that this dose results from a medium-speed film-screen system, we can state that with a 400-speed system, the uterine dose can be reduced to about

$$100/400 \times 123 \text{ mrads} = 31 \text{ mrads}$$

Dose Reduction in Radiography

We have at least eight ways to minimize patient exposure during radiography. They are extremely important and should be learned and used by every radiologic technologist. We shall now discuss them individually, and then summarize them.

1. *Beam Filtration.* Patient exposure can be greatly reduced by the insertion of an aluminum (Al) filter in the beam to remove the lower energy photons which would otherwise be absorbed by the patient and not contribute to the radiographic image. In other words, a suitable aluminum filter decreases unnecessary exposure of the patient. Naturally, this requires a filter of the proper thickness. Recommendations for filter thickness have been expressed either as (a) minimum total thickness in mm aluminum equivalent (Al eq), including inherent filtration by the glass window of the x-ray tube and the cooling oil in the tube housing, or as (b) half-value layer (HVL) of the x-ray beam in mm Al.

Table 23.9 shows the minimum Al eq filtration needed for each of three useful kVp ranges according to *NCRP Report No. 33* (1968). You can see that above 70 kVp, the *total* filtration must be no less than **2.5 mm Al equivalent.**

It turns out that *21 CFR* (see Table 23.8) also recommends a *half-value layer* of nearly 2.5 mm Al for x-ray beams with energy above 70 kVp. Since there may be uncertainty as to the inherent filtration of a particular x-ray tube, it is much simpler for the physicist to add sufficient aluminum filtration until the correct HVL has been attained. Thereafter, the filter is permanently fixed in position and used for the entire range of radiographic and fluoroscopic kVp. (The *U.S. Code of Federal Regulations—21 CFR*

TABLE 23.8
REQUIREMENTS FOR FILTRATION IN RADIOGRAPHY°

Operating kVp	Minimum Total Filter° (inherent plus added) in mm Al eq	Minimum HVL°° in mm Al
Below 50	0.5	0.3
50–70	1.5	1.2
Above 70	2.5	2.3

°From NCRP Report No. 33 (1968).
°°*From U.S. Code of Federal Regulations — 21 CFR 800.10, April 1983.*

800.10, April 1983 — specifies minimum HVLs for various kVp ranges on all new equipment manufactured after December 1, 1983.)

How much dose reduction can we expect from beam filtration? Table 23.9 shows that a 3 mm Al filter (added) reduces exposure to the testes to about 1/3 to 1/2 that with an unfiltered beam at 65 to 68 kVp (Ardran and Crooks). Another study with a pelvic phantom showed that a 3 mm Al filter reduces skin exposure to about 1/5 that without filtration at 60 kVp other exposure factors (mAs) being adjusted to maintain equal radiographic density (Trout and others). Furthermore, in this latter study it was found that at 60 kVp the mAs had to be increased to compensate for the removal of x-ray photons by the filter, but this did not increase the actual exposure of the patient. At 130 kVp, filters up to 3 mm thick did not require an increase in mAs. (With modern higher kVp technique and rare earth screens, and prevailing gonadal shielding, testicular doses have been markedly reduced.)

TABLE 23.9
REDUCTION IN EXPOSURE OF MALE GONADS (TESTES)
WHEN 3 mm Al FILTER IS ADDED°

Examination	Added Filtration	kV	Testes Uncovered	Testes Covered 1 mm Sheet Lead
			mR	mR
Pelvis, AP	0	65	2,000	42
15 × 12 in. film	3 mm Al	65	670	24
Lumbar Spine, AP	0	68	24	5
10 × 12 in. film	3 mm Al	68	6	2

°Data of Ardran, G. M., and Crooks, H. E. (Courtesy of authors and *British Journal of Radiology*.)

2. **Beam Limitation ("Collimation").** As we have already explained (see pages 386–395), a decrease in the cross-sectional area of the beam not only avoids unnecessary exposure of tissues outside the area of interest, but it also reduces the amount of scattered radiation generated within the patient. We achieve this most efficiently by means of beam-limiting devices that are loosely called collimators. The variable shutter type is preferred. Federal regulations require that all new equipment must have *positive variable beam-limiting devices* interlocked with the bucky tray so that the beam size is automatically restricted to the particular cassette size. A manual override may be provided for use, with certain limitations, in the event of system failure. Neither the length nor the width of the x-ray field in the plane of the film (or other image receptor) may differ from the corresponding film dimension by more than 3 percent of the FFD. Thus, at a 40-in. (100-cm) FFD, the x-ray field dimension at the film must be within 3 cm of the corresponding film dimension.

3. **Gonadal Shielding.** Although the beam should be so restricted that direct exposure of the gonads does not occur unnecessarily, additional precautions should be taken to use specific gonadal shields when these do not obscure the area of interest. In the majority of radiographic examinations, specific shielding of the testes should be used when they lie within the beam, or within 5 cm beyond the edge of the beam. The shield must have a lead equivalent of at least 0.5 mm. Ovaries should be shielded whenever possible, although this is limited by uncertainty as to their exact location (see Table 23.10 for thickness data on Pb).

TABLE 23.10

THICKNESS OF LEAD REQUIRED TO REDUCE
THE INCIDENT RADIATION TO 0.5 PERCENT
OF THE INITIAL EXPOSURE RATE,
WITH RADIOGRAPHIC BEAMS*

kV_p	Half Value Layer	Thickness of Lead for 0.5% Transmission		Usual Thickness of Lead Available
		mm.	in.	
60	0.5 mm Al	0.21	0.008	1/100 in.
100	1.0 mm Al	0.68	0.026	3/100 in.

*From data of Trout ED, Geiger RM. (Courtesy of *American Journal of Roentgenology.*)

4. *Modified Radiographic Projection.* In radiography of girls for scoliosis one should use a posteroanterior projection. This reduces the breast dose by at least 98 percent without appreciable loss of radiographic quality.

5. *High-speed Image Receptors.* A great reduction in patient exposure results from the use of the highest speed intensifying screens and films consistent with satisfactory image quality. Rare earth screens having a speed of 400, that is, about four times that of medium speed screens, decrease patient exposure to *about* one-fourth relative to medium speed without significantly impairing recorded detail. Therefore, high-speed film-screen systems should prevail in routine radiography, although slower systems still play an important role in special situations where fine recorded detail is needed.

6. *Optimum Film Processing.* Automated film processing has become universal in modern radiology departments. Processors require careful maintenance for optimal function. This assures consistently high radiographic quality in properly exposed films and helps avoid repeat examinations.

7. *High Kilovoltage.* As we pointed out earlier (see pages 159–160), higher kVp entails more efficient tube operation. Also, for equal radiographic density, higher kVp with lower mAs delivers a smaller absorbed dose to the patient. Therefore, we should use the highest kVp consistent with optimum image quality. With modern equipment capable of a wide range of mAs values, an optimum kVp can be selected for a particular anatomic part. The most modern radiographic equipment is provided with automatic exposure control (phototiming) which selects optimum kVp and mAs for various anatomic regions.

8. *Careful technique* to minimize repeat examinations.

Based on the preceding discussion, we may summarize as follows the important factors in reducing patient exposure during radiography — so-called *minimum dose radiography:*

1. *Beam filtration* to provide an *HVL of 2.5 mm Al* (see also pages 528–529).

2. *Beam limitation ("collimation")* to the smallest possible dimensions by a variable beam-limiting device. Equipment of the most modern type is provided with positive (automatic) variable beam-limiting devices with manual override.

3. *Lead shielding of the gonads* by at least 0.5 mm lead equivalent (see Table 23.10 for attenuation of diagnostic x rays in lead). Omit the shield if it hides the area of interest.

4. *PA projection for scoliosis examination* of girls to minimize breast dose.

5. *Highest speed screen-film systems* with good image quality; 400-speed systems using rare earth screens and appropriate films in general radiography.

6. *Optimum film processing* for consistent results.

7. *Highest practicable kVp* that still yields good image quality.

8. *Careful technique* to minimize repeat examinations.

In any procedure using ionizing radiation, whether radiography, fluoroscopy, or nuclear medicine, the patient should be questioned about previous exposures. The total exposure, especially to gonads, embryo, and bone marrow, should be kept at the lowest level consistent with medical needs.

Radiographic examinations that expose the gonads or uterus in fertile women should be performed only when absolutely necessary, since the patient may not be aware that she is pregnant. It is during the first few months that the embryo is most susceptible to fatal injury or the induction of serious congenital anomalies. Some authorities recommend the so-called *10-day rule:* such examinations should be performed only during the first 10 days following the onset of a menstrual period (ie, the "safe" period) because ovulation and pregnancy are much less apt to occur during this time than later in the menstrual cycle. However, if a radiographic examination is medically necessary, it should be carried out with strict adherence to all available protective measures as described above.

Protection in Mammography

As stated earlier on pages 344–347, skillful technique minimizes breast dose while producing mammograms of superb quality. To achieve these goals requires dedicated mammographic machines with special mammographic tubes having molybdenum targets and filters; low-dose mammographic screens and films, with or without grids having ratios of 3:1 or 4:1; and an efficient breast-compression device.

Except for extremely tender breasts, compression must be used in all mammographic examinations to reduce breast thickness and make it more uniform. The important advantages include:

1. Decreased exposure factors, with reduction of breast dose.

2. Diminished amount of scattered radiation, thereby improving contrast.

3. Improved recorded detail by bringing the breast closer to the image receptor.

Depending on the number of views per breast (usually two for screening and three for diagnostic) and on the use of a grid, the glandular dose to the breast at a depth of about 4 cm ranges between about 0.05 to 0.15 rad (cGy) per examination. In xeromammography the breast dose is about 0.75 rad (cGy). These values are, of course, approximate since breast dose varies widely with thickness and density of the glandular tissue. To put these doses in perspective, recall that the natural background dose in the Colorado Plateau ranges up to 0.14 rem per year!

Computed Tomography Scanning

In general, the tight collimation for each slice with relatively small side scatter to adjacent slices engenders about the same dose as a conventional radiographic examination. For example, a complete CT scan of the head delivers about the same dose as a full radiographic skull series.

One can determine the dose in CT scanning by measuring the absorbed dose at the *center* of one "slice" with a small dosimeter in a water phantom, while scanning this slice and three adjoining slices on both its sides (see Figure 23.2). Under these conditions, the dosimeter will record the doses from the direct beam through the center slice, as well as the scattered radiation from the adjoining slices (Curry and others).

Noise, which degrades image quality, can be reduced by increasing the exposure, but this obviously raises patient dose. Thus, techniques are set to use the smallest exposure that will produce optimal contrast resolution. The collimator should also be checked periodically to assure its proper function; if too wide, scattered radiation increases and acts as another noise factor.

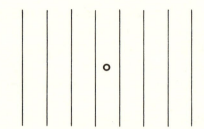

Figure 23.2. Method of measuring absorbed dose in CT scanning. A small dosimeter is placed in the center of the middle "slice" in a water phantom, remaining there while all seven slices are scanned. The dosimeter measures the absorbed dose of the primary beam through the middle slice, as well as that of the scattered radiation from the adjoining slices.

Patient Protection in Fluoroscopy

Fluoroscopy presents a considerable hazard, so before a patient is subjected to this type of examination, a careful history should be taken to ascertain the recency and number of previous fluoroscopic studies. X-ray exposures produce cumulative effects, and further exposures must be governed by what has gone before.

Several precautions should be observed to reduce exposure during fluoroscopy:

1. *Intermittent Fluoroscopy.* It is good practice to use intermittent activation of the fluoroscopic tube to decrease patient exposure and prolong tube life.

2. *Restriction of Field Size.* The size of the fluoroscopic field must be limited by suitably collimating lead shutters placed between the tube and the patient. As we have seen, the skin and depth doses, as well as the scattered radiation into the surroundings, decrease as the area of the radiation field is decreased. Therefore, the fluoroscopic beam should be restricted to the smallest field that includes the area of interest, to reduce exposure of both the patient and personnel. According to the Federal Register Code *21 CFR 800.10,* neither the length nor the width of the useful beam at the image receptor shall exceed that of the visible image area by more than three percent of the source-image receptor distance. In this case, the image receptor is the input screen of the image intensifier.

3. *Correct Operating Factors.* An increase in kVp produces an

increase in fluoroscopic image brightness. If this is accompanied by an increase in filtration and a compensating decrease in mA, exposure rate to the patient is actually decreased with no change in image brightness. In other words, for a given image brightness, x-ray exposure of the patient decreases as kVp is raised and mA is lowered. Since further improvement becomes insignificant above 100 kVp, the recommended operating factors are 90 to 100 kVp, 2 to 3 mA, and 2.5 mm Al equivalent filter (or beam HVL 2.3 mm Al) for bright (image intensified) fluoroscopy. Image resolution can be improved by increasing the exposure rate, but this is strictly limited by maximum permissible exposure of the patient. For children the kVp should be reduced to a value depending on body thickness, just as in radiography. The source-skin distance must be at least 15 in. (38 cm) with stationary and 12 in. (30 cm) with mobile fluoroscopic equipment according to the federal code cited above. Distances longer than these, with compensating increase in exposure factors, do not significantly reduce patient exposure.

4. *Filtration.* We can increase the hardness of an x-ray beam by using a suitable filter. This removes relatively more soft than hard x rays, thereby removing a large portion of non-image-producing radiation which would otherwise be absorbed in the patient's skin. In fluoroscopy, the same filtration is required as in radiography: total of 2.5 mm Al equivalent, or HVL of 2.3 mm Al.

5. *Exposure Limits.* Although there is the hazard of cumulative small doses of radiation during one's lifetime, there is as yet no official limit to x-ray exposure of patients in diagnostic radiology. Obviously, this should be reduced to a minimum consistent with medical needs. According to the federal code cited above, the exposure rate at the table top (ie, patient's entrance exposure rate) must be no greater than 10 R/min for fluoroscopic equipment with automatic exposure rate control and no greater than 5 R/min without automatic exposure rate control. This does not apply to the exposure rate in the radiographic spotfilm mode. In addition, a timer must be provided to preset the on-time of the fluoroscopic tube for a cumulative total of no more than 5 min. At the end of this time, an audible signal must indicate that the time limit has been reached and must continue to sound as long as fluoroscopy continues until the timer is reset.

Some fluoroscopic units have an optional high-level control which temporarily increases x-ray output above the 5 R/min limit at the patient's skin entrance surface. This control shall be subject to continuous manual operation only, and be accompanied by an audible signal. Extreme caution should be exercised in the use of the high-level control, as radiation injuries have been reported by failure to deactivate this control.

6. ***Primary Protective Barrier.*** This must be provided to protect the fluoroscopist by intercepting the entire cross section of the useful beam at any SID (ie, distance between x-ray source and input screen). The fluoroscopic tube must not produce x rays unless the barrier is in its proper position. This barrier must limit the fluoroscopist's exposure from transmitted radiation, essentially, to no more than 2 mR/hr for each R/min of the patient's entrance exposure rate.

PROTECTION FROM ELECTRIC SHOCK

On initial installation, all radiographic equipment must be properly **grounded** to prevent electric shock. Ordinarily, secure connection by a suitable conductor to an underground metal water pipe should be adequate. In some instances, with very high energy x-ray units, grounding may have to be made through a metal rod that has been driven into the ground until it reaches permanently wet ground; this may have to be evaluated by actually measuring the resistance of the grounding circuit.

The building codes of most cities require all electric wall outlets, wherever located, to be grounded in all newly constructed buildings.

Personnel should always observe the ***one-hand*** rule: never touch an electric device with one hand while the other hand is submerged in the processing tanks or is touching a conductor.

High-voltage, low-amperage shock tends to "throw," whereas low-voltage, high-amperage shocks tends to "hold" the victim. In the latter instance, do not grasp the victim directly, but first either open the main switch or remove him/her by means of a dry board, a dry wad of newspaper, or a dry rope. Immediately call for medical aid, in the meanwhile applying cardiopulmonary resuscitation (CPR).

PROTECTION IN NUCLEAR MEDICINE

The handling of radionuclides requires certain precautions because of the harmful effects of chronic exposure to ionizing radiation, as already described. But the problems associated with radionuclides differ in some respects from those prevailing in the diagnostic x-ray department; hence, the need of a separate section dealing with health physics in nuclear medicine. Medical compounds containing radionuclides are called *radiopharmaceuticals.* In this section we will consider protection as it relates to the diagnostic use of radionuclides.

Types of Radiation. Before discussing the harmful effects of radiation emitted by radionuclides and the methods of protection, let us review the types of radiation involved. radioactive substances give off three types of radiation: alpha particles, beta particles, and gamma rays. Since radiopharmaceuticals emit only gamma or beta rays, our discussion will be limited to them. Not only do gamma rays have much greater penetrating ability than beta particles, but they also differ from them in character. Thus, gamma rays are electromagnetic waves or photons, whereas beta particles consist of high-speed electrons. Consequently, protection must differ for these two kinds of radiation.

Harmful Effects. These have already been described in detail on pages 513–517. The same effects are produced by beta and gamma radiation as by x radiation for equal absorbed doses. However, betas are much less penetrating, their effects being limited to the skin from external sources, or in their immediate vicinity from internal sources. The effects of x and gamma rays (photons) are manifested deep in the body even from external exposure.

Dose Equivalent Limit. Since the quality factor $\overline{Q}$ for gamma rays is the same as that for x rays, the dose equivalent limit is also the same, especially since exposure is low, in the diagnostic range.

Principles of Radiation Protection. There are basically four factors, suitable combinations of which are used to reduce or eliminate radiation hazards in nuclear medicine. These include: (1) *distance,* (2) *shielding,* (3) *time of exposure,* and (4) *limitation of activity* of source to smallest required amount. In this section, we shall assume that the radioactive sources are sealed in the form of capsules or needles. Liquid preparations of radionuclides, except for the hazard of spill, ingestion, or inhalation, may also be regarded as sealed sources insofar as external protection is concerned.

1. **Distance.** This is often the easiest and least expensive method of reducing the exposure rate to the permissible level. In the case of gamma rays, the inverse square law applies with reasonable accuracy for the purpose of protection in storage and handling. Long forceps should be used for picking up encapsulated radio-pharmaceuticals; capsules should be used whenever possible to avoid spillage of liquids. Storage containers that emit gamma rays should be as remote as possible from areas habitually occupied by personnel and general public.

2. **Shielding.** Protective materials placed between a radioactive source and its surroundings to reduce the exposure rate are called **shields** or **barriers.** Proper shielding contributes materially to the reduction of **gamma-ray exposure rate,** especially when combined with distance. The shield may be incorporated in the container itself, or it may be placed as a barrier around the source in the storage or working area. **Lead** of proper thickness serves as the most satisfactory barrier material for storage of radiopharmaceuticals because of its efficiency as an absorber of gamma rays. As a general rule, gamma-ray emitting radiopharmaceuticals should be stored in their shipping container and surrounded with lead bricks 5 cm (2 in.) in thickness, on all sides that might constitute radiation hazard. Routine monitor surveys should then establish the adequacy of the barrier.

Lead-shielded syringes should always be used for intravenous injection of radiopharmaceuticals, as it has been shown that appreciable cumulated doses to hands result from the use of unshielded syringes. In addition, disposable gloves should be worn to protect against contamination by radionuclides, and by infection as well.

Shielding of **beta radiation** differs from that of gamma rays. Since the range of beta particles in soft tissues is usually limited to several mm, the external hazard can be readily controlled. The range of a beta particle depends on its energy; therefore low energy beta particles (ie, less than 0.3 MeV) are absorbed by the container, by the cornified layer of the skin, or by several feet of air. On the other hand, the medical radionuclide ^{32}P (maximum electron energy 1.7 MeV) requires special shielding. Plastic such as polystyrene is preferred because its low atomic number reduces the intensity of brems radiation produced when the beta particles interact with the atoms of the shield.

3. *Length of Exposure.* Since the total exposure is equal to the exposure rate times the time, it is obvious that the faster a given procedure is carried out, the smaller is the exposure incurred by the operator. Those responsible for injecting diagnostic radionuclides should therefore be experienced in venesection.

4. *Limitation of Activity of Stored Radiopharmaceuticals.* The exposure rate is directly proportional to the activity of the source. Furthermore, the exposure rates from all sources must be added to obtain the total exposure rate. Hence, the quantities stored should be kept at a minimum consistent with the requirements of the department.

Radiation Monitoring. Besides requiring that all personnel wear film badges (see pages 512–513), every Nuclear Medicine Department should conduct periodic surveys to determine the exposure rates from stored radionuclides from possibly contaminated work areas and waste receptacles, especially when liquid preparations are used; and from possibly contaminated clothing, shoes, instruments, and hands. Such surveys are advisable at least weekly, and even more often when deemed necessary. Whenever patients receive radionuclides in quantities other than tracer doses, clothing, sputum cups, bedpans, urinals, and other utensils that may have become contaminated must be surveyed before they can be declared "safe." All contaminated articles must be stored in special bins with appropriate lead shielding and at a permissible distance until the activity has decreased to a level a few times that of background.

The two most popular survey instruments include (1) the *survey meter,* a portable battery operated *GM counter* with a ratemeter, and (2) the *cutie pie,* a direct-reading *ionization chamber* (see Figures 23.2 and 23.3). Both are calibrated to read in mR/hr (milliroentgens per hour). In addition, the survey meter can indicate counts per minute and is usually provided with earphones for audible detection of counts. A survey meter operating continuously as a GM counter should be placed at the exit of any laboratory where high level activity prevails, so that a constant check can be maintained on personnel leaving the laboratory.

Unsafe Practices. The following list of unsafe practices serves as a useful check list of the safety of one's own nuclear medicine department.

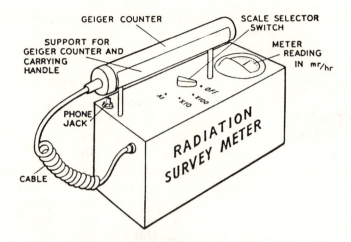

Figure 23.3. Battery-operated portable survey meter with GM counter. A phone jack is included for audible detection. This type of meter is inaccurate because it is very energy dependent.

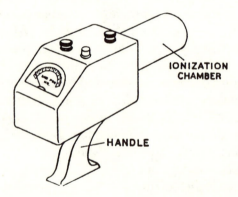

Figure 23.4. Battery-operated ionization-chamber type survey meter known popularly as a "cutie pie."

1. Inadequate planning—excessive time required for carrying out procedures.

2. Improper monitoring.

3. Inadequate shielding.

4. Inadequate use of trays and paper covering of work areas when radioactive liquids are used.

5. Pipetting solutions by mouth instead of remote apparatus.

6. Poor work habits, such as smoking or eating in the laboratory.

7. Improper waste disposal.

8. Failure to keep adequate radiation safety records of radionuclides received, waste disposal, and personnel exposure.

9. Failure to post special signs to mark storage and disposal areas.

Waste Disposal. The Nuclear Medicine Department must provide for disposal of radioactive waste products such as unused radiopharmaceuticals, contaminated disposable materials, cleanup papers, patients' excreta, etc. Excreta may be discharged directly into the sewage system, but other materials should be placed in a clearly marked storage bin until monitored activity decays to a few times background. Fortunately, the low activity of the radiopharmaceuticals used in diagnosis does not ordinarily pose a problem. In any case, requirements of the State Health Department must be fulfilled.

RADIATION HYGIENE

Assuming that all the above safety measures have been in place and the technologist has been conscientious in observing all the rules, can we be certain that radiation hazards have been eliminated? Insofar as we know, there have been no reported harmful effects from exposures approximating the recommended dose equivalent limit. Periodic blood counts, while they may detect early radiation injury, are a poor substitute for preventative monitoring by film badges or other devices. If the dose equivalent limit is exceeded, prompt corrective steps should be taken.

Upon accepting a new job, the technologist must bring records of previous exposure and undergo a complete blood count and physical examination. Upon terminating employment, the technologist must be given a complete summary of the incurred radiation exposure.

The danger from ionizing radiation in the well-controlled Radiology Department is virtually nil, provided all personnel make full use of all protective devices and equipment and constantly bear in mind the ALARA principle.

QUESTIONS AND PROBLEMS

1. State the causes of background radiation. What is the average individual background exposure per year from natural and from artificial sources?

2. State how dose equivalent (H) and effective dose equivalent (H_E) are derived. Why may we use H instead of H_E in diagnostic radiology?

3. Describe two methods by which one may determine the amount of whole-body radiation being received in the Radiology Department.

4. Discuss the effects of excessive whole-body exposure to ionizing radiation. What are the locally damaging effects of overexposure to ionizing radiation?

5. Name the materials widely used in protective barriers. What material (and thickness) is required for the walls of ordinary radiographic rooms?

6. State the annual occupational dose equivalent limits for radiologic technologists. Why may this be used instead of effective dose equivalent?

7. What are the dose equivalent limits before and after pregnancy is known? For the general public annually?

8. What special precautions should nuclear medicine technologists take when injecting radiopharmaceuticals?

9. Where should the personnel monitor (film badge or TLD) be worn during fluoroscopy?

10. What is ALARA? Should it be enforced regardless of cost?

11. Discuss minimum exposure radiography in (1) fluoroscopy and (2) radiography, including the important factors in each.

12. Assume that the conditions of your x-ray department are "safe," from the standpoint of ionizing radiation. State three rules of conduct on your part that will help keep your exposure below the occupational dose equivalent limit.

13. Name and justify the four main protective measures against whole body exposure in nuclear medicine.

14. Why must x-ray equipment be grounded? Is a gas pipe satisfactory? A water pipe?

Chapter 24

NONRADIOLOGIC IMAGING

In this chapter we shall deal with two important diagnostic devices that do not employ x rays. These include magnetic resonance and ultrasound imaging systems.

MAGNETIC RESONANCE IMAGING

To a casual observer, a magnetic resonance imaging (MRI) unit bears a resemblance to a computed tomography (CT) scanner. However, they operate on radically different physical principles. Whereas CT requires a rotating x-ray beam, MRI depends on the behavior of positively charged particles—atomic nuclei—in an external magnetic field that undergoes modification in the process of creating an MR image. Note again that MRI is a nonradiologic process in that the radiation it uses, while electromagnetic in nature, has a much lower frequency (longer wavelength) than x rays and depends on continually-spinning atomic nuclei within the body, rather than on x rays generated by electron interactions with an x-ray tube target. A physical phenomenon called *nuclear magnetic resonance* (*NMR*) underlies MRI.

Nuclear Magnetic Resonance

Although relatively new in its application to the imaging of atomic structures, nuclear magnetic resonance (NMR) has been in use for about 50 years in spectrography to identify atoms and molecules and to study their properties. When applied to imaging (MRI), it provides a much wider range of diagnostic information, in much greater detail, than with any other prevalent imaging system. In particular, MRI achieves *superb low-contrast resolution* without bone or air artefacts, and does not employ ionizing radiation. Low contrast resolution with MRI is about 20 to 40 times greater than with x rays (Bushong).

543

The physical basis of NMR is extremely complex, but it fortunately lends itself to some degree of simplification for our purpose. A review of magnetism should be helpful at this point.

Nuclear Spin. A magnet consists ultimately of atomic-size magnets, each with a north and south pole, induced by the intrinsic continual spinning of atoms, nuclei, protons, neutrons, and electrons, on their individual axes. These spinning particles are called *magnetic dipoles,* the strength of which is unique to each kind of chemical element. An ordinary magnet contains innumerable iron atomic dipoles, arranged in domains and lined up in the form of bulk magnetization, with a north pole at one end and a south pole at the other (see pages 86–95).

Since MRI is based on NMR, which involves spinning nuclei, and since hydrogen nuclei—protons—are the most abundant in the body and have strong magnetic dipoles, we shall base our description of NMR primarily on protons. The fields of the spinning proton dipoles point in a particular direction and are therefore vector quantities indicated by the *right hand rule.* If you allow your right fingers to curl in the direction of proton spin (see Figure 24.1), your thumb will point in the direction of the magnetic field, that is, toward the dipole's north pole. The direction and intensity of this field define the proton's *magnetic dipole moment.* The term moment refers to the tendency of a dipole to rotate in an external magnetic field due to the torque (twisting force) generated by interaction between the two fields—dipole and external. Since forces have direction, they are *vector* quantities.

Alignment of Magnetic Dipoles. In the body, the hydrogen nuclear (proton) dipoles normally arrange themselves in random directions (helter skelter) like the iron atoms in a piece of soft iron. This randomness results from thermal agitation of the particles, in the absence of an external directional force. Therefore, the algebraic sum of all the magnetic moments is zero and there is no net magnetization of the body. The earth's magnetic force is too weak to align the dipoles to any appreciable degree. But when a person is placed in a strong external magnetic field, B_0, the resulting torque causes some of the proton magnetic moments to line up with B_0 (ie, "parallel"), while an almost equal number remain directed opposite to the field ("antiparallel"). It so happens that only a tiny excess (about 1 per million) align themselves parallel

SPINNING PROTON

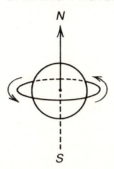

Figure 24.1. Inherent spinning of protons, which behave as miniature magnets. They therefore possess magnetic moment, the tendency to turn in an external magnetic field due to the force between them.

with the field. By convention, the parallel orientation is designated the "up", and the antiparallel the "down" position (see Figure 24.2). The up proton dipoles are in the ground (lowest) energy state, with their north poles in the direction of the south pole of the external field. As shown on Figure 24.3, a small, freely-moving magnet like a compass needle lines up similarly when placed in an external magnetic field.

Nuclear Precession. At the same time the protons spin on their axes, these axes also rotate (see Figure 24.3), an effect known as

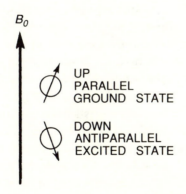

Figure 24.2. Precession of spinning protons around their axes, when situated in an external magnetic field. The "up" position represents a lower energy state, and the "down" position an upper energy state.

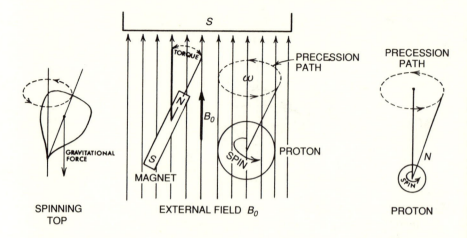

Figure 24.3. Precession of protons compared with a small bar magnet in a magnetic field, and with a spinning top which precesses on account of the force of gravity.

precession. This is caused by the torque generated between the proton dipole moment and the external field. Although the precessing protons are said to line up parallel with external field B_0, they actually have their axes inclined at a small angle with B_0. An analogy is the precession or wobble of a toy top under the influence of gravitational force. In summary, then, we have proton magnetic dipole moments induced by spontaneous, continual spin of protons on their axes and, at the same time, precession of the axes due to the torque of the external field.

Precession has a specific *frequency,* that is, number of rotations per second, symbolized by ω (Greek *omega*) called the **Larmor frequency;** this is proportional to the intensity of the external field B_0 and related to it by a constant γ (Greek *gamma*) that is unique to the specific particle. Gamma is known as the **gyromagnetic** or **magnetogyric ratio.** Table 24.1 gives values of for a few typical nuclei that lend themselves to NMR; it so happens that only nuclei with odd numbers of protons *and* neutrons are subject to NMR (Curry and others). Note the high value of γ for ordinary hydrogen -43 MHz/tesla (Hz is the unit of frequency, which is 1 cycle per sec, and MHz is 1 million Hz). Tesla (T) is the SI unit of magnetic field strength, in this case, of B_0.

Larmor frequency ω is expressed by

TABLE 24.1
APPROXIMATE GYROMAGNETIC RATIOS*

Nuclide	Gyromagnetic Ratio
	MHz/T
Hydrogen 1 (^{1}H)	42.6
Hydrogen 2 (^{2}H) (deuterium)	6.5
Carbon 13 (^{13}C)	11
Sodium 23 (^{23}Na)	11
Phosphorus 31 (^{31}P)	17

*Modified from Curry TS III, Dowdey, JE, Murry RC Jr: Christensen's Physics of Diagnostic Radiology, 4th ed., Philadelphia, Lea & Febiger, 1990. By permission.

$$\omega = \gamma \, \boldsymbol{B_o} \text{ MHz (megahertz)} \tag{1}$$

You can see that γ is in MHz/T and B_o is in T, so $\dfrac{MHz}{T} \times T =$ MHz. Also, the precessional frequency or speed of axis rotation of the nuclear dipole can be increased by using a stronger magnet B_o. The field strengths of clinical MRI imagers range from 0.15 to 1.5 T and are constant for each imager.

Considering only the precessing protons in the parallel, up, or ground state position (lowest energy level), we find that their small excess over those in the down position creates a resultant net magnetization M superimposed on B_o (see Figure 24.4). In other words, M represents the algebraic sum of the individual precessing nuclear magnetic moments (ie, difference between up and down protons), and precesses with the same frequency ω. As will be shown later, M gives rise to an NMR signal under certain conditions.

Resonance. By mechanical analogy, resonance is exemplified by the sympathetic vibration of one object when its natural vibrational frequency matches that of another vibrating object. For example, striking the middle A key on a piano will transfer energy to the air in the form of sound waves of the same frequency, 435 Hz, which will then set the A string of a nearby violin into vibration with the same frequency—the violin A string is said to be tuned to the piano A, or the two strings are in *resonance.* Note that the energy of the vibrating piano A has been transferred to the air as sound waves which, in turn, have transferred energy to the violin A, all these waves having the same frequency.

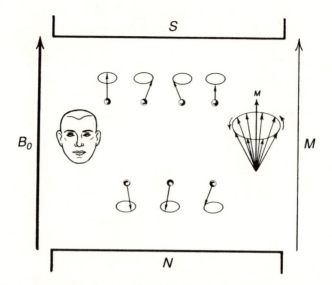

Figure 24.4. Basic relations in magnetic resonance imaging. On the left is a frontal view of a patient's head in the supine position within the external, main magnetic field B_0 directed according to the thick arrow. The field also contains precessing protons, in the lower energy state with their N-poles directed toward S-pole of the magnet; and precessing protons in the higher energy state with their N-poles directed toward the N-pole. Actually, the lower-state protons are in extremely slight excess. On the right is a figure (taken from Figure 24.5) representing the summation of the protons in their lower energy state, with a net magnetization M parallel to B_0. Note that whole figure is oriented as it would appear to one looking vertically downward at the MRI couch.

Similarly, a condition of resonance is created by application of a radiofrequency (RF) electromagnetic pulse with the same Larmor frequency as the precessing net magnetization M, if the RF pulse is directed perpendicular to M or B_0. In this process, energy is transferred to some of the precessing protons (and M), which abruptly turn down to a higher energy state. If the RF field designated as B_1 lasts one-fourth cycle (90°), these protons "flip" down 90° together with their net magnetization onto the transverse plane (ie, perpendicular to B_0) where they continue to precess, as shown in Figure 24.5. In the transverse plane, net magnetization is labeled M_1. Note that the rotation of M_1 in the transverse plane has the Larmor frequency of the protons. When the RF was applied, it caused the precessing "up" protons to get into step so as to assume identical point-for-point positions in their rotational paths; the precessing protons are said to be *in phase* in the transverse plane. While rotating in the transverse plane, the net magnetization energy sets up an alternating voltage (AC), whose radiofrequency matches the Larmor ω, in a nearby receiver coil; this *signal* in the receiver coil eventually contributes to the MR image.

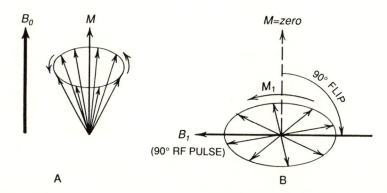

Figure 24.5. In A is shown a group of precessing protons creating a net magnetization vector M; the diagram is oriented as explained in Figure 24.4, with the net magnetization vector being the summation of the precessing protons. M rotates with the same Larmor frequency as the precessing protons.

In B, a 90° radiofrequency (RF) magnetic pulse B_1 has been applied at 90° to M (ie, transversely across the MRI couch), causing the net magnetization to "flip" (spiralling) down to the patient's *transverse* plane, where M continues to rotate, initially at Larmor frequency. Now M is zero and M_1 at maximum.

When the RF turns off at the end of one-fourth cycle, the RF signal fades and disappears. This is called *free induction decay* (*FID*) of the signal because it occurs spontaneously in the absence of the RF field (see Figure 24.6). You can see that FID is a damped oscillation wave with exponentially declining amplitude, lasting only a fraction of a second. FID represents so-called T_2 relaxation, to be explained below.

Factors Governing NMR Signal. Three conditions determine the intensity of an NMR signal, and the brightness of the resulting MR image: (1) proton density, (2) T_1 relaxation, and (3) T_2 relaxation. These will now be explained.

1. *Proton density.* This is simply the number of mobile hydrogen nuclei — protons — per unit volume of tissue in the NMR field. This translates into the number of protons in a given voxel

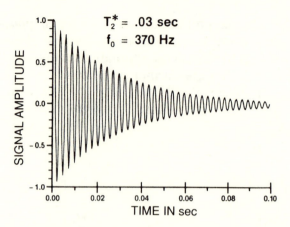

Figure 24.6. Computed free-induction decay (FID) wave form due to a single radiofrequency of $f_0 = 370$ Hz. The amplitude of this signal decays exponentially with time constant T_2^*. * represents the effects of inhomogeneities in the external magnet. Without *, the arbitrary time constant T_2 here is 0.03 sec. (*From Wehrli FW, by permission of the author and GE Medical Systems.*)

(corresponds to voxel or unit tissue element in CT). Obviously, the greater the proton density, the greater will be the number of precessing protons contributing to net magnetization and the resulting signal. Note that proton density inheres in each tissue and differs from tissue to tissue.

2. T_2 *Relaxation.* On delivery of the 90° RF pulse perpendicular to B_0, the protons in the net magnetization in the transverse plane precess in phase (in step) as noted above. Then cessation of the RF pulse causes the precessing protons to *dephase* (get out of step) due to three factors: thermal molecular agitation, "spin-spin" interactions with neighboring protons, and inhomogeneity (nonuniformity) of the external field B_0. As a result, transverse magnetization decays exponentially (FID) with time, according to the time constant T_2^* (like exponential decay of a radionuclide, mathematically) (see Figure 24.7). T_2^* may be defined as the time it takes for transverse magnetization to decay to 37 percent of its maximum (initial) value. The asterisk in T_2^* indicates the inclusion of B_0 inhomogeneity. T_2 denotes *spin-spin relaxation* because it results only from interactions among spinning protons; it is also called *transverse relaxation* because it occurs perpendicular to B_0. As T_2 relaxation or demagnetization occurs, the transverse net magnetization M_1 spirals upward along a path with decreasing

diameter (like the upper part of a beehive) until M_1 reverts to M parallel to B_o. This corresponds to FID mentioned earlier. It must be emphasized that RF signals are emitted *only* during transverse relaxation.

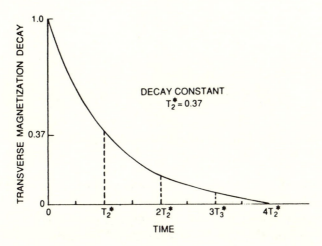

Figure 24.7. Exponential decay curve of transverse relaxation T_2 (*hypothetical*).

3. *T_1 Relaxation.* This is longitudinal remagnetization as the protons return to the lower energy state, with their net magnetization M parallel to B_o. While M_1 is spiraling upward as described under T_2 relaxation, it also has a vertical component which ends up as M. T_1 relaxation occurs exponentially with time, but this is a **growth** of magnetization from M = 0, to M = M (see Figure 24.8); when M turned down 90° due to the RF pulse, its value declined to zero along its initial axis; and during relaxation it returned to its maximum value along its initial longitudinal axis.

T_1 is the time constant and is unique to various tissues. The protons absorbed energy from the RF pulse during excitation at the resonant frequency, to precess in the transverse plane. They return this energy to the tissues during relaxation, so T_1 relaxation is also called **spin-lattice relaxation** (tissues are a lattice or network of structures). Exponential buildup of net magnetization M means that the time constant T_1 represents the time it takes for M to return to 63 percent of its original equilibrium value parallel to B_o (see Figure 24.8). T_1 is also called **longitudinal relaxation** because it occurs along the axis of the longitudinal field B_o. Again note that

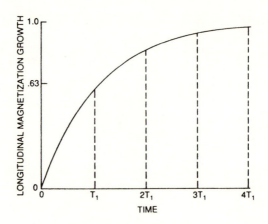

Figure 24.8. Exponential growth curve of longitudinal relaxation T_1 (hypothetical).

T_1 relaxation *does not emit an RF signal;* as will be shown later, it must first be converted to a T_2 signal.

Relaxation Times. Relaxation times and proton densities have a strong effect on the MR images produced by various normal and pathologic tissues. In water, $T_1 = T_2$, but in soft or solid tissues T_1 is much longer than T_2; T_1 relaxation therefore occurs more slowly than T_2. As it turns out, T_2 can be changed to a larger extent than T_1 by various tissue abnormalities such as cancer, resulting in a greater likelihood of detection.

Several factors influence relaxation times, especially T_2. For example, the rate at which complex molecules "tumble": the slower the tumble, the faster the relaxation. Thus, fatty molecules such as the lipids in the white matter of the brain tumble relatively slowly, so their protons relax faster than water molecules.

Spin-Echo. This is a technical procedure in which a sequence of 90° and/or 180° RF pulses at selected time intervals are used for one or more of the following reasons: (1) minimizing the effects of B_o inhomogeneity on T_2 relaxation, (2) changing T_1 relaxation to T_2 signals, and (3) preferentially increasing the intensity of signals from T_2 relaxation and processed T_1 relaxation, and (4) proton density. Spin-echo technique involves the production of so-called "echoes" which are simply enhanced selected signals. An example of this process is the one that intensifies the signal from transverse relaxation—T_2—to increase brightness and improve contrast in the resulting MR image. The following steps are involved (see Figure 24.9):

1. 90° RF pulse generates free induction decay according to time constant T_2^* which includes B_0 inhomogeneity, followed by a

2. 180° RF pulse after a selected interval TR (about 400 to 6000 msec), followed by the spin-ech signal at time TE—this is the time interval between the first RF pulse of 90° and the middle of the spin-echo, also called *echo time.*

3. Additional pulses may be applied.

4. Echoes decay with time constant T_2, since B_0 inhomogeneity has been counteracted by the spin echo process. Image brightness depends on the intensity of the echo signal.

Spin-echo may be explained by the effect of B_0 inhomogeneity on protons as they precess in the transverse plane after a 90° pulse. Owing to different local values (nonuniformity) of B_0, precession of the protons will occur at different frequencies (recall that Larmor frequency ω is proportional to B_0). As shown in Figure 24.9, if one thinks about the simple case of two different precessional frequencies in the transverse plane, one must be faster than the other. After the first 90° RF pulse, dephasing occurs as the faster precessions get farther and farther ahead of the slower ones. Following the 180° pulse, transverse magnetization moves down and around 180° and comes to lie on the opposite (mirror image) side of the transverse plane, with the faster precession now *behind* the slower; they are still dephased. When the faster precession catches up with the slower one, they rephase and generate an echo—a relatively intense signal. The TR value (time between the pulses) has a pronounced effect on the signal intensity and the resulting MRI brightness and contrast.

From NMR to MRI

What has been said thus far describes the signals generated by the basic NMR processes. In recent years, the principles of NMR have been developed into magnetic resonance imaging (MRI) to produce exquisitely detailed images of anatomic structures, both normal and pathologic.

Although MR images bear a superficial resemblance to CT images, MRI: (1) requires no ionizing radiation; (2) provides much better low-contrast image resolution; (3) yields images directly in

SPIN ECHO (90° –TR– 180°)

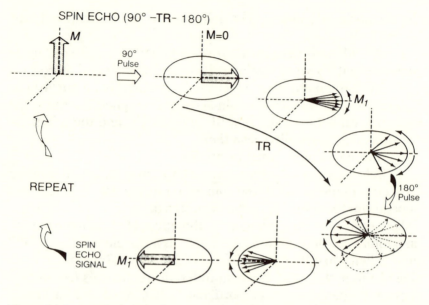

Figure 24.9. A typical spin-echo sequence. First a transverse 90° RF pulse flips net magnetization down 90° to rotate in the transverse plane, where spinning protons get out of step, some rotating behind the others. After a time period TR (repetition time) a 180° RF pulse is applied transversely, which now causes net magnetization M_1 to rotate 180° down and around into the horizontal plane on the mirror image side of the 90° plane (arrows pointing to the left). Now the lagging protons catch up with those ahead, and magnetization becomes focused (thick arrow toward the left), during which a spin-echo signal is emitted. Note that this is how T_1 relaxation, which by itself does not emit a signal, has been converted to transverse relaxation to emit a signal. Other sequences may be used to vary the kind of MRI weighting. (*Courtesy of Harms SE, et al, and Radiographics.*)

multiple sections; (4) is not subject to artefacts from high and low density materials such as bone and air, although magnetic materials such as iron, cobalt, and nickel introduce artefacts and may cause injury due the strong external magnetic field; (5) does not require injection of organic iodides for image enhancement, although gadolinium injection improves contrast with much less hazard than organic ioidides; and (6) serves as an excellent method of measuring blood flow, but this will not be included here.

Let us now apply and extend the principles of NMR to MRI— the actual imaging of the body in multiple slices. But first we must clarify some basic terminology related to the patient's and MRI axes. Figure 24.10 shows the position of a patient on an MRI couch. The *longitudinal axis* corresponds to the *z-axis*, as well as the axis of the external field B_0 and net longitudinal magnetization M. The *x-axis* runs from side to side and is the transverse axis, so it lies in the transverse plane. Finally, the *y-axis* is anteroposterior and, with the x-axis, defines the *transverse plane* (recall that application of a 90° RF pulse generating a field perpendicular to the longitudinal or z-axis flips net magnetization onto the transverse or xy plane).

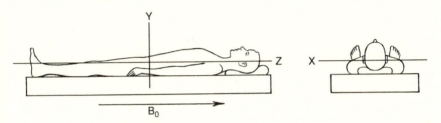

Figure 24.10. Orientation of standard axes relative to the patient's position and external field B_0.

The next three sections deal with MR image brightness and contrast and how they are affected by T_1 and T_2 relaxation, and by proton density.

Image Brightness. The brightest images are produced for tissues with a short T_1—fast remagnetization after a 90° RF pulse. As already explained, tissues with a high fat or lipid content, such as the white matter in the brain, have a low tumbling rate and therefore a short T_1 and T_2. Aqueous molecules such as water in cysts and cerebrospinal fluid have a rapid tumbling rate and therefore a long T_1 and T_2 (see Table 24.2).

T_1-weighting preferentially intensifies the T_1 contribution to the MR image, but T_1 relaxation does not generate an RF signal, so it must be changed to transverse magnetization and T_2 relaxation by sequential pulsing based on spin-echo technique. A short TR (time between pulses) stops T_1 remagnetization early, along the *lower* part of the longitudinal magnetization growth curve (see Figure 24.8). A 90° RF pulse now would convert the relatively small longitudinal magnetization to a correspondingly small transverse magnification giving rise to a weak signal. But a long TR allows more time for the growth of longitudinal magnetization, say in the upper part of the growth curve, resulting in a stronger transverse

TABLE 24.2
APPROXIMATE T_1 AND T_2 RELAXATION TIMES FOR VARIOUS TISSUES

Tissue	T_1	T_2
	msec	msec
Cerebrospinal fluid	2000	150
Fat	220	80
Brain		
White matter	650	90
Gray matter	750	100
Muscle	600	40
Liver	480	50
Kidney		
Medulla	680	140
cortex	360	70
Blood	800	180

*Averaged from sources in the literature. Values differ among various authors. Data here are for a 1-T magnet.

magnetization after a 90° pulse and yielding a stronger signal and image brightness. Thus,

SHORT TR and WEAK MAGNETIZATION
$\rightarrow$ WEAK TE SIGNAL
LONG TR and STRONG MAGNETIZATION
$\rightarrow$ STRONG TE SIGNAL

Optimal brightness of image occurs when TR (time between pulses) exceeds a $3T_1$ interval for a particular tissue. Note that in the process just described, T_1 (longitudinal) relaxation has been changed to T_2 (transverse) magnetization to acquire an RF signal. However, this must be modified to achieve optimum contrast resolution with spin-echo technique, as described below.

With *T_2 relaxation,* which also follows an exponential decay curve, a T_2-weighted RF pulse sequence causes tissues with a longer T_2 to give a stronger echo and a brighter image.

Proton density is an inherent property of various tissues and has a strong effect on signal intensity and eventual image brightness. As indicated above, proton density is simply the number of protons per unit volume of tissue—in practice, per *voxel.* (It is assumed that the protons are mobile so that they can participate in magnetization and demagnetization.) In and of itself, proton density has a major influence on image brightness because more protons are available to contribute to signal intensity.

Image Contrast. Besides image brightness, image contrast has a major impact on image quality. As in radiography, contrast in an MR image, as recorded on film, denotes variation of image density from tissue to tissue. But pulse sequence, TR, and TE present white matter as white, and gray matter as dark, but manipulation of these factors can alter the shades of contrast, and even invert the images so that white becomes dark and dark becomes white. In any event, we must be able to distinguish easily between various tissues by producing optimum contrast and brightness.

Contrast is the resultant of three basic NMR factors: T_1 relaxation, T_2 relaxation, and proton density. These can be manipulated to optimize contrast.

1. *T_1 Contrast.* Growth of longitudinal magnetization after an RF pulse occurs at different rates for different tissues. To obtain optimum contrast we must find a TR (time after an RF signal) which yields an echo at the instant that the ratio between the magnetization of both tissues is at maximum (see Figure 24.11). However, this may occur at points on the curves where signal intensity is too low. In general, a TR = $0.63T_1$ will provide a satisfactory combination of brightness and contrast (analogous to density and contrast in a radiograph). For T_1-weighting, one should select an intermediate TR and short TE (time between echoes) because a short TE decreases T_2 contrast more than T_1 contrast.

2. *T_2 Contrast.* T_2 relaxation can be used to enhance image contrast because different tissues have different relaxation times according to their T_2 constants. Comparison of the decay curves of transverse demagnetization for two hypothetical tissues (see Figure 24.12) shows how contrast can be maximized. Contrast is the ratio of the residual transverse magnetization for the two tissues at a particular time, for example, A/B. For maximum contrast an intermediate TE should be selected (TE is spin-echo time, between the first RF pulse and the center of the following spin-echo). Recall that spin-echo (see pages 552–553) is used to intensify RF signals form transverse magnetization, to produce a bright image. But if the selected TE turns out to be too long, image brightness suffers on account of the low residual transverse magnetization. A long TR interval (time between successive pulses) minimizes T_1 contrast, thereby favoring T_2 contrast. Thus, by adjusting the spin-echo sequence to a long TR and intermediate TE, we obtain a T_2-weighted image.

3. *Proton-Density Contrast.* Proton density is the number of

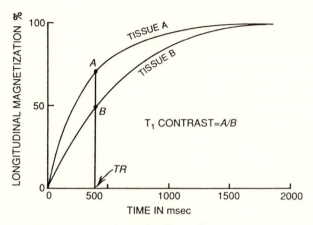

Figure 24.11. Growth curves of longitudinal magnetization for tissues having different relaxation constants T_1. Contrast is derived from the curves reaches a maximum at A/B. Spin-echo repetition time TR here is 500 msec. (*Adapted from Keller PJ, by permission of the author and GE Medical Systems.*)

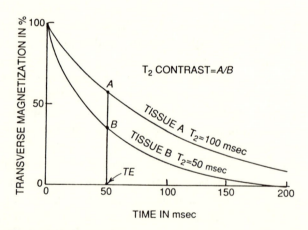

Figure 24.12. Decay curves of transverse magnetization for tissues having different relaxation constants T_2. Contrast reaches a maximum at A/B; spin-echo TE (time to echo) is here 50 msec. (*Adapted from Keller PJ, by permission of the author and GE Medical Systems.*)

mobile protons (ie, capable of producing a signal), per unit volume of tissue—in practice, per voxel. A high proton density creates a bright image (pixel) of the voxel. Proton density contributes to contrast, which depends on the ratio of proton density in one voxel to that in another.

Figure 24.13 shows how proton density contributes to image contrast. Just as with T_1, growth of longitudinal magnetization after a 90° RF pulse occurs at different rates for different tissues, but when based on different proton densities, the magnetization curves have a different shape. In Figure 24.13 you see the longitudinal magnetization curves for two voxels that have the same T_1 values but different proton densities; the upper curve represents the voxel with higher proton density. Contrast due to proton density is the ratio of the longitudinal magnetization of the two voxels at a particular instant. The ratio A/B in the figure is the contrast at time TR. Note that as you follow the curves upward, contrast progressively increases to a maximum at TR, after which it remains constant up to full magnetization M (equilibrium values). Since maximum contrast is reached late in the magnetization process (high on the curves) as compared to T_1, a long TR is desirable; this also minimizes T_1 contrast. At the same time, a short TE minimizes T_2 contrast. Therefore, proton-density weighting requires a long TR (as in Figure 24.13) and a short TE.

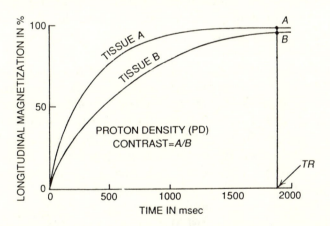

Figure 24.13. Growth curves of longitudinal magnetization resulting from tissues having different PD (proton density) relative values. Maximum contrast is reached at A/B; repetition time (TR) here is 1800 msec, although it may be much longer. (*Adapted from Keller PJ, by permission of the author and GE Medical Systems.*)

Table 24.3 presents the weighting factors, rationale, and effect on white and gray matter of the brain as they appear on the MR image.

Conversion of RF Signals to MR Images. This requires a proce-

TABLE 24.3
MR IMAGE CONTRAST PRODUCED BY WEIGHTING

			Brain Matter	
Weighting	TR	TE	White	Gray
Proton Density	long (min° T_1)	short (min T_2)	dark	light
T_1	intermediate (enh°° T_1)	short (min T_2)	gray	gray
T_2	long (min T_1)	intermediate (enh T_2)	dark	light

°minimizes
°°enhances
Note: Cerebrospinal fluid has long T_1 and T_2, so it appears hyperintense (white) in both T_1- and T_2-weighted images.

dure called **signal encoding** which makes it possible to create images from RF signals. A signal's amplitude (strength) is proportional to the degree of transverse magnetization elicited by the resonant RF pulse which, in turn, is proportional to the number of excited protons in the voxel. Signals can be enhanced by weighting of T_1, T_2, and proton density to produce spin-echoes. These signal differences may be small from tissue to tissue, but by MRI manipulation, they may become distinguishable—*low contrast resolution.* In fact, MR image contrast resolution exceeds that of CT by a factor of 30 or so.

Signal-to-image conversion is highly complex, requiring the use of **auxiliary coils** to generate RF field gradients across the main field B_0. A **field gradient** causes a change of field strength along a particular axis. Wehrli uses a simple device to demonstrate the basic principle (see Figure 24.14). Two cylindrical holes are drilled in a teflon block and filled with water; they have been aligned with the *z-axis,* parallel to B_0, at the same distance *along* the *y-axis,* but at different distances *along* the *x-axis.* Without a field gradient along the *x-axis,* both samples experience the same field B_0, so free induction decay (FID) displays a single frequency, as in Figure 24.14A. But with a field gradient applied along the *x-axis,* the samples experience different fields and the resulting FID consists of two superimposed signals (see Figure 24.14B). By a mathematical process called Fourier (foory-ay) transformation (FT), the plot of amplitude *vs* time is changed to amplitude *vs* frequency, which now shows a single frequency spike without the field gradient, and two spikes with the field gradient (see Figure 24.15A and B). Thus, with this simplified example, application of a field gradient and Fourier transform make it possible to recognize the two samples.

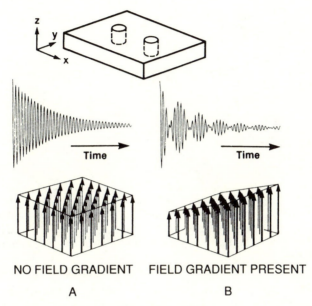

NO FIELD GRADIENT   FIELD GRADIENT PRESENT

A   B

Figure 24.14. *A.* In the absence of a field gradient ($G_x = G_y = G_z = 0$) both samples experience the same field, resulting in a simple free-induction decay (FID) wave form. *B.* In the presence of a field gradient G_x, the two samples experience different fields, resulting in a complex FID wave form. (Figure 24.14A *from Wehrli FW, by permission of the author and GE Medical Systems.*)

In practice, three sets of field gradient coils are required. They cooperate with the main field B_0 to produce the electromagnetic image by defining (1) slice thickness and (2) signal encoding, the location and intensity of the various spin-echoes emitted by the excited voxels in the selected slice. The position and function of these field gradient coils will receive a brief description (see Figure 24.16).

1. *Slice Thickness.* RF field gradient coils surround the patient, with the gradient field along the *z-axis* (parallel to B_0). Then a 90° pulse tuned to the Larmor frequency of the precessing protons within the selected slice will establish its position, and its band width (range of frequencies in the pulse) will determine its thickness (usually 5 to 10 mm). The margins of the slices are not sharply defined — some of the RF energy spills over to adjoining tissue to produce "noise." A gap of about 30 percent between

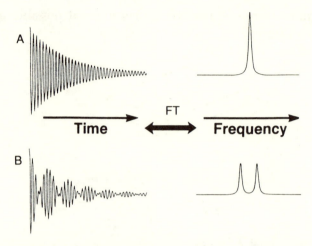

Figure 24.15. By a mathematical process called Fourier transformation (FT) applied to the FID wave forms in Figure 24.14, the time line is changed to a frequency line showing a single frequency signal in *A*, and two signals in *B*. (*From Wehrli FW, by permission of the author and GE Medical Systems.*)

slices reduces this effect (Curry and others). Note that slices can be selected for other orientations—oblique, longitudinal, etc, depending on the orientation of the *z* coils.

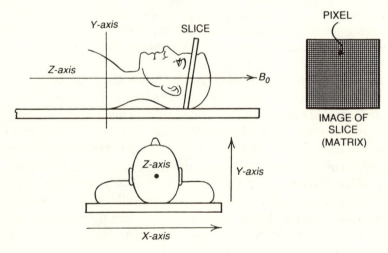

Figure 24.16. Matrix obtained by computer processing of innumerable bits of position and signal-intensity data accumulates during MR imaging.

2. *Signal Encoding.* Signal intensities and their precise location must be identified so that each voxel will have a distinctive pattern that is sent to the MRI computer network for processing. This requires two kinds of encoding of the information: *phase* and *frequency.*

a. *Phase encoding* involves the use of *field gradient coils* which cause a shift in RF wave phase along the *y-axis* while the signals are being generated within the slice (described in [1] above). In essence, phase encoding divides the slice into horizontal rows of voxels, each now being identified by a particular wave phase.

b. *Frequency encoding* employs field gradient coils that generate an RF field gradient along the *x-axis* to give each voxel in the slice a unique frequency, thereby dividing the slice into vertical columns.

The intersections of the rows and columns characterize the voxels, each with its distinctive information. This is digitized by the computer, processed, and converted to the final MR image.

Noise in MRI

In assessing image quality, we must take into account the amount of *noise* present. As noted before, noise in any system has a random distribution in both time and space, and impairs image quality by covering up information. Any factor that improves the signal-to-noise (S/N) ratio (ie, signal intensity relative to noise intensity) will improve image quality. The important elements that enhance the S/N ratio include:

1. *Voxel volume.* As this is increased (smaller matrix), the S/N ratio improves and the time needed to acquire an image decreases, but this occurs at the cost of some decrease in image resolution.

2. *TE Value (time-to-echo).* A short TE results in a stronger signal, increasing the numerator of the S/N ratio without contributing to noise. Thus, a short TE improves the S/N ratio.

3. *TR Value (repetition time).* Signal intensity increases with a longer TR interval because it allows more time for longitudinal remagnetization, producing a stronger signal. Therefore, a longer TR improves the S/N ratio, but it also increases the time needed to create a slice.

4. *Signal Averaging.* As with digital fluoroscopy, smoothing of signal and noise intensities decreases the effect of noise. In MRI,

this is accomplished by computer averaging of the signal and noise intensities of *each* voxel several times. The signal intensities are more consistent over time than the noise intensities because the latter are random events and have greater fluctuations. Thus, the average noise intensity is lower than that of the signals, resulting in a higher S/N ratio. At the same time, averaging increases the time needed to acquire an image.

External Field Magnets for MRI

There are three types of magnets available to produce the external B_0 field: (1) superconducting, (2) resistive, and (3) permanent. These will be briefly described.

1. *Superconducting Magnets.* This type is the one most widely used at present. Its strong, constant magnetic field is generated by coils within the gantry surrounding the patient; the magnetic field (lines of force) parallel the patient's cephalocaudal axis. *Superconductivity* refers to the ability of certain metals to have virtually no electrical resistance at extremely low temperatures of about $-270°C$! A magnetic field intensity of 2 T can be obtained, although most clinical models operate at 1.5 T.

The magnet is actually a *solenoid,* composed of a copper coil containing particles of a superconducting metal such as niobium and titanium and operating at a temperature of $-270°C$. This temperature is maintained by concentric chambers, the inner one containing liquid helium and the outer one liquid nitrogen. Between and around them are vacuum chambers to serve as insulators from heat. The liquid helium and nitrogen must be replenished periodically because of gradual vaporization.

Superconducting magnets provide the strongest and most homogeneous fields available. In operation, a current is supplied to the magnet and then turned off. The current, however, flows steadily because of the superconductivity of the coil, and generates a strong magnetic field by electromagnetic induction.

2. *Resistive Magnets.* A resistive magnet consists of a series of four solenoids within the gantry surrounding the patient. The name *resistive* means that the solenoids must be continuously supplied by electric current during MRI. Field intensity and homogeneity are reasonably good. Power consumption is extremely high, but purchase and maintenance are much less than with

superconducting magnets. Resistive magnets generate a field intensity of about 0.5 T.

3. **Permanent Magnets.** These resemble ordinary permanent bar magnets, but are fabricated from special ceramics that can be magnetized. In the form of building blocks, they can be stacked to form magnets large enough to surround a patient. Although inexpensive in purchase and maintenance, field intensity is low—about 0.1 to 0.15 T—and homogeneity less than that of the other types of magnets.

Surface Coils

When especially good recorded image detail is required for a small structure, such as an arm or leg, *surface* receiving coils can be placed directly in contact with the zone of interest. They provide a limited field of view with smaller pixels than ordinary MRI, with resulting enhancement of detail. Moreover, the S/N ratio is improved. Surface coils must be very carefully positioned with relation to the main magnet to avoid loss of signal intensity.

A Typical MRI Unit

At the outset, it was stated that an MRI unit resembles externally a CT scanner. As shown in Figure 24.17, the *gantry* (outer rectangular shell) contains the field magnet. Within the circular opening lies a *shim coil,* which improves the uniformity of the main magnetic field B_0. To prepare for the examination, the technologist slides the patient head-first into the gantry opening by means of the movable couch top. To the left of the gantry you can see the control system.

Hazards of MRI

The strong magnetic stationary field can be hazardous because many magnetic objects can become severely damaging projectiles. This includes such objects as iron tools, surgical instruments, and even metal carts and oxygen tanks. Internally, any iron-containing implants may malfunction (eg, pacemaker) or be displaced (metallic foreign bodies). Especially dangerous are intracranial surgical clips

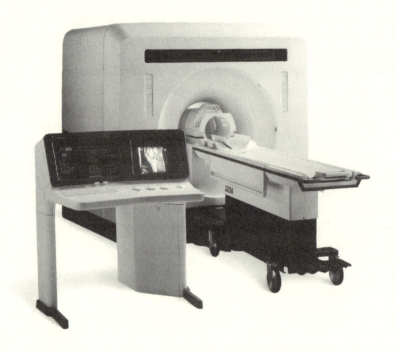

Figure 24.17. A typical MRI unit. The external magnet is contained within the massive gantry. The patient is slid into the gantry opening head-first. Note that the entire body is not contained within the opening at any one time. (*By courtesy GE Medical Systems*)

which could tear brain tissue, although surgical clips elsewhere in the body pose no significant hazard.

When a patient requires resuscitation during MRI, he or she must be moved elsewhere, unless special, nonmagnetic resuscitation equipment is available.

Damage can occur outside the body to calculators, watches,

credit cards, and electronic equipment, among others, unless warranted to be immune to strong magnetic fields.

So-called "fringe" fields around the magnet may extend for some distance — about 11 m from a 0.5-T field — although proper shielding will minimize it.

RF gradient fields may cause mild muscle contractions and cardiac arrhythmias, or light flashes during orbital MRI, but these have been temporary and without serious consequences. Examination of pregnant women with MR has not been proven safe.

ULTRASOUND IMAGING

This section will concentrate on the use of ultrasound waves in medical diagnosis. But first we must acquire an understanding of the physical nature of sound in general, and then turn to the special properties and application of ultrasound.

Nature of Sound

As a form of mechanical energy with wave characteristics, nevertheless, sound differs from electromagnetic radiation in a number of ways. Electromagnetic radiation, as described earlier on pages 150–151, consists of transverse waves that can be transmitted in a vacuum as well as in various media, depending on the energy of the radiation. Sound waves, on the other hand, are longitudinal in nature and require a material medium (solid, liquid, or gas) for their transmission; they will *not* pass through a vacuum.

Sound must be generated mechanically by an oscillating (vibrating) body of matter. The bell is a familiar example (see Figure 24.18); when struck by the clapper, the entire bell is set into vibration, a rapid, periodic back-and-forth motion of its wall. This transfers mechanical energy to the surrounding air in all directions (isotropic). To simplify matters, let us look at only the right side of the bell: the outward component of its vibration compresses the adjacent air on the right (molecules move closer together). The right wall of the bell instantly moves to the left (inward component of vibration), now rarefying the air on the right (molecules move farther apart). This rapid, periodic, alternating compression and rarefaction of the air creates a wave form which moves outward as a *longitudinal wave*. It is important to note that the air molecules do *not* move continuously in the direction of the wave, but rather in a to-and-fro motion in the alternating zones of compression and rarefaction; in fact, they move over a distance of only a few μm (millionths of a meter). The to-and-fro motion has the same frequency as the vibrating bell. As you may recall, *frequency* is the number of cycles per second. The resulting wave form, representing the summation of the alternate compressions and rarefactions, also has the same frequency. When it reaches an ear, the eardrum is set into vibration, again with the same frequency and perceived as sound. This process is summarized in Figure 24.18.

The above description of sound waves also applies to the transverse waves in water relative to a fishing float: throwing a pebble into the water nearby generates visible waves that pass outward in all directions, but the float only bobs up and down. This means that the water molecules also bob up and down; it is only the wave form that moves out over the water.

Properties of Ultrasound

Like sound generally, ultrasound obeys the *wave equation,* which also applies to electromagnetic radiation:

$$v = \nu\lambda \tag{2}$$

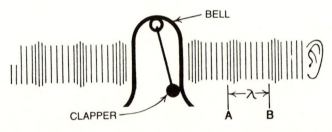

Figure 24.18. An example of audible sound-wave production by a bell. When struck by the clapper, the bell rings, sending out sound waves in all directions. Shown here are only a portion of the laterally transmitted waves in air. The waves proceed in trains of alternating zones of compression and rarefaction of air, symbolized by the closeness of the vertical lines. Wavelength λ is the distance between the beginning of two successive compressions (A to B), which pass outward ahead of zones of rarefaction. Upon entering the ear, the waves set up vibrations in the typanic membrane (eardrum) with the same frequency. These vibrations set up electrical impulses in the auditory nerve, which carries them to the auditory center for processing.

where v is the velocity of sound in m/sec (meters per second), ν the frequency in Hz (hertz = 1 cycle/sec), and λ the wavelength in m. In Figure 24.18, λ is the distance between two corresponding points in successive cycles, such as A and B. Simply put, one cycle equals the distance between the start of one compression and the start of the next compression. Frequency ν is the number of compression-rarefactions per sec, measured in Hz. The relation of units in the wave equation (2) is revealed by:

$$\frac{m}{sec} = \frac{1}{sec} \times m$$

so,

$$m/sec = m/sec$$

The components of equation (2) will now be explained under separate headings, with the addition of amplitude and direction.

1. *Frequency.* Sound waves with frequencies ranging from about 15 Hz to 20,000 Hz are audible to the normal human ear (dogs can hear higher frequencies than humans). The mechanism for hearing is the transfer of sound wave energy to the eardrum, which goes into vibration with the same frequency as the sound. Thus, if one strikes an A note (frequency 435 Hz), this excites a frequency of 435 Hz in the air in the form of sound waves which, in turn, produce the same frequency in the ear. Medically useful ultrasound involves frequencies of 1 to 10 MHz (1 MHz [megahertz] = 1 million Hz).

2. *Velocity.* The term velocity refers to speed in a given direction. If direction is along a straight line, velocity and speed are the same. Velocity of any object is defined by

$$\text{velocity} = \text{distance/time} \qquad (3)$$

For example, in ordinary units, we drive to Dallas over a distance of 100 miles in 2 hours; what is the velocity or speed? Substituting these values in equation (3),

$$\text{velocity} = 100 \text{ mi}/2 \text{ hr} = 50 \text{ mi/hr}$$

The same equation applies to sound waves as follows: the velocity of sound waves equals the distance between any two points such as A and B in Figure 24.18 divided by the time it takes the wave front to move through this distance (like the front end of our car going from Tyler to Dallas).

The velocity of sound, and ultrasound in particular, depends on the density (g/cm³) and the compressibility of the conducting medium. Since various soft tissues are essentially liquids and have about the same density and compressibility, they conduct ultrasound with about the same velocity—1540 m/sec. The most rapid conduction occurs in bone (very dense and rigid), whereas slowest conduction occurs in gas (low density and high compressibility); if gas were not highly compressible, how could it be stored under pressure and used to inflate a tire?

3. *Wavelength.* Since velocity changes insignificantly between various soft tissues, you can see from equation (2) that a higher frequency ultrasound must be accompanied by a proportional decrease in wavelength (doubling ν goes with halving λ). According to the wave equation (2), an ultrasound frequency of 2 MHz and a velocity of 1540 m/sec gives rise to a wavelength found by

$$1540 \text{ m/sec} = 2 \text{ MHz} \times \lambda$$

Rearranging, and changing 2 MHz to $2 \times (10)^6$ or 2 million Hz,

$$\lambda = 1540/2,000,000 = 0.0008 \text{ m, or } 0.8 \text{ mm}$$

With a higher ν of 5 MHz, λ would be shorter by inverse proportion:

$$2/5 \times 0.8 \text{ mm} = 0.32 \text{mm}$$

4. *Amplitude or Intensity.* Although all forms of sound can be measured by physical methods, ordinary sound is unique in that its intensity can be perceived as *loudness.* Amplitude increases with more rapid back-and-forth movement and greater excursion of the molecules producing increased compression, resulting in increased intensity. The more vibrational energy available for transfer to the medium, the greater will be the intensity or amplitude of the ultrasound waves. In fact, the dimension of amplitude is *power/unit area,* expressed in *watts/cm.²*

5. *Direction.* At the outset, it was stated that ordinary sound waves are emitted in all directions from a vibrating bell; this also applies to other sources such as vocal cords, musical instruments, and so on. But ultrasound waves are directional, that is, they move straight forward or backward from the source, at least in their initial path.

Production of Ultrasound

A special device called a **transducer** generates ultrasound waves for medical imaging. At its heart is a type of material known as a **piezoelectric crystal.** Piezoelectricity (studied intensively by Pierre Curie in the 1890s) is a property peculiar to certain natural and artificial crystals which, when deformed or compressed, generate a voltage; and, conversely, when subjected to a voltage, becomes stressed or deformed. Application of a high-frequency AC voltage causes the crystal to vibrate with its resonant frequency, thereby converting electrical energy to mechanical energy. A sufficiently high frequency will generate ultrasound waves. Figure 24.19 shows a typical transducer, described in the legend.

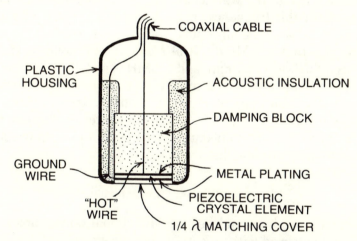

Figure 24.19. Idealized drawing of an ultrasound transducer, which is capable of both sending ultrasound waves and receiving their reflected waves. (Modified from Curry TS III, Dowdey JE, Murry RC Jr: *Christensen's Physics of Diagnostic Radiology,* 4th ed., Philadelphia, Lea & Febiger, 1990. By permission.)

The piezoelectric crystal now widely used consists of *lead zirconate titanate* or **PZT**® of which there are at least two varieties. (Natural quartz has piezoelectric properties, but is not so efficient as PZT.) The molecules of a piezoelectric crystal are polarized, one end positive and the other end negative. When the high frequency voltage pulse is applied across the crystal, it alternately thickens and thins along its short axis, at its resonant frequency; and generates ultrasound waves as a beam in the air in front of the crystal face, and also back toward the damping block. Note that both faces of the crystal vibrate synchronously, thereby increasing the intensity of the ultrasound. In the usual B-mode operation (to be described later), the damping block must stop the crystal vibration within a microsecond or so because the transducer must be ready immediately to receive reflected waves (echoes) from tissue interfaces within the body.

The **damping block** usually consists of powdered rubber and tungsten blended with an epoxy resin. The proportion of these ingredients depends on the transducer's frequency and is chosen to achieve optimum suppression of the backward-directed ultrasound.

Crystal **resonant frequency** depends on its thickness, which should be one-half the wavelength of the ultrasound waves being produced. At this frequency, both faces of the crystal vibrate in synchrony and generate ultrasound waves with maximum intensity. It turns out that a 2-MHz crystal element composed of PZT-4® requires a thickness of 1 mm, and a 1 MHz crystal a thickness of 2 mm (Curry and others). A trick called **quarter-wave matching** improves the efficiency of ultrasound energy transfer from the crystal to the skin—this requires a layer of material of appropriate thickness and acoustic impedance (see below) to be equivalent to one-fourth of a wavelength.

Ultrasound Beam Characteristics

Although an ultrasound beam is directional (anisotropic), it is not entirely focused without certain modifications.

1. **Unfocused Beam.** An ultrasound beam leaving a flat crystal element has an initial cylindrical segment, followed by a diverging conical portion (see Figure 24.20). The length of the former and the divergence of the latter depend on ultrasound frequency and crystal diameter, as follows:

a. *Frequency.* The higher the frequency, the longer will be the cylindrical segment or **near field** (Fresnel [fre'nel] zone). At the same time, the *far field* (Fraunhofer zone) becomes less divergent at higher frequencies (compare Figures 24.20A and B). The best **lateral resolution** exists at the junction of the near and far fields (ie, ability to display two closely spaced points in the same plane, as two separate images—see also pages 318–319). **Depth resolution**

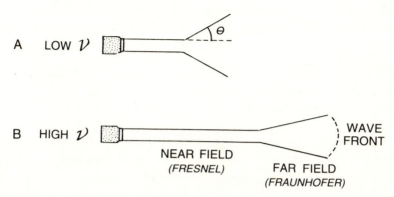

Figure 24.20. Transducer crystal diameter is the same in A and B. At higher-frequency ultrasound, the near field is longer (ie, reaches deeper into the body), while the far field spreads less (ie, has a smaller divergence angle θ). The zone of maximum lateral resolution lies at the junction of the near and far fields.

improves at higher frequencies (points closely spaced in depth, displayed as two separate images). But note that as frequency is increased, greater absorption of sound energy occurs in the tissues, weakening beam intensity.

b. *Crystal Diameter.* Increasing crystal diameter increases near zone length, but worsens lateral and depth resolution (see Figure 24.21). The effect of crystal diameter on near-field length and far-field beam spread can be obtained from the following equations (Sprawls):

$$NFL = D^2/4\lambda \text{ cm} \qquad (4)$$

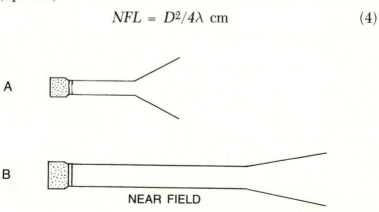

Figure 24.21. The crystal in B has a larger diameter than in A. With the wider crystal, the near field is longer and the far field has a smaller divergence angle.

where NFL is near-field length, and λ wavelength, both in cm.

The divergence angle DA may be derived from

$$DA = 70\lambda/D \qquad (5)$$

where DA is defined in Figure 24.21, and D is the crystal diameter. As expected, beam intensity in power/cm² decreases as the divergence angle increases. Therefore, the far field in A is less intense than in B.

2. *Focused Beam.* By focusing the ultrasound beam, one can narrow it at a particular depth, thereby improving both lateral and depth resolution. This can be effected in two ways:

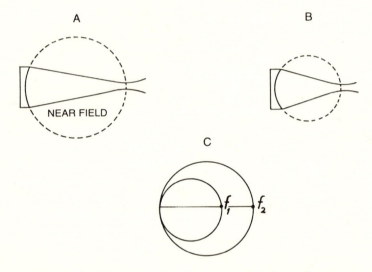

Figure 24.22. On the left in A and B, in contact with the circles, are concave-faced scintillation crystals. The crystal face in B has a greater degree of curvature (smaller radius) than in A, so its near field is shorter, that is, its focal zone is closer to the skin and is narrower than in A. Lateral resolution is maximum at the focal zone and is obviously better than with an unfocused beam. In C is shown how the degree of curvature affects the focal distance f_1 and f_2. Recall that a circle of smaller diameter has a greater curvature.

a. *Concave crystal face.* The crystal element must be fabricated with a *concave face;* the greater the concavity (crystal curvature inward), the nearer the beam will focus relative to the crystal. Stated technically, a shorter focal length goes with a more highly concave crystal. This is shown in Figure 24.22.

b. *Acoustic lenses.* Plastic *acoustic lenses* having a flat surface on one side and a concave surface on the other are placed with the flat side against the crystal face (see Figure 24.23). The shorter the focal length (greater curvature), the nearer the lens will focus the beam relative to the lens face.

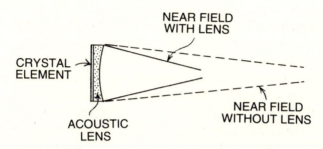

Figure 24.23. A concave acoustic (plastic) lens in front of a flat crystal focuses the ultrasound beam in the same manner as a concave crystal.

The **quality** or **Q-factor** of an ultrasound beam comprises its **bandwidth** (total range of frequencies in the beam) and its pulse duration. Thus, the narrower the bandwidth, the purer will be the sound because of the fewer frequencies present. Also, the shorter the duration of the pulse, the more time the transducer can be in the "receive" mode for reflected (echo) signals, since the transducer serves alternately as *both* a sender and receiver of ultrasound. For optimum depth resolution, the pulse should contain only a few wavelengths; the shorter the distance they occupy, the closer two points can be in depth and still be imaged as two separate points.

Reception of Ultrasound

As already stated, the transducer has two functions: not only does it generate an ultrasound beam when stimulated by a high-frequency electric pulse with the selected frequency, but it also receives the beam (echo) after its reflection from the boundary between dissimilar tissues within the body. Ordinarily, the applied pulse lasts for about 1 μsec (microsecond), then abruptly goes into the "receive" mode for about 999 μsec. Thus, the total duration of the send-receive sequence is about 1000 μsec = 1 msec. However, a transducer cannot, with a single crystal element, simultaneously send and receive.

Behavior of Ultrasound in Matter

Upon entering the body, an ultrasound beam may interact with tissues in one or more of the following processes: (1) reflection, (2) refraction, and (3) absorption (attenuation). Each of these will be explained.

1. **Reflection.** All of us have seen the reflection of light from a mirror or other smooth surface. Although this process involves transverse (electromagnetic) waves, the longitudinal sound waves also undergo reflection, in the case of ultrasound, from interfaces or boundaries between different tissues.

Both ultrasound and light obey the **law of reflection,** namely, the angle of incidence and the angle of reflection are equal (see Figure 24.24). Thus, sound and light "bounce off" an appropriate surface in a particular direction.

The factor that determines the percent of the incident beam undergoing reflection is a property, peculiar to various tissues,

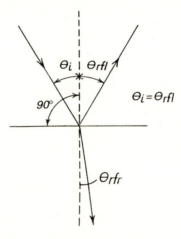

Figure 24.24. Above the horizontal line (reflecting tissue interface), the left arrow shows the direction of an ultrasound beam making an angle of incidence θ_i with the perpendicular (dashed line). The beam is reflected at the same angle θ_{rfl} as shown by the arrow upward and to the right. Thus, the angle of incidence equals the angle of reflection, a law which also applies to a light beam. The line directed downward below the interface indicates the direction of the refracted beam which does not follow the same law; it bends toward the perpendicular at an angle θ_{rfr}, the angle of refraction. Refracted ultrasound waves serve no useful purpose and create noise.

called **acoustic impedance.** The greater the difference in acoustic impedance, the greater will be the percentage reflection. A simple relation for acoustic impedance is given by:

$$\mathbf{Z} = \rho \boldsymbol{v} \text{ rayls} \tag{6}$$

where, for a given tissue, Z is its acoustic impedance, ρ its density, and v the speed of sound in it, in cm/sec. The unit of acoustic impedance is the **rayl** (named after the famous British physicist Lord Rayleigh). As the velocity of sound in all soft tissues is virtually the same—154,000 cm/sec (1540 m/sec)—Z is nearly proportional to the density of the particular tissue. The velocity of sound in bone is very high, and so is density of bone, so it has a high Z (= 7.8 rayls) and a high percentage of reflection. Air has an extremely small Z, 0.0004 rayl, with virtually *100 percent reflection.* Note that it is the **difference** in Z between two tissues, not their position relative to the boundary, that determines the degree of reflection; the same percent reflection occurs whether the ultrasound beam goes from soft tissue to air, or from air to soft tissue.

A simple and convenient way to calculate the percent reflection at an interface employs the following equation (Curry and others):

$$\mathbf{R} = \left(\frac{\mathbf{Z}_1 - \mathbf{Z}_2}{\mathbf{Z}_1 + \mathbf{Z}_2} \right)^2 \times 100 \tag{7}$$

where R is the percent of the incident beam reflected when the incident beam is perpendicular (90°) to the surface, and Z_1 and Z_2 are the acoustic impedances of the two tissues (see Table 24.4). For example, at the interface between fat and muscle, Z_1 for fat = 1.38, and Z_2 for muscle = 1.70. What is the percent reflection at the interface?

$$R = \left(\frac{1.70 - 1.38}{1.70 + 1.38} \right)^2 \times 100$$

$$= \left(\frac{0.32}{3.08} \right)^2 \times 100 = 1\% \text{ (approx)}$$

Note the very small percent reflection. Yet, it suffices to produce an echo that can be amplified by the system to produce a satisfactory image. It also leaves a large percent of transmitted beam for reflection (echo production) deeper in the body.

For an interface between the transducer crystal (Z_1 = 30) and air (Z_2 = 0.0004),

TABLE 24.4
APPROXIMATE VALUES OF ACOUSTIC IMPEDANCE FOR
AIR, WATER AND VARIOUS ORGANS*

Material	Acoustic Impedance
	rayls
Air	0.0004
Fat	1.38
Water	1.54
Brain	1.58
Kidney	1.62
Liver	1.65
Muscle	1.70
Eye lens	1.84
Skull	7.8
PZT-4 (piezoelectric crystal)	30.0

*Modified from Curry TS III, Dowdey JE, Murry RC Jr: *Christensen's Physics of Diagnostic Radiology*, 4th ed., Philadelphia, Lea & Febiger, 1990. By permission.

$$R = \left(\frac{0.0004 - 30}{0.0004 + 30} \right)^2 \times 100$$

The value 0.0004 is so tiny relative to 30, that it may be ignored, leaving

$$\left(\frac{-30}{30} \right)^2 \times 100 = -1^2 \times 100 = 100\%$$

This means that the entire ultrasound beam has been reflected back to the transducer at the crystal-air interface! Incidentally, the fact that the right side of equation (7) is squared means that the value of R is always positive, and that the direction of the beam across a boundary does not change R. It also confirms the need for coupling oil between the transducer and skin to eliminate the chance of air being trapped between them. Table 24.4 gives the acoustic impedance of air, water, and a number of tissues of interest in medical ultrasound studies.

The *intensity* of the reflected beam relative to the incident beam is important in diagnostic ultrasound. Relative amplitude is expressed as the *logarithmic ratio* of the reflected intensity to the incident intensity, by analogy to the sensitivity of the human ear to sounds of various degrees of intensity. When one sound is 100 times as intense as another, it is perceived by the normal ear as twice as loud; if 1000 times intense, it is perceived as 3 times louder. It turns out that 2 is the logarithm of 100, and 3 is the logarithm of 1000. At this point you must review pages 23–24 for an explanation of logarithms, to enable you to follow the discussion of relative amplitude in ultrasound.

In evaluating the intensity or power of a reflected beam relative to an incident ultrasound beam, one obtains an extremely small ratio, requiring the use of negative powers of ten. Since the unit for the ratio of audible sound intensity—the decibel—was already in use, it was only natural for it to be adopted for ultrasound. Logarithms, being small numbers for ultrasound, are much more convenient to use.

The following equation defines *relative amplitude (RA)*:

$$RA = log\ \frac{A_r\ watts/cm^2}{A_i\ watts/cm^2}\quad bels\ (B)\qquad(8)$$

where A_r is the amplitude of the reflected ultrasound beam, and A_i that of the incident beam. Since the *decibel* (*dB*) is already widely used, we convert bels to decibels: 1 B = 10 dB. Suppose the numerator in equation (8) is 0.001 watts/cm^2 and the denominator 100 watts/cm.^{2} Substituting these values in the equation,

$$log\ \frac{0.001}{100} = log\ \frac{10^{-3}}{10^2} = log\ 10^{-5} = -5\ B, or\ -50\ dB$$

The negative answer indicates that the reflected intensity is much weaker than the incident intensity. (If we invert the fraction, it gives the reciprocal value—how much greater the incident amplitude is than the reflected amplitude; in this case, 50 dB.)

As a general rule, a 50 percent loss of intensity of the reflected beam relative to the incident beam represents a loss of 3 dB (ie, the amplitude ratio −3 dB).

The *angle of incidence* of an ultrasound beam has an important bearing on imaging. As the transducer serves both to send and receive ultrasound energy, you can see from Figure 24.25A that when the transducer is aimed toward the skin at a large angle of incidence, the reflected beam (echo) will return at some distance from the transducer and be missed. With a small angle of incidence as in Figure 24.25B, typically less than 3 degrees, the echo will enter the transducer during its reception mode. At a particularly large angle of incidence known as the *critical angle*, total reflection occurs at the skin.

2. *Refraction.* This occurs when an ultrasound beam passes, at an angle other than 90 degrees, from one tissue into another with

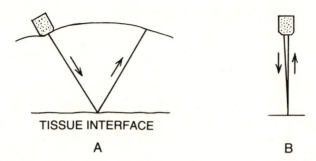

TISSUE INTERFACE

A   B

Figure 24.25. In *A* the transducer is directed at a large angle to the vertical, so it misses the returning ultrasound echo. In *B* the transducer is directed at a small angle, so it detects the ultrasound echo.

a change in velocity. The frequency remains constant, so the wavelength has to change according to the wave equation (2). Refraction involves only a fraction of the beam, increasing with the angle of incidence, and is manifested by a change in direction (see Figure 24.24). The refracted ultrasound contributes nothing useful to the image, but passes deeper into the body where it gives rise to artefacts. In practice, however, with the angle of incidence of the beam relatively small, less than 3 degrees, very little ultrasound is refracted.

3. *Absorption.* Due to friction among molecules in their back-and-forth motion, reduction in intensity of the ultrasound beam occurs as it traverses matter. Friction results in degradation of a part of the molecules' kinetic energy to heat. Analogous to x rays, **attenuation** of ultrasound beam intensity is exponential with depth in the tissue, governed by the **attenuation constant.** Therefore, we can apply the concept of **half-value layer** (**HVL**) to ultrasound. HVL in this case depends on the frequency of the ultrasound waves: the greater the frequency, the greater the attenuation coefficient. This means that a **high-frequency beam shows less penetration that a low-frequency beam.** In general, the attenuation of ultrasound per cm of tissue is approximately proportional to frequency so that doubling the frequency doubles attenuation, and halving frequency halves attenuation (both per cm). Bushong states that attenuation in soft tissue is about 1 dB/cm/MHz.

Ultrasound Image Displays

There are four main categories of ultrasound image display: (1) A mode, (2) T or TM mode, (3) B mode, and (4) Doppler. These will be considered under separate headings.

1. ***A-Mode Display.*** At one time the only method of displaying echoes received by the transducer and converted to electric signals, the A mode shows the signals as spikes along a horizontal axis (see Figure 24.26). The height of a spike represents the amplitude of the reflected signal; A mode stands for *a*mplitude mode. The position of a spike along the horizontal axis is proportional to the return-time of the echo, which depends on the depth of the relevant tissue interface. Because the signal loses amplitude by virtue of the attenuation of its reflected beam along its path (dB/cm), a specially designed amplifier automatically corrects for this lost amplitude, taking into account the distance the echo has traveled. Early on, the A mode was used to determine the possible lateral shift of the midplane "complex" of the brain by a contralateral subdural hematoma (see Figure 24.26). A good deal of subjective input by the operator impaired the reliability of this procedure. Modern A-mode equipment has greatly improved the accuracy of ***depth*** measurement of structures within the body, but signal amplitude is largely ignored.

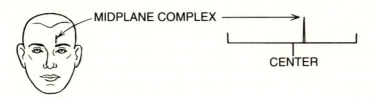

Figure 24.26. An ultrasound A-mode display used some years ago to determine the displacement of a "midplane complex" in the brain. It is here identified as a spike along the horizontal depth-line. Measuring the distance of the spike from the actual midsagittal plane of the skull indicated pressure from a subdural hematoma or other mass on the side opposite the shift. A-mode is still used for depth measurement in other conditions, but with more sophisticated equipment.

2. ***TM or M Mode.*** The TM (time-motion) mode has its main application in echocardiography to show the motility of the cardiac valve leaflets and muscle wall. In this mode, the spikes representing ultrasound echoes are electronically converted to dots on a

single B-mode display. The scanner incorporates a paper recording strip, the vertical axis of which records time, and the horizontal axis depth of the structure under study (see Figure 24.27). The motion of a valve leaflet or heart wall proceeds back and forth along this depth axis. If the dots (signal "blips") are moved electronically to the top of the recording strip and allowed to drop to the bottom, they will describe a series of transverse waves, which denote the amplitude of motion of the structure. Thickened leaflets or scarred muscle will show diminished motion. A radionuclide added to the process with appropriate imaging will enhance the visibility of heart wall motion and improve diagnostic capability.

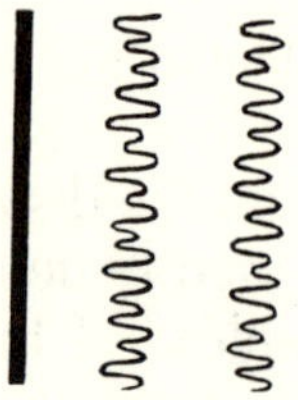

Figure 24.27. TM-mode ultrasound display. This shows the mobility of an internal structure such as a heart valve leaflet, recorded on a moving strip.

3. **B-Mode Display.** We use ultrasound most frequently in the *brightness* or *B mode* because of its extensive diagnostic application, especially in the abdomen and pelvis. The B-mode display involves the *pulse-echo sequence* with the buildup of an image from innumerable echo blips. The dot elements in the image correspond to the pixels in CT and MR images, and to dots in a half-tone print. You can see the latter by viewing a newspaper photo with a magnifying glass. In the B mode, the brightness of the dots is proportional to the intensity of the echoes they represent.

The imaging process in B mode requires that the technologist move the transducer over the skin in a scanning fashion, the transducer crystal being "coupled" to the skin by a layer of oil (see page 278). This is why it is called "contact scanning." The transducer may be moved straight across the skin as in a *linear scan,* but much more frequently as a combination of linear and rocking motion in a *compound scan.*

In a compound scan, the operator must time transducer motion to coincide with the image on the screen. The innumerable dots have to be assembled and processed by the computer to build up an intelligible image. To accomplish this, the computer is linked to a jointed arm, which consists of three segments connected by two joints (see Figure 24.28). Thus, the assembly has a total of three joints, each of which contains a sensor. The three sensors contain variable resistors (potentiometers) whose resistance varies with the position of the arm and its attached transducer. This continually "informs" the computer about the sizes of the angles at the joints as the transducer scans the area of interest. From this information, the computer determines the position and depth of the echoes. In addition, motion of the linked arm is limited so as to image a single slice at a time: the slice is made in one direction, so the computer "turns" the image, allowing the observer to see the slice in face view.

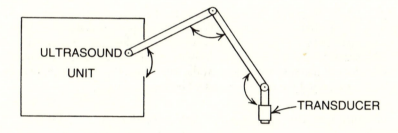

Figure 24.28. The triple-jointed arm of an ultrasound unit. These joints contain electrical resistance sensors that "tell" the computer the position and depth of the tissue interface, as the operator scans the region with the transducer.

Until 1972, B-mode scanners produced images directly on storage cathode ray tubes, giving extremely high-contrast images (short scale) with no shades of gray (similar to high-contrast radiographs). Such images were difficult to interpret. Marked improvement, notably in long-scale contrast, resulted from introduction of the *gray-scale imaging* that is in general use today. This was achieved by the use of a *scan converter,* which discards superfluous echoes from the same point and allows only the strongest echo blips to be stored for imaging. The final image is displayed on a TV monitor.

4. *Doppler Scanning.* We have all experienced the Doppler Effect—the increasing frequency (higher pitch) of a fire engine siren as it approaches, and the decreasing frequency (lower pitch) as it recedes. This is caused in the first instance by compression of

sound waves (higher frequency) ahead of the siren and, in the second instance, by rarefaction of sound waves (lower frequency) behind the siren (see Figure 24.29). The name **Doppler shift** has been applied to this phenomenon, which provides the basis for Doppler scanning. Doppler, an Austrian physicist, discovered this shift about 150 years ago for both sound and light.

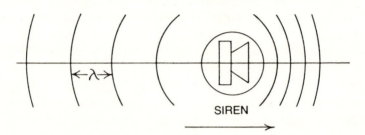

Figure 24.29. Simplified diagram to show the Doppler effect. Motion of a sound source like a siren causes the sound waves to be pushed together in front and pulled apart behind. Therefore, in the direction of the siren's motion, the wavelength shortens and the frequency increases, so the sound pitch increases (like high notes compared to low notes). At the same time, the wavelength behind the siren increases and the frequency (pitch) becomes lower.

Doppler scanning employs **two transducers** side-by-side in a single unit, one to send and the other to receive ultrasound waves, either in pulsed or continuous mode. **Pulsed Doppler** uses individual pulses to determine the depth and velocity of moving internal structures by controlling the space between the pulses according to the depth of a given structure.

The **continuous** mode allows us to evaluate the velocity of blood flow in an artery or vein by reflecting ultrasound waves off blood cells. A Doppler shift occurs first when the moving cells receive the incident beam, and again when the echo returns from the moving cells. For this procedure, the transducer unit is hand-held at a relatively wide angle to the skin surface because of the double-transducer design, in which the receiving transducer is separate from the sending one. If blood motion is slow enough, the Doppler shift gives rise to frequencies in the audible range—the operator can assess the approximate velocity of blood flow by listening to the pitch of the sound: the lower the pitch, the slower the flow. In addition, the shift signal can be processed by an appropriate electronic device to yield a frequency distribution which can be recorded. This gives a more accurate estimate of blood flow through the vessel.

Continuous Doppler scanning has contributed greatly to non-invasive assessment of heart valve mobility, and adult and fetal cardiac wall motion, as well as the patency of blood vessels on the basis of blood flow pattern.

Question of Biohazard

Although diagnostic ultrasound has been in use for a number of years, no biologic injury has thus far been reported at the prevailing dose levels. Still, there is ongoing investigation of possible harmful effects, especially with regard to embryonic and fetal injury, since ultrasound is widely and frequently used to diagnose fetal abnormalities.

High-intensity ultrasound, in the range of about 10 or more watts/cm^2 in experimental animals, has revealed tissue changes such as release of free radicals, tissue disruption at interfaces, and production of tiny cavities. None of this has been found in the diagnostic range with about one-tenth the amplitude (Bushong).

Although opinion may change in the future, it is generally held that the benefit/risk ratio is so high that the major contribution of ultrasound to medical diagnosis warrants its continued use.

QUESTIONS

1. What are magnetic dipoles? Which one is used in MRI, and why?
2. Describe the effect on the magnetic dipoles when a person is placed in a strong external magnetic field.
3. Define and explain the precession of magnetic dipoles. What is meant by Larmor frequency and its relation to resonance?
4. Give an example of resonance, in your experience.
5. State and describe the factors contributing to the MRI signal.
6. Define "free-induction decay" and what it represents.
7. How is T_1 relaxation converted to an MR signal?
8. State the factors contributing to MRI contrast.
9. What are gradient coils? State their purpose.
10. Briefly discuss the spin-echo process.

11. Explain the purpose of signal encoding and its relation to voxel creation.
12. List the sources of noise in MRI. How can it be suppressed?
13. Describe the types of external field magnets in MRI. Which one is the most advanced? What is meant by tesla (T)?
14. What are some of the hazards of MRI?
15. Describe sound waves. How do they differ from, and resemble, electromagnetic waves?
16. How are ordinary sound and ultrasound waves generated?
17. State the wave equation and explain the terms.
18. What is a piezoelectric crystal? On what does its resonant frequency depend?
19. Describe a transducer with the aid of a simple diagram. What are its functions?
20. Why does a transducer crystal have to be coupled to the skin by oil?
21. Discuss the characteristics of an unfocused ultrasound beam. How can such a beam be focused? Compare the shapes of a focused and unfocused beam.
22. Discuss the fate of an ultrasound beam in the body. How are ultrasound waves reflected and received?
23. What happens to the depth of the "near field" as ultrasound frequency is increased?
24. What percentage of an ultrasound beam is reflected from the boundary between soft tissue and air? Explain acoustic impedance.
25. Discuss relative amplitude and its applicable units.
26. Describe three modes of ultrasound image display.
27. Explain the principle of Doppler imaging.
28. Briefly outline the biologic hazards of ultrasound.

APPENDIX—ANSWERS TO PROBLEMS

Chapter 1.

1. (a) $\frac{1}{2}$ (b) $\frac{3}{4}$ (c) $\frac{1}{15}$ (d) $\frac{4}{5}$
2. (a) $1\frac{4}{5}$ (b) $17\frac{1}{84}$ (c) $1\frac{29}{30}$
3. 25 bu
4. (a) $1\frac{1}{3}$ (b) $1\frac{1}{63}$ (c) $1\frac{13}{27}$
5. b = ac c = b/a
6. (a) 8 (b) -3 (c) $2\frac{1}{3}$ (d) $4\frac{4}{9}$ (e) 44
7. (a) $\frac{2}{3}$ (b) 5 (c) $10\frac{1}{2}$ (d) 15
8. 20 cm
9. tripled
10. 12.56 cm^2
11. 600 sq ft
12. about $3\frac{3}{4}$ min
13. 3,600,000
14. 4.24×10^5
15. 1.6×10^6
16. 5
17. 10^4 or 10,000
18. 11
19. 3
20. 10^9
21. 7

Chapter 2.

11. 20° C
12. 104° F

Chapter 4.

7. 12
8. At. no. 6
   Mass no. 8

Chapter 6.

8. 2 amps
9. 150 volts

587

11. 1.8 Ω; 4, 2, and $\frac{2}{3}$ amps in branches; $6\frac{2}{3}$ amps in main lines.
13. 24 watts

Chapter 9.

8. 56.6 kV
9. 42.4 A

Chapter 10.

3. 110,000 volts = 110 kV
10. 100/240 or 5/12

Chapter 13.

11. $\frac{1}{10}$ sec

Chapter 14.

20. Underexposed
21. Correctly exposed

Chapter 16.

9. 10 gal (38 liters).

Chapter 18.

13. 80 R/min 5R/min
14. 300 mA
16. 20 mAs
31. Image diameter 10 in. (25.4 cm). Percentage magnification 11% Magnification factor 1.11

Chapter 19.

18. 16 in. (40.6 cm)
19. 7 mm
21. 8 in. × 10 in. (20.3 cm × 25.4 cm)

Chapter 22.

9. 7.7%
10. 435 cps, corrected, with GM counter. No correction needed for scintillation counter.

Chapter 24.

24. 100%

BIBLIOGRAPHY

Batson OV, Carpentier VE. Stereoscopic depth perception. *Am J Roentgenol* 51:202, 1944.

Bergeron RT. Manufacturers' designation of diagnostic x-ray tube focal spots. *Radiology* 111:421, 1974.

Bernstein H, Bergeron RT, Klein DJ. Routine evaluation of focal spots. *Radiology* 111:427, 1974.

Bloom WL, Jr, Hollenbach JL, Morgan JA. *Medical Radiographic Technic*, 3rd ed. Springfield, Thomas, 1969.

Bookstein JJ, Voegeli E. A critical analysis of magnification radiography. *Radiology* 98:23, 1971.

Bucky GP. A grating-diaphragm to cut off secondary rays from the object. *Archives of the Roentgen Ray*, June 1913.

Bureau of Radiological Health (BRH), U.S. Department of Health, Education, and Welfare. Rockville, MD 20852.

> *Population Exposure to X rays U.S., 1970.*
>
> *Pre-release Report: X-ray Exposure Study (XES) Revised Estimates of 1964 and 1970 Genetically Significant Dose, 1975.*
>
> *Gonadal Shielding in Diagnostic Radiology, 1974.*
>
> *Handbook of Selected Organ Doses for Projections Common in Diagnostic Radiology, 1976.*

Bushong SC. *Radiologic science for technologists. Physics, Biology, and Protection,* 4th ed. St. Louis, Mosby, 1988.

Chamberlain WE. Fluoroscopes and fluoroscopy. *Radiology* 38:383, 1942.

Code of Federal Regulations (21 CFR 800 to 1299). Federal Register, U.S. Government Printing Office. April 1, 1983.

Curry TS III, Dowdey JE, Murry RC Jr. *Christensen's Physics of Diagnostic Radiology,* 4th ed. Philadelphia, Lea & Febiger, 1990.

Doi K, Rossman K. Effect of focal spot distribution on blood vessel imaging in magnification radiography. *Radiology* 114:435, 1975.

Drew PG. The NMR phenomenon: putting spinning nuclei to work. *Diagnostic Imaging*, August 1983.

Eastman Kodak Company, Rochester, N.Y.:

> *The Fundamentals of Radiography,* 1980.
>
> *Radiography in Modern Industry.*
>
> *Sensitometric Properties of X-ray Films.*

Ettinger A, Fainsinger MH. Zonography in daily radiological practice. *Radiology* 87:82, 1966.

Feig SA. Can breast cancer be radiation induced? In *Breast Carcinoma, the Radiologist's Expanded Role.* Logan WW (Ed.) New York, John Wiley, 1977.

Gaulden ME. Possible effects of diagnostic X-rays on the human embryo and fetus. *J Arkansas Med Soc, 70:* 424, 1974.

Gould RG, Hale J. Control of scattered radiation by air gap techniques: application to chest radiography. *Am J Roentgenol 122:* 109, 1974.

Grigg ERN. *The Trail of the Invisible Light.* Springfield, Thomas, 1965.

Harms SE, Morgan TJ, Yamanashi WS, Harle TS, Dodd GD. Principles of nuclear magnetic imaging. *RadioGraphics 4:26,* 1984 (Special Edition).

Humphries RE. Personal communication.

Jaffe C, Webster EW. Radiographic contrast improvement by means of slit radiography. *Radiology 116:* 631, 1975.

Jans RG, Butler PF, McCrohan JL, Thompson WE. The status of film-screen mammography: results of the BENT study. *Radiology 132:* 197, 1979.

Johns HE, Cunningham JR. *The Physics of Radiology,* 4th ed. Springfield, Thomas, 1983.

Keller PJ. *Basic Principles of Magnetic Resonance Imaging.* General Electric Medical Systems, 1991.

Liebel-Flarsheim Company. *Characteristics and Applications of X-ray Grids.* Cincinnati, 1968.

Littleton JT, Rumbaugh CL, Winter FS. Polydirectional body section roentgenography: a new diagnostic method. *Am J Roentgenol 89:* 1179, 1963.

Mettler FA, Guiberteau MJ. *Essentials of Nuclear Medicine Imaging.* 3rd ed. New York, Grune & Stratton, 1990.

Milne ENC. Characterizing focal spot performance. *Radiology 111:* 483, 1974.

National Council on Radiation Protection and Measurements (NCRP) Reports: NCRP, P.O. 7910 Woodmont Ave., Suite 800, Bethesda, MD 20814.

66 *Mammography* (1980)

68 *Radiation Protection in Pediatric Radiology* (1981)

71 *Operational Radiation Safety — Training* (1983)

74 *Biological Effects of Ultrasound: Mechanisms and Clinical Implications* (1983)

82 *SI Units in Radiation Protection and Measurements* (1985)

85 *Mammography — A User's Guide* (1986)

86 *Biological Effects and Exposure Criteria for Radiofrequency Electromagnetic Fields* (1986)

91 *Recommendations on Limits for Exposure to Ionizing Radiation* (1987)

93 *Ionizing Radiation Exposure of the Population of the United States* (1987)

94 *Exposure of the Population in the United States and Canada from Natural Background Radiation* (1987)

95 *Radiation Exposure of the U.S. Population from Consumer Products and Miscellaneous Sources* (1987)

99 *Quality Assurance for Diagnostic Imaging* (1988)

102 *Medical X-Ray, Electron Beam and Gamma-Ray Protection For Energies Up To 50 MeV (Equipment Design, Performance and Use)* (1989)

103 *Control of Radon in Houses* (1989)

104 *The Relative Biological Effectiveness of Radiations of Different Qualities* (1990)
105 *Radiation Protection for Medical and Allied Health Personnel* (1989)
107 *Implementation of the Principle of As Low As Reasonably Achievable (ALARA) For Medical and Dental Personnel* (1990)

Pfalzner PM. Attenuation coefficient: finite or infinitesmal? *Physics in Med & Biol 11:* 132, 1966.

Potter HE. The Bucky diaphragm principle applied to roentgenology. *Am J Roentgenol 7:* 292, 1920.

Potter HE. History of diaphragming roentgen rays by the use of the Bucky principle. *Am J Roentgenol 25:* 396, 1931.

Quimby EH, Feitelberg S, Silver S. *Radioactive Isotopes in Clinical Practice.* Philadelphia, Lea & Febiger, 1968.

Ridway A, Thumm W. *The Physics of Medical Radiography.* Reading, A–W, 1968.

Rosenstein M. *Handbook of Selected Organ Doses for Projections Common in Radiology.* HEW/FDA 76-8031, 1976.

Rossmann K, Lubberts G. Some characteristics of line spread-function and modulation transfer function of radiographic films and screen-film systems. *Radiology 86:* 235, 1966.

Rossman K. Point spread-function, line spread-function, and modulation transfer function. *Radiology 93:* 257, 1969.

Seemann HE. *Physical and Photographic Principles of Medical Radiography.* New York, John Wiley, 1968.

Selman J. *The Basic Physics of Radiation Therapy,* 3rd ed. Springfield, Thomas, 1990.

Selman J. *Elements of Radiobiology.* Springfield, Thomas, 1983.

Spiegler P, Breckenridge WC. Imaging of focal spots by means of the star test pattern. *Radiology 102:* 679, 1972.

Sprawls P Jr. *Physical Principles of Medical Imaging.* Rockville MD, Aspen, 1987.

Ter-Pogossian M. *The Physical Aspects of Diagnostic Radiology.* New York, Hoeber, 1967.

Theron WO, Moore R, Amplatz K. The evaluation of high-speed screen-film combinations in angiography. *Radiology 114:* 449, 1975.

Trout ED, Gager RM. Protective materials for field definition in radiation therapy. *Am J Roentgenol 63:* 396, 1950.

Trout ED, Kelley JP, Larson VL. A comparison of an air gap and a grid in roentgenography of the chest. *Am J Roentgenol 124:* 404, 1975.

INDEX